D0734917

Marine Invertebrates
And Plants of the Living Reef

- How to identify most reef invertebrates from the Gulf of Mexico, Caribbean Sea, Florida and the Tropical Atlantic.
- For the Miniature Reef Aquarist • SCUBA Diver • Snorkler • Ichthyologist
- Illustrated with 432 color photographs

Dr. Patrick I. Colin

H-971

Photographs were taken by the author unless credited otherwise.

This book is dedicated

to my parents,

Helen Louise and Charles William

Colin,

for their encouragement

Distributed in the UNITED STATES by T.F.H. Publications, Inc., One T.F.H. Plaza, Neptune City, NJ 07753; in CANADA to the Pet Trade by H & L Pet Supplies Inc., 27 Kingston Crescent, Kitchener, Ontario N2B 2T6; Rolf C. Hagen Ltd., 3225 Sartelon Street, Montreal 382 Quebec; in CANADA to the Book Trade by Macmillan of Canada (A Division of Canada Publishing Corporation), 164 Commander Boulevard, Agincourt, Ontario M1S 3C7; in ENGLAND by T.F.H. Publications Limited, Cliveden House/Priors Way/Bray, Maidenhead, Berkshire SL6 2HP, England; in AUSTRALIA AND THE SOUTH PACIFIC by T.F.H. (Australia) Pty. Ltd., Box 149, Brookvale 2100 N.S.W., Australia; in NEW ZEALAND by Ross Haines & Son, Ltd., 18 Monmouth Street, Grey Lynn, Auckland 2, New Zealand; in SINGAPORE AND MALAYSIA by MPH Distributors (S) Pte., Ltd., 601 Sims Drive, #03/07/21, Singapore 1438; in the PHILIPPINES by Bio-Research, 5 Lippay Street, San Lorenzo Village, Makati Rizal; in SOUTH AFRICA by Multipet Pty. Ltd., 30 Turners Avenue, Durban 4001. Published by T.F.H. Publications, Inc. Manufactured in the United States of America by T.F.H. Publications, Inc.

CONTENTS

Preface

This book had its origin in the frustration of the non-specialist in attempting to identify any but the most common invertebrate animals or plants occurring on Caribbean reefs. There was simply no single source having photographs in color taken in the natural habitat plus information concerning the natural history of these organisms. The situation has been different for the fishes of the area since the publication in 1968 of two books, *Caribbean Reef Fishes* by John E. Randall (T.F.H. Publications) and *Fishes of the Bahamas and Adjacent Tropical Waters* by James E. Boehlke and Charles C.G. Chaplin. The present effort is designed to fill the need for a guide to reef invertebrate and plant identification and to serve as a companion volume to Randall's book.

This book is designed to serve first as an identification guide for the professional marine scientist and amateur snorkler or SCUBA diver. Marine research in tropical Atlantic areas is increasing rapidly and resort accommodations catering to divers have proliferated in the last ten years, resulting in the need for a starting point concerning reef organism identification. Beyond simple identification the present volume is intended to provide information on the natural history of Caribbean reefs and their organisms, drawn from the scientific literature and the author's personal observations. In addition, this book can also act as an introduction to the technical literature on Caribbean reefs for both amateur and professional students of marine biology.

The Caribbean region is of specific concern in this book, but since much of this fauna and flora also occurs outside the Caribbean proper, it is useful in areas such as the Bahama Islands, Turks and Caicos Islands, southern Florida and tropical portions of the Gulf of Mexico. To a lesser extent the organisms covered occur at Bermuda and along the tropical

Atlantic coast of South America. No species are included which occur in the peripheral areas but not in the Caribbean proper.

In attempting to cover the entire spectrum of macroscopic reef organisms, exclusive of fishes, any attempt to be comprehensive would be futile. Very few groups of reef organisms are completely known, much less photographed in nature, and in most cases only representatives of specific groups have been included. The entire invertebrate fauna and flora of the Caribbean is much greater than just those occurring on reefs, and again many species have not been considered. The coverage of specific groups within this book is roughly proportional to the abundance, distinctiveness and ease in observing of the elements in that particular group. Knowledge of Atlantic reef organisms is increasing rapidly; although the references cited are the most recent available in most cases, many may be superseded by the time this volume is published.

Various persons and institutions have contributed to the completion of this book. Charles Cutress of the University of Puerto Rico has freely offered information, much unpublished, from his superb knowledge of reef invertebrates and assisted in identifying some photographs. Ronald Larson, Ileana Clavijo and Dennis Opresko have also assisted in identifying and describing certain groups. Field work, during which some photographs were taken, was carried out on vessels from the University of Miami School of Marine and Atmospheric Sciences and the Department of Marine Sciences, University of Puerto Rico, Mayaguez. A large number of people offered assistance in the form of discussion, identifications or aid in field work, and the following are gratefully acknowledged: Walter H. Adey, Charles and Deborah Arneson, Frederick M. Bayer, James E. Boehlke, James N. Burnett-Herkes, Bruce Chalker, Nick and Anne Chitty, Richard T. Davies, Walter M. Goldberg, Martin F. Gomon, Robert Gore, Edwin J. Gould, Eileen Graham, Roger T. Hanlon, Willard Hartman, Jeremy B.C. Jackson, Lynton S. Land, Judith C. Lang, Charles Messing, Pamela Oatis, John C. Ogden, Francisco Palacio, Shirley Pomponi, Tony Rae, C. Richard Robins, Paul Sammarco, William F. Smith-Vaniz,

Jon C. Staiger, Dennis L. Taylor, Lowell P. Thomas, Ronald Thresher, Richard K. Wallace, Ernest H. and Lucy B. Williams, and Robert C. Work.

Any errors in the identifications and information contained in this volume are the responsibility of the author.

Patrick L. Colin
La Parguera, Puerto Rico

Deep-reef specimens of *Montastrea cavernosa* and *M. annularis* growing in their plate-like forms.

Introduction

The Caribbean Sea, occasionally termed the American Mediterranean, is a marginal sea separated from the Atlantic Ocean by the chain of islands comprising the Greater and Lesser Antilles. It is bounded on the south by the northern coast of South America and on the west by the Central American continent. The Caribbean measures a maximum of 1500 km from north to south and 3000 km from east to west, covering an area of over 2,600,000 square km. While the passages between the islands bordering the Caribbean are not deep by oceanic standards, the basins of this sea reach a depth of 7000 m. Besides the islands bordering the Caribbean, a large number of other islands also exist on submarine ridges crossing portions of this sea, separating its four deep-water basins.

The conditions prevailing throughout the Caribbean are tropical, and coral reefs abound in this area. These tropical conditions are also found outside the Caribbean proper, and the coral reefs of the western Atlantic Ocean constitute a rich assemblage ranging from as far north as Bermuda (32° N) to near Rio de Janeiro, Brazil (23° S). The reefs of the Bahama Islands, the Turks and Caicos Islands, portions of the Gulf of Mexico and southern Florida are quite similar in their species composition to the Caribbean and will be discussed in the present volume as they are actually inseparable biologically from the Caribbean proper. Areas of marginal reef growth, such as Bermuda, will also be discussed although they are often fairly distinctive biologically from the Caribbean.

The reefs of this region encompass nearly the entire range of reef morphologies. The classically designated fringing reefs (shallow and occurring near shore) and barrier reefs (shallow but occurring some distance from shore with an intervening lagoon) occur here. True barrier reefs exist in several western Atlantic locations. The longest Atlantic barrier reef flanks the coast of Belize (British Honduras) for a distance of some 240 km at a distance of 12 to 40 km from the coast (Thorpe and Stoddart 1962, Matthews 1966), being interrupted only occasionally by deep-water passages of modest width. Only lengthy barrier reef structures exist at Andros Island, Bahamas and Isla de Providencia (Milliman 1973).

Fringing reefs with shallow depths and occurring on or very close to continental and insular margins occur in much of the tropical western Atlantic area. The point of distinction between fringing and barrier reefs is difficult to determine, and combinations of these two types can occur on a single reef. The width, depth and agency responsible for the development of the shallow crest of the fringing (as well as the barrier) reef can vary widely.

Large, often rocky masses of reef occurring on open bottoms are usually termed "patch" or knoll reefs. These include variations on the general form, such as algal cup reefs ("boilers"). Mound-like patch reefs of considerable size and extensive cavern development occur. Such reefs can occur behind (landward) or in front of barrier or other offshore reefs.

Reefs may occur on offshore banks that do not break the surface or as a slightly elevated area which does not reach near the surface at the edge of the shelf surrounding islands or continents. These represent what could respectively be termed "bank reefs" and "submerged reefs" and are well documented in the Caribbean area (MacIntyre 1967, 1972).

Atolls also occur in the Caribbean and adjacent areas. Basically an atoll is a ring of shallow reef, often having gaps in its leeward side, rising out of oceanic depths with a lagoon at its center. The basement rock is usually volcanic but is often deeply submerged under thick layers of calcareous substrate. Bryan (1953) listed 26 atolls in the Caribbean and adjacent regions, while in a somewhat more refined estimate Milliman (1969) estimated 15 reef structures in this area which could be classified as atolls. He also reported on four southwestern Caribbean atolls. Stoddart (1962) examined three atolls off the coast of Belize. Outside the Caribbean

An area with entrances to buttress caves, Spring Gardens, Jamaica, depth 10 m.

Large head of *Montastrea annularis* abutting on the *Acropora palmata* zone at Discovery Bay, Jamaica, depth 7 m. A colony of *Millepora complanata* is growing in one of the crevices of the *M. annularis*.

Reef view, Discovery Bay, Jamaica. 15 m. Visible in the photo are stony corals, sponges and algae.

proper, atolls exist in the Bahamas (Milliman 1967) and the Campeche Bank area of Yucatan, Mexico (Kornicker and Boyd 1962, Logan 1969, Logan *et al.* 1969).

Various terms are used to delimit physiographical areas on reefs. Reef crest or reef flat is used for the often broad zone of barrier and near-shore fringing reefs which is extremely shallow or at times emergent. This zone can also be defined biologically. A similar, but quite distinctive biologically, shallow platform on some reefs is an algal ridge, produced by coralline algae in areas of high wave energy. The zone leeward and usually landward of the reef crest is the back reef, while that to seaward is the fore reef. Typically the fore reef slopes quite steeply to seaward, often at depths below about 20-30 m; such deeper areas are often termed the fore reef slope or deep reef. Within these broad zones of the reef there are easily identifiable sub-habitats, usually based on environmental conditions.

The reef flat is more subject to variable conditions than any other reef habitat. Light intensity is high, so high that it is probably detrimental to many organisms that could occur there. Wave action can be strong, even though much of the energy of waves has been expended on the fore reef. This is particularly true during storms when the increase in wave height over normal causes waves to break farther up onto the reef flat. Low tidal levels often result in aerial exposure of reef flats. In combination with calm conditions reducing splash from breaking waves, this exposure can be a limiting factor for some organisms and under proper conditions can result in mass mortalities of reef flat organisms (Glynn 1968). The effect of intense sunlight on the often limited volumes of water above reef flats must not be ignored as this can cause considerable heating and subsequent death of organisms. Precipitation in combination with reef flat emergence could also cause mortality due to lowered salinity. Some of the corals occurring on reef flats are limited to depressions which become pools during low tide emergence (Glynn 1973c). Other species such as *Siderastrea radians*, *Porites asteroides* and *Diploria clivosa* can survive moderate exposure (Glynn 1973c).

When the reef flat is submerged there is often a direct

movement of water from the fore reef to back reef caused by wind-driven surface currents or wave propagated transport. The speed of these currents across the reef flat can be greater than the surface currents of surrounding areas (Glynn 1973a) and there can be a large decrease in the phytoplankton and zooplankton populations in water crossing the reef (Glynn 1973b).

The back reef environment poses somewhat different challenges to organisms. Often there is no expedient means for transport of sediment—often very fine—away from the environment. It cannot be moved seaward through sloping channels to be sent over the drop-off to the depths due to the presence of the reef flat. Consequently much of the rocky substrate may be buried and exposed hard substrate may be the factor limiting certain stony corals (Kissling 1965) and gorgonians (Kinzie 1973). Back reef areas receive relatively little wave action and are often more turbid than nearby fore reef environments. Due to their shallow depth, light intensity is high.

The fore reef environment is usually heterogeneous, divisible into zones on the basis of the dominant benthic organisms. This distribution is largely controlled by wave action. On reefs exposed to moderate wave action a zone of *Acropora palmata* occurs immediately seaward of the reef flat where most of the energy of breaking waves is expended. Shinn (1963) described the orientation of branches of *A. palmata* with respect to waves. A zone of mixed coral growth occurs somewhat deeper. Even in this environment (about 8 m deep) sea fans and gorgonians are oriented perpendicular to the direction of incoming waves. Water temperature is more stable than in back reef areas and light intensity decreases with depth.

Deeper (greater than 10 m) on fore reefs either a mixed coral community or one dominated by a single species, *Acropora cervicornis*, exists. Even within the mixed communities certain corals, usually *Montastrea cavernosa* and *M. annularis*, are more common than others.

Channels serve as conduits for the transport of sediments down slope toward deep water. The depth of these channels is controlled by various factors including the height the coral has grown to around them (Goreau and Land 1974). Buttresses have coral masses rising as much as 10 m above the surrounding terrain, with a steep face at their outer ridge and sand channels in between them; these are an extreme development of vertical relief by corals (Goreau 1959). The corals of adjacent buttresses can grow together, forming a cavern in what was previously an open sand channel. The buttress caves can still maintain the function that the channel served, sediment transport, but open up a new habitat with its own fauna.

The cave habitat is occasionally abundant on reefs and produces protection from wave action, intense light and major currents. This environment has been discussed by various authors (Hartman and Goreau 1970, Goreau and Goreau 1973). A similar environment, at least in terms of illumination, is that occurring underneath large plate-like colonies of coral (Jackson *et al.* 1971).

Many factors determine the structure of a stony coral community. A colony initially forms by the settlement of a larva, the planula, in an available location. The selection of this location is determined by the presence of other species which the planula can outcompete, the presence of conspecifics and the proper environmental conditions. Often the larva selects a less than optimal location and is programed for an early death or at best a stunted growth form. Tiny coral colonies are open to predation by a variety of organisms such as sea urchins or fishes, and it takes some time before a colony grows beyond the stage where a single predatory incident can easily eliminate the entire colony. As colony size increases, other predatory or destructive forces become more important in determining the fate of the individual colony. Adjacent corals may attack via the extracoelenteric feeding response or other colonial animals and some plants may attempt to overgrow the living coral surface. Boring organisms begin their attack on the coral skeleton, and the coral must constantly be producing more calcareous structure to account for this weakening. The physical factors present at the location the larvae settle are also crucial in survival of the colony. A sizeable percentage of larvae must make "mistakes" in site selection and are doomed to be eliminated by biological or physical factors of the location selected. One adaptation around this problem is to have the planulae move only a short distance by crawling from the parent colony and settle in a location close to the successful parent. These are only a few examples of the factors affecting a single colony and this, considering the thousands or more of colonies, is what determines the character of the coral community.

Coral reefs are also affected by long term phenomena. Within the geologically recent past the major oceanographic and climatic events have been the repeated glaciations with glacial-interglacial cycling of conditions. Atlantic reefs were rather drastically affected by glaciation in two different ways. First, there may have been a lowering of sea surface temperature by several degrees Centigrade, although authorities still disagree about whether this occurred and how much was the lowering if it did. Secondly and perhaps more importantly, sea level was alternately lowered (as much as 135 m during the last glaciation 17,000 years ago) and raised by the deposition of large amounts of water in the ice caps during glaciation and its release during interglacial periods. This cycling of sea level went on throughout the Pleistocene and for perhaps longer.

What are today the reefs of the Caribbean were only 17,000 years ago dry land. What reefs existed at that time? Around most Caribbean islands today at -135 m (the depth of past sea level) occurs a vertical or near-vertical wall continuing down often a considerable distance. Reef development was probably not good along this ancient shoreline. There simply was little substrate within the zone of sufficient light for stony corals, and what was available was not particularly suitable, being steeply sloped. The broad, nearly flat island margins of the Caribbean, where most reefs occur today, were gone, elevated over 100 meters above the sea and replaced by vertical cliffs. The area suitable for reef development during glacial periods was probably 10% or less of that which is suitable today.

Sea level lowering also changed the character of various restricted bodies of water and perhaps allowed reef development on offshore banks presently below coral growth limits. In the Bahamas the Tongue of the Ocean and Exuma Sound are deep basins connected by both a relatively narrow deep-water opening and more widely by shallow water connections to adjacent areas. Lowered sea level cut off these shallow connections, causing these two basins to become nearly enclosed deep-water bays with no shallow margins.

Obviously sufficient populations of Caribbean reef-dwelling organisms were present to repopulate the previously exposed areas after sea level rose. Only a few authors have dealt with this problem. Goreau (1969) discussed various aspects of this subject and felt chance factors might have been important in determining present day distribution of certain reef forms. Adey (1975) analyzed the growth of a reef in St. Croix in relation to rising sea level.

Typical view of the reef at 15 m depth on the leeward (west) side of Barbados. Large colonies of *Siderastrea* and *Montastrea* are visible with broad sediment areas.

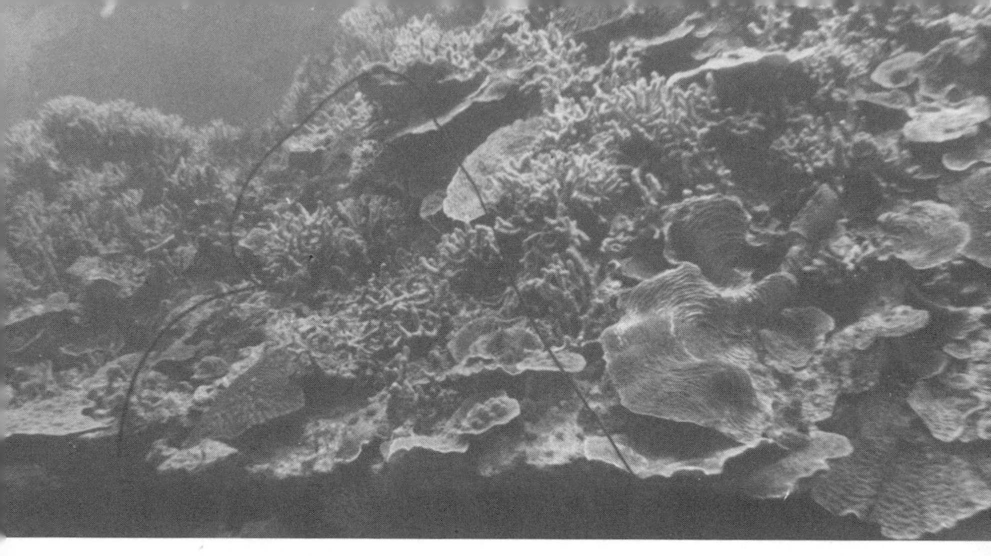

General view of a deep-reef coral slope at Spring Gardens, Jamaica, 30 m depth. Large plates of *Agaricia* are visible with clumps of *Madracis* very common. The whip-like antipatharian, *Stichopathes lutkeni*, is seen in the center of the photo.

Flattened plates of *Montastrea annularis* as are typically found in deep-reef environments where light is dim and coral adapted to take advantage of what light is available. A few small clumps of *Acropora cervicornis* are visible. Photo taken at Discovery Bay, Jamaica at 30 m depth.

Reefs in the Caribbean Sea

Although numerous reports exist on the reefs of this region, the Caribbean as a whole cannot be described as well studied. The coral reefs can vary considerably over both short and long distances and are profoundly influenced by local conditions and their history. In approaching the descriptive work done in this area, it is most useful to consider the reefs of the Caribbean by subregions. The present discussion will include information not previously published; while relevant publications will be cited, they will not all be recapitulated for the sake of space.

The Greater Antilles (Cuba, Jamaica, Hispanola, Puerto Rico) are the largest islands of the Caribbean and, except for one instance (Jamaica), have both an Atlantic (northern) and a Caribbean (southern) shore. The Atlantic shores of these islands will be included in the present discussion. Between the islands are relatively narrow passages or channels which may also contain some islands.

Cuba is the largest island in the West Indies, being over 1100 km in length, and much of its coasts are fringed by reefs. Kuhlmann (1970, 1971, 1974) and Duarte-Bello (1961) have reported on these reefs, but for such a large area the amount of published information is small.

The best known Caribbean reefs occur around the island of Jamaica, particularly on its north coast. Work by T.F. Goreau published by himself (1959, 1963, 1969, and others) and with co-workers such as N.I. Goreau, E. Graham, W.D. Hartman, L.S. Land, J.W. Wells and C.M. Yonge (see bibliography) has greatly contributed to knowledge of Jamaican reefs and more significantly to the ecology and geology of coral reefs in general. More species of stony corals have been recorded from Jamaica (at least 64 species) than from any other Atlantic locality, and conditions on

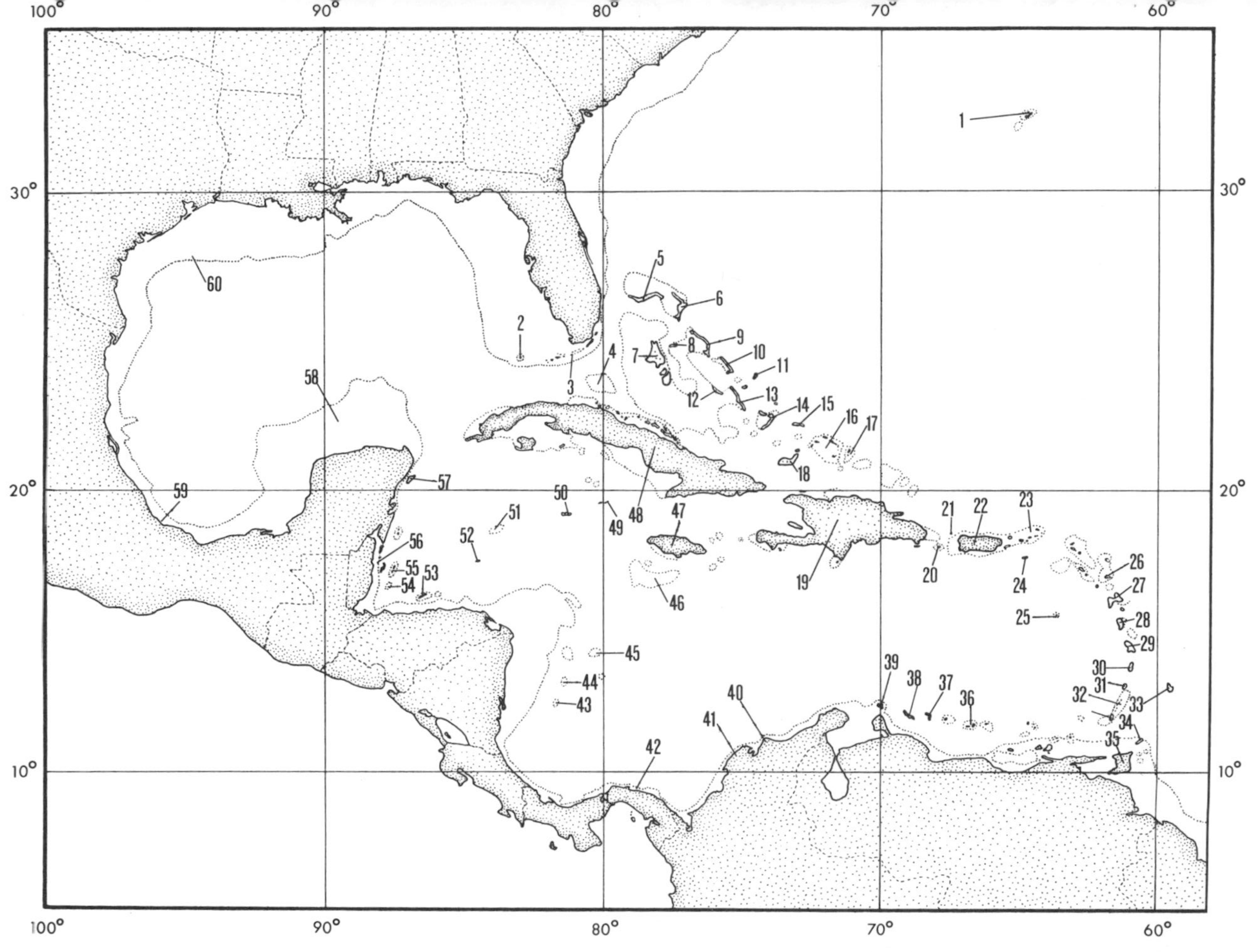
100°
90°
80°
70°
60°
30°
20°
10°
1
2
3
4
5
6
7
8
9
10
11
12
13
14
15
16
17
18
19
20
21
22
23
24
25
26
27
28
29
30
31
32
33
34
35
36
37
38
39
40
41
42
43
44
45
46
47
48
49
50
51
52
53
54
55
56
57
58
59
60

View of general reef on the west coast of Barbados, 15 m depth.

Localities in the tropical western North Atlantic Ocean mentioned in the text. 1—Bermuda; 2—Dry Tortugas, Florida; 3—Florida Keys; 4—Cay Sal Bank, Bahamas; 5—Grand Bahama Island; 6—Great Abaco Island; 7—Andros Island; 8—New Providence Island (Nassau); 9—Eleuthera Island; 10—Cat Island; 11—San Salvador; 12—Great Exuma Island; 13—Long Island; 14—Acklins Island; 15—Mayaguana Island; 16—Caicos Bank; 17—Grand Turk Island; 18—Great Inagua Island; 19—Hispanola (Haiti and Dominican Republic); 20—Mona Island; 21—Desecheo Island; 22—Puerto Rico; 23—Virgin Islands; 24—St. Croix (Virgin Islands); 25—Aves Island; 26—Antigua; 27—Guadeloupe; 28—Dominica; 29—Martinique; 30—St. Lucia; 31—St. Vincent; 32—Grenadines; 33—Barbados; 34—Tobago; 35—Trinidad; 36—Los Roques; 37—Bonaire; 38—Curacao; 39—Aruba; 40—Santa Marta, Colombia; 41—Cartagena, Colombia; 42—San Blas Islands, Panama; 43—San Andres; 44—Old Providence Island; 45—Serrana Bank; 46—Pedro Bank; 47—Jamaica; 48—Cuba; 49—Cayman Brac and Little Cayman; 50—Grand Cayman Island; 51—Misteriosa Bank; 52—Swan Island; 53—Roatan; 54—Glovers Reef; 55—Lighthouse Reef; 56—Belize (British Honduras) barrier reef; 57—Cozumel; 58—Campeche Bank; 59—Veracruz, Mexico; 60—Flower Gardens Bank, Texas.

much of the north coast seem near optimal for reef development.

The structure of the fringing reefs along Jamaica's north coast varies somewhat although this variation is well documented. Buttressed fore reefs (Goreau 1959, Goreau and Goreau 1973) occur in a number of locations, but a true buttress zone is lacking in areas such as west of Discovery Bay. Reefs in areas of high sediment deposition also occur in Jamaica, usually in the somewhat enclosed bays such as Discovery Bay or near the mouths of rivers.

The next island of the Greater Antilles moving east is Hispanola, which contains the Dominican Republic and Haiti. Its reefs are poorly known, not having been described in the recent scientific literature. The only account, dealing with a small area of reefs in the Gulf of Gonave, is that of Beebe (1928). Lying about midway between Haiti and Jamaica is small Navassa Island, whose reefs have not been examined.

The Mona Channel, between Hispanola and Puerto Rico, contains three islands where many of the photographs in this book were taken. Mona Island and its small satellite island, Monito, sit near the middle of the channel approximately 80 km from the Puerto Rican coast and are bathed by oceanic water. The entire northern and eastern coasts of Mona Island (about 15 km in length) and all the coast of Monito (about 1-2 km) are vertical cliffs rising from the water surface to heights of 45 to nearly 100 m. These cliffs continue vertically below the surface to depths of 23 to 27 m at Mona and to 37 m at Monito. Below this submarine cliff the bottom quickly becomes nearly level and continues as a gradually deepening rocky plain for some distance offshore. This vertical escarpment has an abundant sponge fauna, both massive and encrusting, and significant amounts of encrusting invertebrates and macroalgae. Stony corals are nearly absent from this face, with only a few non-reef building or minor reef constructing species such as *Tubastrea aurea* and *Madracis decactis* present. Where a nearly level bottom exists, such as on infrequent blocks that have fallen from the cliff face and come to rest on the rocky shelf, some of the more important reef building corals may occur.

The southern and western shores of Mona Island are quite different from the areas of submarine cliffs. A sloping shelf extends out from shore and reef development is fairly extensive. Some areas have shallow fringing reefs, probably extensively of algal origin, including Playa de Sardinera and Playa de Pajaros. The rocky shelf is narrowest (less than 0.5 km wide) at Piedra de Carabinero where an abrupt transition from a nearly level to nearly vertical bottom occurs at only 10 to 15 m depth. Stony coral development is much greater on this vertical wall than that occurring on the northern and eastern coasts.

The marine communities around Monito Island are similar to those of the northern coast of Mona Island, except the vertical escarpment goes somewhat deeper and the fauna and flora are of a similar nature, but perhaps better developed than at Mona Island.

The third island of the Mona Channel is Desecheo Island. It lies about 20 km off the western coast of Puerto Rico and, although steep and rugged, it lacks the vertical cliffs of Mona and Monito Islands. At the shore the rocky bottom drops quickly to 10-12 m and extensive submarine caves, often going 10 to 30 m inland, are developed in this area. These caves have an abundant colorful encrusting sponge fauna and numerous ahermatypic corals, hydroids and bryozoans. Below the rocky shore a sandy band exists on the southwest side of the island at 13 to 15 m and a gradually sloping reef begins at 15 m. This reef has its most extensive development at 23 m, with tremendous masses of *Montastrea cavernosa* and large, extremely abundant colonies of *Eusmilia fastigiata.* Below 30 m the coral development decreases and the bottom remains only slightly sloping to 45 m.

Puerto Rico presents a dramatic contrast in marine conditions and resultant reef development with the islands of the Mona Channel. Sizable rivers flow into the sea from this island and poor soil erosion practices have increased the already substantial sediment load these rivers carry. Much of the southern and western coasts have a broad shelf as much as 20 km in width with fine sediments which are easily suspended. As a result water transparency is much lower around Puerto Rico than at the islands of the Mona Channel.

A series of well developed shallow reefs exist off the southwestern part of the island. In the La Parguera region there are series of shallow reefs progressively further offshore with moderate (to 20 m) depths between them. These reefs have been dealt with by Almy and Carrion-Torres (1963), Glynn (1973a) and Glynn *et. al.* (1964). On the windward (seaward) sides of these reefs exist lush growths of *Acropora palmata* which slope rapidly to about 7 m, where *Montastrea cavernosa* and *M. annularis* become dominant. At the lower limit (about 10 to 18 m) of these reefs a steep escarpment is found with extensive vertical development of coral knolls and moderate cavern development. Below 15 to 20 m a nearly level sandy to muddy bottom is found.

A very shallow or emergent narrow reef flat occurs; on the leeward side of the reef a *Porites porites* community is often found (Glynn 1973a). In many instances the bottom slopes quickly on the leeward side and consists of sand below about 2 m. At 8 to 12 m a series of rocky outcrops or knolls usually occurs. The knolls are in an area of high sediment deposition, and only a small percentage of the surface is covered by living corals. The corals that occur here are those species able to withstand the high sedimentation, such as *Cladocora arbuscula* and the species of *Siderastrea*. The large solitary coral *Scolymia lacera* is common at depths of only 10 m (33 feet) in this habitat.

A well developed "submerged reef" exists at the shelf edge along the southwestern Puerto Rican coast in an area of much less turbidity. The top of this reef is shallower than further inshore areas, and at a depth of 17 to 25 m a sharp break in the nearly level bottom occurs, dropping away at an angle of up to 45° into the Caribbean. Sand channels exist on upper portions of this reef (Glynn 1973c) but are quickly lost below the break in slope. Corals of the genus *Agaricia* dominate below 30 m and coral growth occurs to at least 71 m in this locality. A number of coral species unrecorded by Almy and Carrion-Torres (1963) from Puerto Rico occur on these outer reefs.

Some interesting reef structures exist on the western coast north of Mayaguez. The coast between Punta Cadena and Punta Higuero has a narrow insular shelf (the narrowest

Reef dominated by knobby heads of *Montastrea annularis* at Buck Island, St. Croix, Virgin Islands. A variety of gorgonians and corals such as *Diploria* and *Acropora palmata* are also visible.

Mixed zone of corals with *Montastrea annularis* dominating. Photographed at Discovery Bay, Jamaica, 15 m depth.

Siderastrea radians with neon gobies on its surface. This large colony has some white patches where the polyps have been grazed away and some evidence of boring damage. Photographed at 15 m depth, Barbados.

Shallow reef at juncture of *Acropora palmata* and "mixed" coral zone at Discovery Bay, Jamaica, depth 8 m. In the background an almost pure stand of *Acropora palmata* occurs where it is slightly shallower. In the foreground occur a variety of corals. Visible are *Montastrea annularis, Diploria strigosa* and *D. labyrinthiformis.*

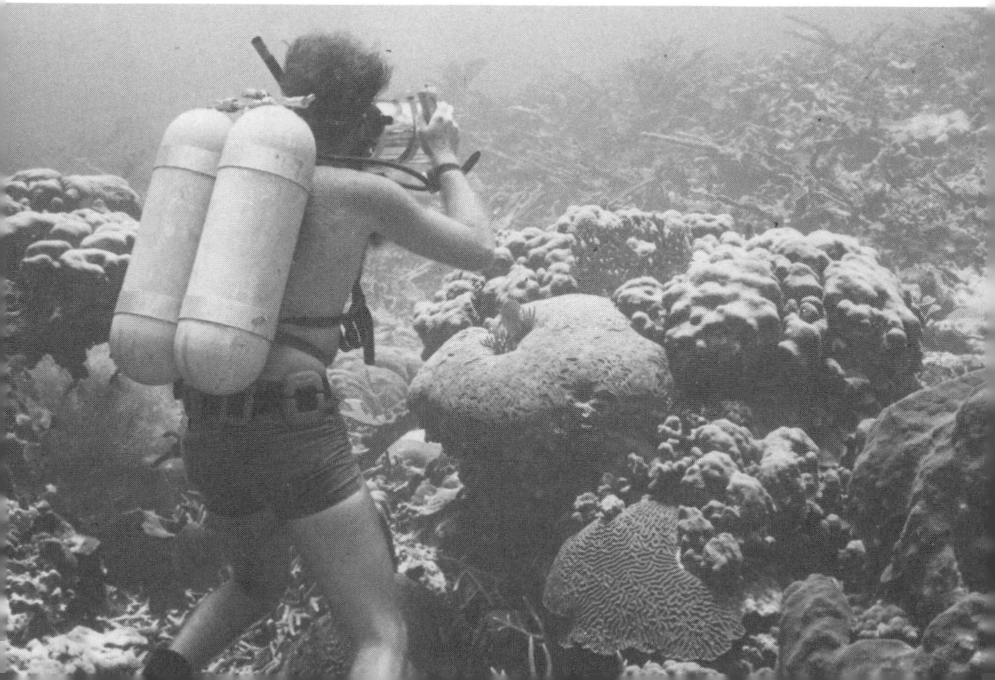

in Puerto Rico) with well developed reefs at its outer edge where the bottom slopes steeply. Stony corals, unusual gorgonians and antipatharians are abundant at depths of 15 to 40 m, but the water clarity is quite variable, being influenced by the local circulation and the discharge of nearby rivers.

Reef growth is reported to be poor along the north coast of Puerto Rico (Almy and Carrion-Torres 1963), probably because of sedimentary conditions and the extreme wave action to which this coast is exposed at times. Isolated coral colonies, such as *Montastrea cavernosa*, do exist on rock substrates in moderate depths, but are not organized into reef structures. Pressick (1970) briefly reported on the distribution of corals in shallow water of a fringing reef at Icacos Island off the eastern coast of Puerto Rico.

The Lesser Antilles are comprised of the smaller islands of the eastern Caribbean, running from the Virgin Islands (east of Puerto Rico) in the north to Grenada in the south and the islands along the Venezuelan coast. These are further divided into the Leeward Islands (Virgin Islands to Dominica), Windward Islands (Martinique to Grenada) and the Venezuelan coast islands (La Blanquilla to Aruba). They are islands of extremely varied marine habitats but lack the conditions associated with land masses of moderate size.

The reefs of the Leeward Islands, with some exceptions, are not really well known. Those of the U.S. Virgin Islands, particularly St. Croix, have been most extensively examined. Work by the West Indies Laboratory in St. Croix (Ogden *et al.* 1972, Multer, 1974, Adey 1974, 1975) has contributed much to our knowledge of this area and reefs in general. Kumpf and Randall (1961) mapped the distribution of shallow marine habitats around St. John, and Clifton and Phillips (1972) have described one small area on that island.

Roos (1971) commented briefly on the reef development around the three Dutch islands in the northern Lesser Antilles (Saba, St. Martin, St. Eustatius). He described coral growth as "scattered around these islands." The flat sandy bottoms of St. Martin and St. Eustatius, as well as the wave swept rocky shores of Saba, were not conducive to luxuriant reef growth.

Tiny Aves Island (distinguished from Islas de Los Aves near Venezuela), lying 200 km west of Guadeloupe, may be the most isolated point of land in the Caribbean. The small sandy islet has a thin veneer of reef to 15 m depth on its leeward (west) side where some protection from the normally high seas exists. Below this depth the bottom becomes a progressively steeper sandy slope reaching near the angle of repose at 45 m. No stony corals were found to exist on this sandy slope, but rocks (possibly not attached to the substrate) with massive sponges and gorgonians were encountered.

The French islands of Martinique and Guadeloupe have received recent attention (Adey and Burke 1976) and the existence of algal ridges on their windward margins is well documented. The deeper reef environments of these and most nearby islands are almost totally unknown.

Kier (1966) examined several localities on the Caribbean (leeward) side of Dominica. Where the bottom sloped steeply close to shore, rocky-coral environments occurred, often reaching 25 m depth only 30 m from shore. Coral development was greatest near 15 m depth and below 22 m the rocky substrate was replaced by a sandy slope. The coral masses, although large, were not thick and barely covered the rocky substrate. Kier (1966) also reported that in some areas, particularly near bays, a gentle sandy slope existed from shore to 25 m depth (a distance of as much as 1 km) without coral reef development.

In the Windward Islands of the Lesser Antilles, Roberts (1972) briefly discussed fringing and patch reefs on the southeastern coast of St. Lucia, the area of most prolific growth on that island. Adams (1968) dealt with reefs on the leeward side of St. Vincent, and Lewis (1975) reported on the shallow reefs around the Tobago Cays of the Grenadine Islands which occur on a shallow bank.

MacIntyre (1972) examined "submerged reefs" off several Lesser Antilles islands by bathymetric profiling, dredging and bottom photography. These submerged reefs appear as a wide ridge rising above adjacent bottom areas at the margin of insular shelves at depths below 15 m. He believed these reefs are recent features, possibly having been formed since the rise of sea level after the last glaciation, but having been submerged sufficiently far by rising sea level that their growth no longer exceeds destruction caused by boring organisms.

The island of Barbados, lying nearly 160 km to the east of the main chain of the Lesser Antilles, has been relatively well studied. Lewis (1960) has dealt with near-shore coral communities and (1965) deeper water reef communities. MacIntyre (1968) charted the distribution of marine communities along a portion of the western coast of Barbados and also (1967) reported on possible submerged reefs off this coast.

The islands off the Venezuelan coast have well developed reef structures, those of the Dutch islands (Curacao, Bonaire and Aruba) being reported on most extensively through work carried out by the Caribbean Marine Biological Institute in Curacao. Various aspects of the marine environments of Margarita Island were discussed by Rodriguez (1959) and by Work (1969) for Los Roques. Roos (1964, 1967, 1971) has discussed distribution of stony corals in the Dutch islands. These islands are characterized by a very narrow island shelf which then slopes away steeply. Corals occur here to considerable depths (to over 60 m) and also fringe much of the coastline.

Along the northern coast of South America the reefs of the Venezuelan portion are poorly developed compared to insular areas offshore. Strong upwelling and the discharge of rivers into this region make conditions less than ideal for reef development. The stony coral fauna is limited, Olivares (1973) listing 22 species. Olivares and Leonard (1971) have also dealt with stony corals in this area.

The reefs of the Colombian coast are somewhat more diverse but are strongly influenced by the discharge of rivers, particularly the Rio Magdalena. Near Cartagena a layer of brackish, turbid water overlies clearer seawater and reef development is surprisingly good although severely limited in depth. Reefs of dense *Agaricia agaricites* occur at 5 to 10 m off the western side of Boca Chica, one of the openings of the Bahia de Cartagena. Low reefs of massive corals occur off Tierra del Bomba on a gently sloping bottom. A distinct change in slope occurs at 10 to 15 m, with the bottom still with some coral cover sloping down at 15-20°. Abundant coral cover occurs on the landward side of Banco de Salmadina, a shallow (5 m) bank located approximately 10 km offshore. On the landward side this bank slopes downward at 20-30° and has coral development on portions to at least 45 m. In other adjacent areas of this bank only a sandy slope occurs below 15 m. The seaward slope of this bank has not been examined.

Reef development is limited at Santa Marta, perhaps more influenced by the Rio Magdalena than the Cartagena area or, as Antonius (1972b) suggests, by occasional cold upwelling. The coverage of the bottom by stony corals is sparse and low, with the limit of coral growth at about 24 m. The bottom slopes steeply from shore along the rocky section of the coast and the reef is hardly more than a narrow, thin veneer on this shore. Antonius (1972b) examined three areas around Santa Marta and reported 33 species of scleractinian corals. Geyer (1969) also listed species from one of these areas.

The best reef development in Colombian waters occurs on the offshore islands, Islas del Rosarios, located west of Cartagena. Pfaff (1969) recorded 48 species of stony corals from these islands. A layer of turbid surface water 2-3 m thick still occurs at times, but the oceanic water below is quite clear. Shallow reefs extend out varying distances from these islands, with the shallow shelf being narrowest (0.5 km) on the leeward side of the islands. Near Isla Grande a sharp dropoff occurs on the leeward side of the island at 15 m and continues nearly vertically to 60 m, at which point the bottom begins to become level on a broad plateau of about 90 m depth on which the islands rest. Limited coral development occurs to at least 60 m, and *Montastrea cavernosa* is most abundant at that depth. Isla Tesoro sits 5 km from the other members of the group and rises separately from the 90 m plateau. On its windward (northeastern) side the bottom slopes progressively steeper, becoming nearly vertical about 1 km offshore at nearly 30 m depth then abruptly leveling at 70 m depth. A number of overhanging ledges of 10 m height occur, but there is no substantial development of caves.

The Caribbean coast of Central America is long and varied, including the shores of Panama, Costa Rica, Nicaragua, Honduras, Guatemala, Belize and Mexico. The reefs of this area vary remarkably from structures which have no equal along other tropical North Atlantic continental shores to a complete lack of reefs in certain areas because of the input of large rivers and unsuitable substrate conditions. While areas have been well examined, much remains to be done along this often desolate coastline.

The reefs of the Panama coast, particularly near the Caribbean mouth of the Panama Canal and in the San Blas Archipelago, have been discussed by Porter (1972, 1974b). A well developed algal ridge occurs at many of the San Blas Islands (Glynn 1973c), an area also of lush coral growth. The number of stony coral species occurring in the San Blas is nearly equal to the richest area known, Jamaica. Surprising is the lack of pillar coral, *Dendrogyra cylindricus*, in the Panama-Colombia area (Porter 1972).

Further west, near the Caribbean entrance to the Panama Canal, reef growth is limited to shallow (less than 15 m) depths by the usually turbid water. Near Portobello Bay antipatharians occur at depths as shallow as 8 m but few stony corals occur below 12 m.

No scientific reports exist for the reefs along the Panamanian coast west of the canal or for the Costa Rican, Nicaraguan or Honduran coasts. Long stretches of this shore lack reefs, such as near the Rio Tortuguero in Costa Rica. The continental shelf off Nicaragua becomes broad, reaching well over 160 km in width for much of the coastline. This shelf is spotted with small islands, none of whose reefs have been described in the technical literature. The Islas de la Bahia off the Honduras coast (Utila, Roatan, Bonacca) have well developed reefs that have never been described.

The reefs of Belize have been discussed fairly thoroughly by Stoddart (1962, 1963, 1965, 1969, 1974), Thorpe and Stoddart (1962), Matthews (1966) and Glynn (1973c). An extensive barrier reef, the longest in the Atlantic, occurs on the continental margin with lush coral growth, and numerous reef structures exist in the broad lagoon. Three island groups, the Turneffe Islands, Lighthouse Reef and Glovers' Reef,

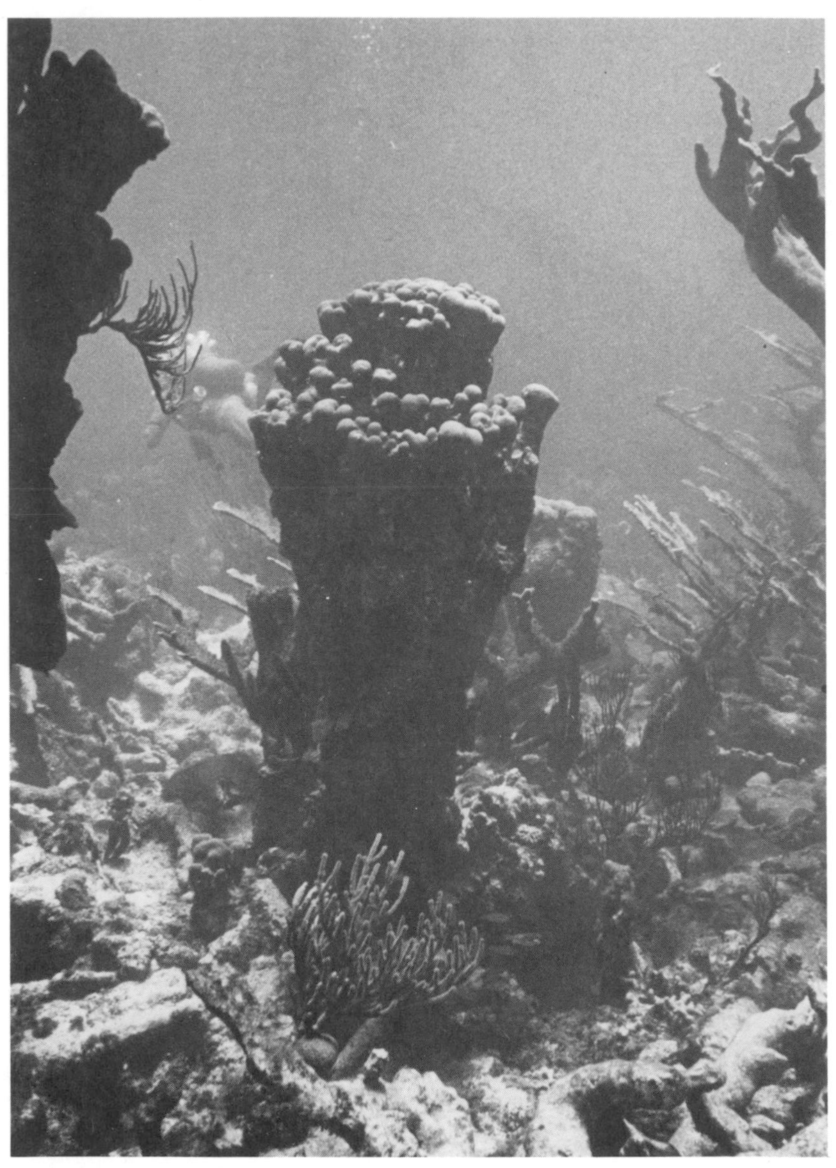

Colony of columnar-knobby *Montastrea annularis* in which the outer columns have been destroyed leaving only the central core of the colony still growing. Photographed at Buck Island, St. Croix, Virgin Islands at 10 m depth.

The east coast of Barbados has a series of beaches and rocky areas exposed to heavy surf from the open Atlantic ocean.

This view was taken looking out the shoreward end of a typical buttress cave on a Jamaican reef. The cave opens in an area of abundant corals and runs seaward a distance of about 100 m. Jamaica, Rio Bueno, 6 m depth.

occur offshore, separated from the continental shelf by deep water. These have been described as atolls by Stoddart (1962). The effect of a hurricane (Hattie - 1962) on reefs and subsequent recovery of those reefs are well documented for this area by reports of Stoddart (1963, 1965, 1969, 1974), the best analysis of such an event yet produced.

Fringing reefs occur along the Caribbean coast of the Yucatan Peninsula north of Belize, but no technical reports have been published regarding this portion of the Mexican coast. The island of Cozumel, a short distance off the Yucatan coast, also has well developed reefs which have not yet been well examined scientifically.

A large number of islands and banks occur in the western Caribbean. Milliman (1969) discussed four atolls (Courtown Cays, Albuquerque Cay, Roncador Bank and Serrana Bank) in the southwestern Caribbean and felt they occurred in similar environmental conditions to those of Pacific atolls. All four had similar ecological zonations across their reef flats.

The only descriptive information on the small island of San Andres is that of Geister (1972), who briefly described an offshore barrier reef on the windward side and a submarine cliff at 10 m depth on the leeward side, and that of Milliman and Supko (1968).

Isla de Providencia has a well developed barrier reef around a northward extension of its shallow bank and numerous patch reefs inside the lagoon, but it has never had its reefs described in the technical literature.

The banks of the western Caribbean region have coral growth of great variability but are little explored. Places such as Serranilla Bank, Bajo Nuevo and Misteriosa Bank are known only as hazards to mariners and their place in the scheme of Caribbean coral reef ecology is a question mark. The tiny Swan Islands north of Honduras have a gently sloping bottom surrounding them with some well developed patches of reef in shallow water.

The final island group on the circuit of the Caribbean is the Cayman Islands, located over 160 km south of Cuba. The group consists of three islands: Grand Cayman, the largest (measuring 31 by 10 km), and two smaller islands, Cayman Brac and Little Cayman (located 6 km apart but over 110 km from Grand Cayman). The reefs of Grand Cayman have been described by Roberts (1971, 1974). He reported the best reef development to exist on the northern shore. Only the protected western coast had patch reefs, while all other shores had a fringing reef.

Numerous small patch reefs do exist inside the shallow lagoon on the eastern end of Grand Cayman. The reefs below 30 m on the northern coast closely resemble many of the deep reefs of Jamaica's northern coast in having a particularly lush growth of coral and sponges at such depths on a steep slope with occasional broad sand channels carrying sediments to the depths.

No information exists in the scientific literature regarding the reefs of Cayman Brac or Little Cayman. The best reef development is again on the long northern shore of these spindle-shaped islands. The bottom slopes away gently from the rocky shore, often with a small irregular fringing reef a short distance offshore. The bottom then drops quickly, usually reaching the "drop-off" within 0.5 to 1 km of shore. Coral development on this outer slope is excellent but somewhat low in profile. Spectacular development of reef caves occurs along the very shallow (10-20 m) "drop-off" of the Bloody Bay - Jackson Point area. These caves contain a typical West Indian cavernicolous fauna of sclerosponges, antipatharians, hydroids, etc. and are only slightly surpassed in extent by the caves of certain sections of Jamaica's northern coast.

A bank reaching to within 30 m of the surface occurs about 20 km west of Grand Cayman, where it rises out of deep water. This bank possesses both rich coral and sponge communities and barren areas consisting of little more than rocky rubble with occasional massive sponges.

Reef offshore from Discovery Bay, Jamaica, with a shallow lagoon behind it. A great deal of the recent work on Caribbean reefs has been carried out on this reef.

Seaward side of the reef crest at Discovery Bay, Jamaica. Abundant *Acropora palmata* can be seen in the shallow water.

A back-reef area at Discovery Bay, Jamaica. Such areas are exposed to high light intensities with little wave action.

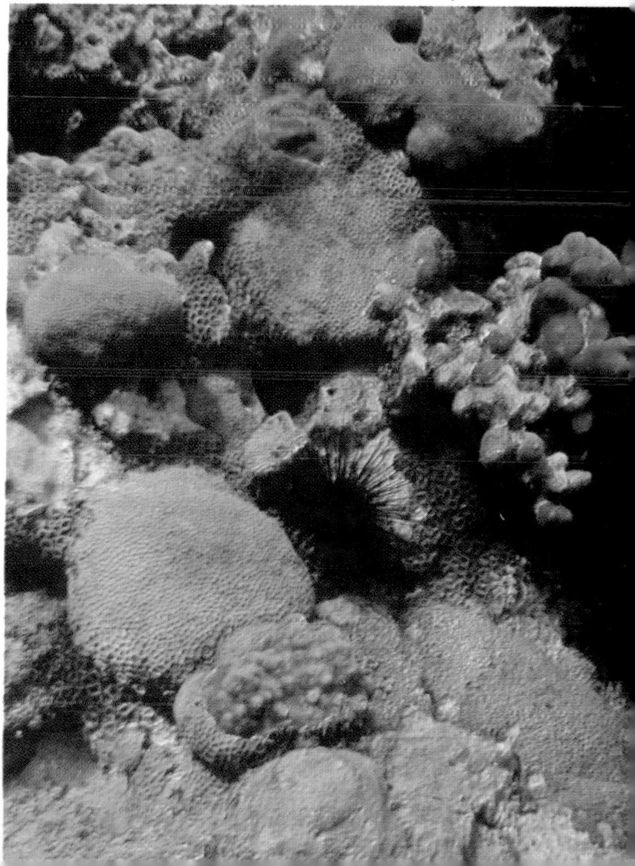

View looking downward onto a typical assemblage of back reef organisms at Discovery Bay, Jamaica. A variety of coral species and the sea urchin *Diadema antillarum* are visible.

Typical view of reef environment on the windward side of Hogsty Reef Atoll, Bahamas, at a depth of 15 m. The coral colonies are relatively low in profile, possibly due to the effect of the occasional stormy seas at this locality.

Reef Areas Adjacent to the Caribbean Sea

THE BAHAMA ISLANDS AND THE TURKS AND CAICOS ISLANDS

While not encompassed by the Caribbean Sea, the Bahamas and the Turks and Caicos Islands are usually considered by geographers to be included in the West Indies. They certainly are part of the Caribbean area in terms of their marine communities and their general environmental conditions.

Stretching from near the southern Florida coast to north of the western end of Hispanola, the Bahama Islands constitute an area covering nearly 900 by 500 km; the Turks and Caicos Islands, a separate political entity, are a continuation of this archipelago off the northern coast of Hispanola. A series of banks, including Silver and Navidad Banks, marks the end of this group near the eastern end of Hispanola.

Only a modest amount of information has been published regarding the reefs of the Bahama Islands. These reefs are rich in their diversity and populations. A number of types are encountered, some of which are distinctive from typical Caribbean reefs.

The islands of the Bahamas occur largely on four shallow-water carbonate banks (the Little Bahama Bank, the Great Bahama Bank, Cay Sal Bank and the Acklins Island Bank) and to a lesser extent as isolated islands separated by deep water without sizable adjacent banks. There are estimated to be nearly 3,000 islands, cays, rocks and other portions of land above water comprising an area of 15,000 square km. The area occupied by the submerged banks (less than 20 m depth) is much greater. Two deep-water basins, the Tongue of the Ocean and Exuma Sound, extend deeply

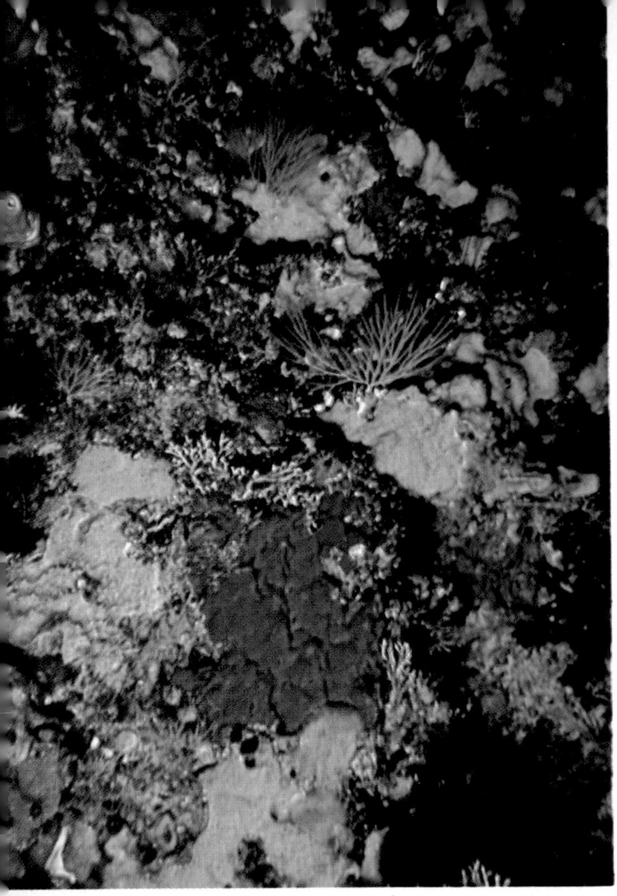

Typical scene over the "dropoff" at Elbow Cay, Cay Sal Bank, Bahamas. A variety of deep-reef gorgonians are visible. Depth 55 m.

The "hole in the wall" at the southern end of the Great Abaco Island is so named from a rock formation on the point. Beyond this spit of land the swells of the open Atlantic pound the eastern shore of the island.

The northern coast of Curacao is exposed to high seas much of the year, and reef development is less extensive than on the more sheltered south coast. A raised platform of rock plunges down 10-15 m to the sea along most of this coast.

Aerial view of the north coast of Curacao with waves beating against the rocky shore and raised shorelines visible further Inland.

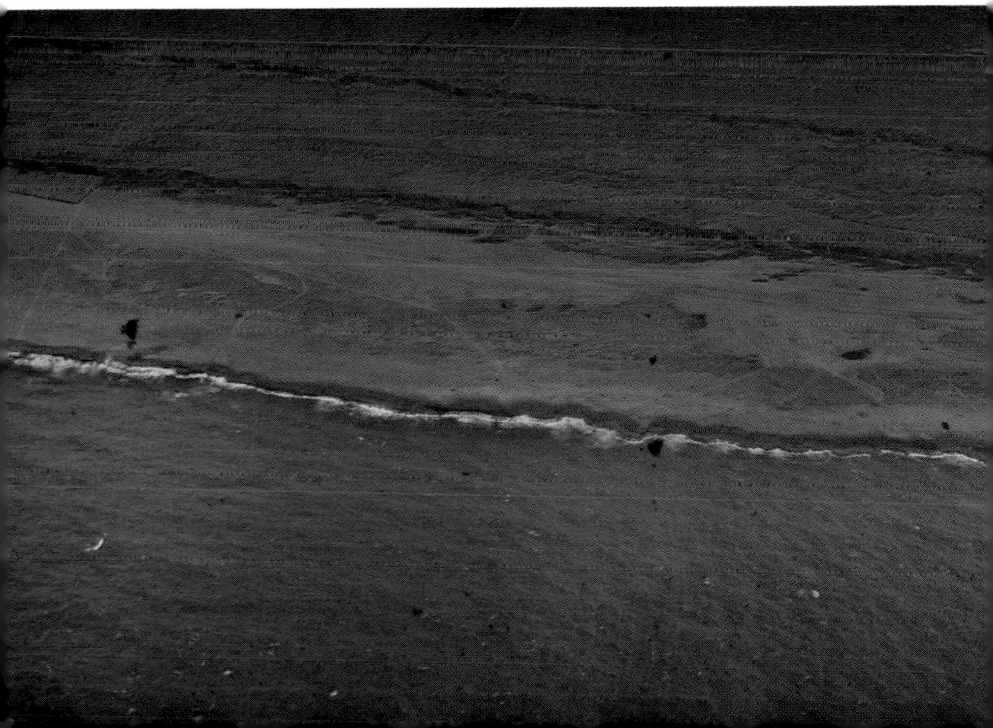

in from the outer margin of the Great Bahama Bank. These deep basins (maximum over 1700 m) greatly increase the shallow-water coast of this bank.

The reefs of the Bahamas typically occur on the margins of these banks, the central portions of the banks comprising basically featureless plains. There is little input of terrigenous (land originated) sediments into the sea in this area and consequently the clarity of the water is generally quite high. These banks were formed by upward growth of reefs on the margins and production of calcareous sediment in their more central portions while a gradual subsidence occurred, producing a layer of carbonate material as much as 3 km thick.

Water temperature conditions are less optimal in parts of the Bahamas for reefs than in many Caribbean locations. At the northern extremes of the Little Bahama Bank (27° N) the winter water temperatures are near 21° C, approaching the lower limits for reef corals.

Only a few studies have dealt in detail with Bahamian reefs. Storr (1964) described reef growth on the eastern side of Abaco Island, Little Bahama Bank. Bunt *et al.* (1972) dealt with an area on the southern side of Grand Bahama Island, also on this bank. Newell *et al.* (1959), Squires (1958), and to a lesser extent Steinberg *et al.* (1965) have described the area around the Bimini group. The U.S. Naval Oceanographic Office (1967) has briefly described the reef tract on the eastern side of Andros Island adjacent to the Tongue of the Ocean. The only area in the southeastern Bahamas that has been reported on is Hogsty Reef, an atoll-like structure (Milliman 1969). These previous studies have largely dealt with the biology of the reef only in shallow water, and practically no published information exists on species composition or abundance below 20 m depth.

The reefs of the Bahamas consist of several different types, some of which are distinctive and perhaps unique to this area. The marine environments from one side to another of an island may vary tremendously. On shores exposed to the open Atlantic, such as along the eastern coast of Eleuthera and Cat Island, reef development is relatively restricted by the force of the ocean swells. These reefs have a more flattened aspect with fewer coral structures rising high above the substrate than those occurring in areas of similar conditions, but more protected from extreme wave action.

The second largest barrier reef in the Atlantic occurs along the eastern shore of Andros Island 1 to 6 km off the coast. U.S. Naval Oceanographic Office (1967) has some detailed drawings of the morphology of this reef.

Patch reefs are abundant in many areas and often reach tremendous size. North of Green Cay (near Nassau, not on the Tongue of the Ocean) the patch reefs rise from 15-20 m depth to reach within 10 m of the surface. They may be 30 m or more in diameter but are separated by broad areas of open sand. Caves are also developed in these reefs.

The reefs occurring on southerly facing shores, such as southern Grand Bahama Island, the southern Berry Islands and south Cat Island, are typically like those described by Bunt *et al.* (1972), a series of rocky fingers with deep, intervening sand channels oriented to seaward at depths of 10 to 27 m. Shoreward groups of rocky patches exist, similar to the fingers in bottom cover but as isolated clumps. The percentage of coral cover is relatively low and a sandy-rubble terrace typically exists below a sharp drop from 21 to 27 m. This deep terrace is variable in width, averaging nearly 100 m, slopes slightly and has a rocky substrate occurring again at nearly 35 m depth. The bottom then slopes away steeply, becoming nearly vertical at 45 to 60 m depth.

An unusual type of reef exists in the areas where deep water occurs near shore on the leeward sides of islands. Such situations occur on the southern end of Eleuthera from Powell Point to Eleuthera Point, the southern end of Long Island and the western edge of the Acklins Island Bank, exclusive of the "bight of Acklins." The "drop-off" usually is within 1 km of shore. A nearly continuous reef, varying in width from 20 to about 150 m, occurs along this edge with a sandy slope with scattered patch reefs occurring landward. The "edge reef" rises above the sand slope as much as 8 m and is reasonably flat on its upper surface at a depth of 20 to 27 m, then drops precipitously at its outer margin to considerably more than 100 m. This "edge reef" seems to dam the

Divers on the deep reef at Discovery Bay, Jamaica. Such areas have lush coral growth and an endlessly fascinating variety of organisms. Depth 30 m.

Another view of the deep reef at Discovery Bay, Jamaica. Corals typically have a plate-like growth form at depths of 30 m and below.

A strongly undercut rocky shore at Elbow Cay, Cay Sal Bank, Bahamas. Waves strike this undercut with tremendous force and cause spray to be forced out of it for considerable distances.

Small Carrie Bow Cay is one of several islands on the barrier reef off the coast of Belize (British Honduras). It sits directly on the reef and has been used by personnel from the Smithsonian Institution to study coral reefs.

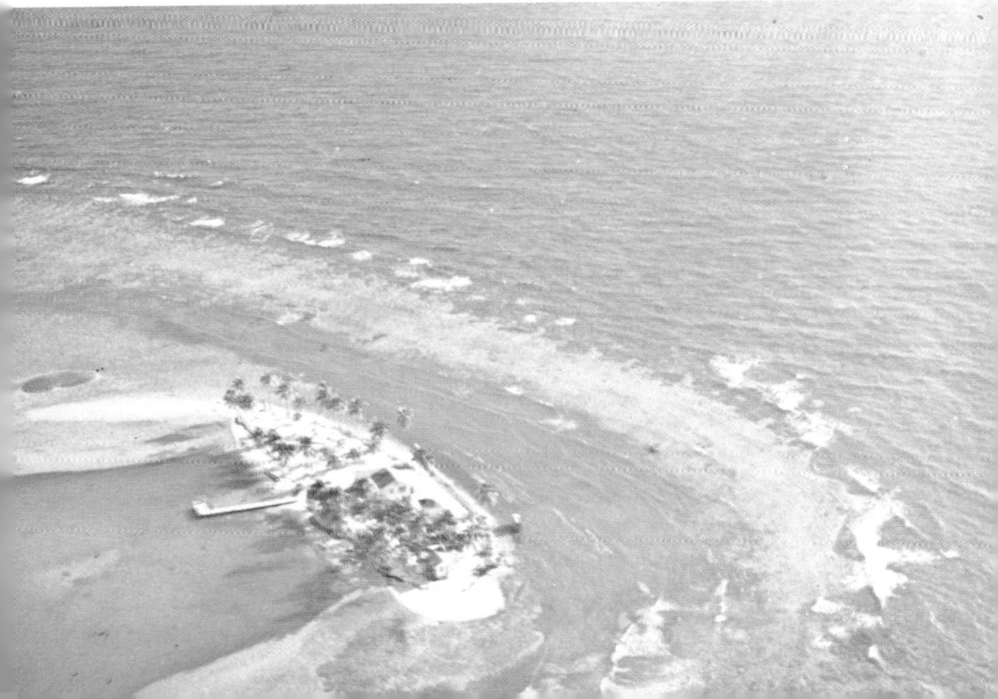

sloping sediment behind it; where deep channels cross it, sizable amounts of sediments are transported through the reef and over the edge. They have only a moderate density of coral on their upper surface, with the numbers decreasing on the landward side. The vertical to nearly vertical face on the outer edge may have abundant populations of deep-water corals and antipatharians.

No reports are available on the reefs of either the Caicos Bank or Turks Bank. A well developed "edge reef" occurs on the western side of Providenciales Island, Caicos Bank, with the shelf edge break occurring at 23 m. On the western side of Grand Turk Island, Turks Bank, no well developed reef structure was found at the anchorage or near the northern tip of the island.

THE GULF OF MEXICO, FLORIDA AND BERMUDA

The southern Gulf of Mexico is totally tropical and has well developed reefs. The structures on the Campeche Bank off Yucatan have been reported on by Kornicker *et al.* (1959), Kornicker and Boyd (1962), Logan (1969) and Logan *et al.* (1969). Alacran Reef is an atoll-like structure but rises only from the depths of the Campeche Bank (about 50 m) rather than from oceanic depths. The fauna and flora of these reefs is West Indian in character.

Rigby and McIntire (1966) reported on reefs near Tampico, Mexico, and some reef-like structures exist off Veracruz (Villalobos 1974, Rannfeld 1972). The stony coral fauna of these areas is limited, perhaps comparable to that occurring in Bermuda.

The northern-most coral communities in the Gulf of Mexico occur some distance off the Texas coast. Bright and Pequeqnat (1973) report on a variety of aspects regarding these communities at the "Flower Gardens Bank," while Tresslar (1973) recorded 17 species of stony corals from this area.

The Florida reef tract extends from the Dry Tortugas, a group of islands and banks west of Key West, to north of Miami. There are considerable changes in the character of these reefs along this tract related to local and general environmental conditions. Numerous reports of the animals and plants of the Dry Tortugas were produced by the Tortugas Marine Laboratory of the Carnegie Institution from 1904 to 1942, but no work covering the distribution and ecology of the reefs at Tortugas was written. In general the reefs are richer in West Indian biota than even the nearby reefs flanking the Florida Keys and in certain instances seem nearly the last area of suitable conditions for certain species. Extensive dense beds of *Acropora cervicornis* occur around some islands and well developed patch reefs are found on the banks. A barrier reef type of formation exists on the southeastern side of the main bank, with abundant coral on its deeper portions at 20-30 m.

Further east the first reefs of the Florida continental shelf occur off the Marquesas Keys, still west of Key West. The shallow fore reef appears similar to most locations along the Florida Keys, but abundant corals and some antipatharians were found on a slightly sloping bottom from 30 to 40 m.

Along the Florida Keys there is a series of shallow outer reefs at 5 to 10 km from shore which parallels the string of islands. These reefs are best developed (particularly near Key Largo) where few or no passages exist between the islands which allow water from shallow Florida Bay to pass over the reefs. The water from Florida Bay is of variable salinity, transparency and temperature and at times is unsuitable for reef corals.

At most locations along the Florida Keys a slope occurs from 20 to nearly 27 m which is steeper than either shallower or deeper slopes and which has an abundant coral fauna. Below this "deep reef" few corals occur and the bottom is generally a gently sloping sediment plain with occasional rocky outcrops.

Inside the offshore chain of reefs are numerous patch reefs, some with lush coral development (particularly species of *Montastrea*) and high vertical relief, in a "lagoonal" area of 10 to 12 m maximum depth called Hawk Channel. Many of these patch reefs are in poor condition due to increased dredging activities in their vicinity.

Various workers have discussed the reefs of this area. Shinn (1963) described the "spur and groove" formations

Two tests of the foraminiferan *Homotrema rubrum* on the undersurface of a piece of reef rubble. Puerto Rico, La Parguera, Laurel Reef, 1 m depth.

The calcareous sponge *Leucosolenia canariensis* on a steep reef face next to the coral *Montastrea cavernosa.* Jamaica, Rio Bueno, 20 m depth.

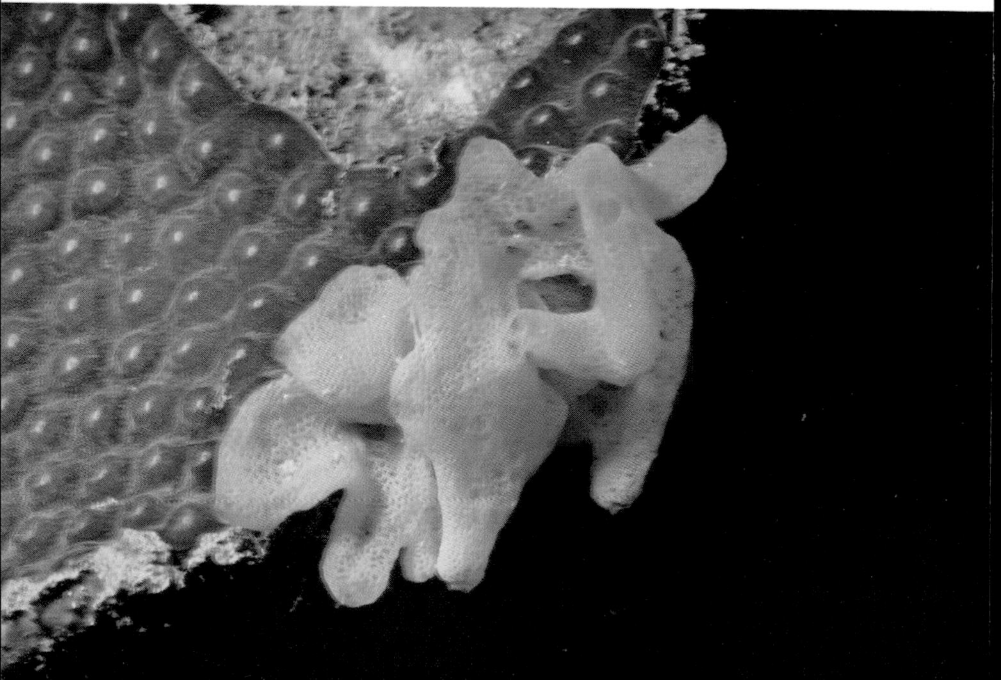

A small group of calcareous sponges, *Scypha barbadensis,* on an overhanging deep-reef face along with the whip-like black coral *Stichopathes lutkeni* and the orange filamentous gorgonian *Ellisella* sp. Bahamas, Elbow Cay (Cay Sal Bank), 45 m depth.

Large examples of *Chondrilla nucula* on a steep rocky face. Puerto Rico, Mona Island, north wall, 15 m depth.

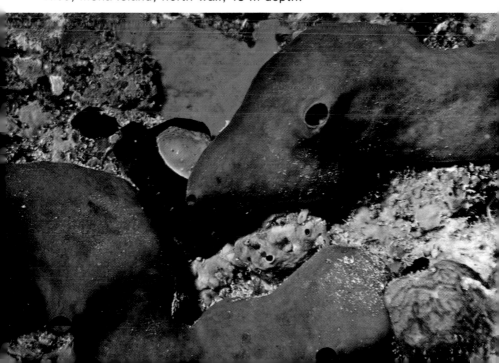

occurring on the outer reefs. Ball *et al.* (1967) discussed the zonation of Florida Keys reefs and the effect of hurricane Donna on that area. Kissling (1965) dealt with coral distribution in an inshore area. Jones (1963) reported on seasonal and daily variation in environmental conditions on a patch reef south of Miami and Opresko (1973) examined the distribution of gorgonians in this same area. Hein and Risk (1975) dealt with bioerosion of coral heads in the Florida Keys.

The shallow offshore reefs cease at Fowey Rocks, 15 km south of Miami, but reef communities continue up the southeastern Florida coast at least to Palm Beach and perhaps farther. Glynn (1973c) reports that Carysfort Reef (50 km south of Miami) marks the northernmost reef with flourishing *Acropora* corals. Goldberg (1973a) reported on such communities off Boca Raton where the bottom slopes gradually from shore and a series of progressively deeper coral communities is found.

Although isolated colonies of a few species of hermatypic corals may occur as far north as North Carolina (MacIntyre and Pilkey 1969) on the continental shore, the reefs of Bermuda (32° N.) occur farther north than any others in the world. These reefs owe their existence to the warming influence of the Gulf Stream current, and the hermatypic coral fauna is limited to only about 20 species. The genus *Acropora* does not occur there, although most other genera which are major reef constructors do occur. A large variety of reef types exist in Bermuda, from algal "cup" reefs or "boilers" to deep, rugged structures. Antipatharian corals are abundant at 40 to 70 m, but coral growth ceases at about 40 m.

The patch reefs of Bermuda possess well developed caverns (Garrett *et al.* 1971) and the shallow water ahermatypic corals have been reported on by Wells (1972).

BELOW THE REEF

The depth at which stony coral growth ceases is a function of many factors, and the communities which occur below these depths in areas of optimal reef growth are just beginning to be examined by methods allowing accurate *in situ* observations. Usually this environment is steep and rugged and nearly impossible to sample by the conventional methods of dredging and trawling. The development of deeper diving techniques and, more importantly, small research submarines has allowed investigators to see and sample directly this habitat.

In the Caribbean area, reefs in Jamaica, Belize and the Bahamas have been examined using submersibles to depths around 300 m. The results of these investigations indicate that there is still a tremendous amount to learn about the "sub-reef" habitat (from 60 or 90 m to 300 m).

Lang (1974), Hartman (1973), Lang *et al.* (1975) have examined the "sub-reef" habitat at Discovery Bay, Jamaica. Ginsburg and James (1973) and Lang *et al.* (1975) reported on various areas in Belize to 300 m depth. Porter (1973) described an area in the Bahamas, and the U.S. Naval Oceanographic Office (1967) has figured the "sub-reef" face to about 100 m depth at Andros Island Bahamas.

Many invertebrates thought to be rare as a result of dredge samples have been found common at depths below 100 m. The large pleurotomarid gastropod *Entemnotrochus adamsoniana* (Yonge 1973a) is such a creature, and the tiny stalkless crinoid *Holopus rangi* can be unexpectedly abundant (Lang 1973).

Macroalgae, particularly members of the genus *Halimeda*, can occur commonly to 70 m and rarely to nearly 100 m. The calcareous plates of *Halimeda* then form a primary component of the sediment occurring in the "sub-reef" environment.

One of the most interesting findings of these investigations is that, at least in Jamaica, the sclerosponges can act as primary framework constructors of reefs at depths below those of stony corals (70-105 m, Lang *et al.* 1975). Hermatypic corals, even in the clearest water, do not form sizable reef structures below 60-70 m and are rare as individual colonies below 80 m.

COLLECTION OF SPECIMENS

It is an intended purpose of this volume to discourage the wholesale collection of reef invertebrates and plants and

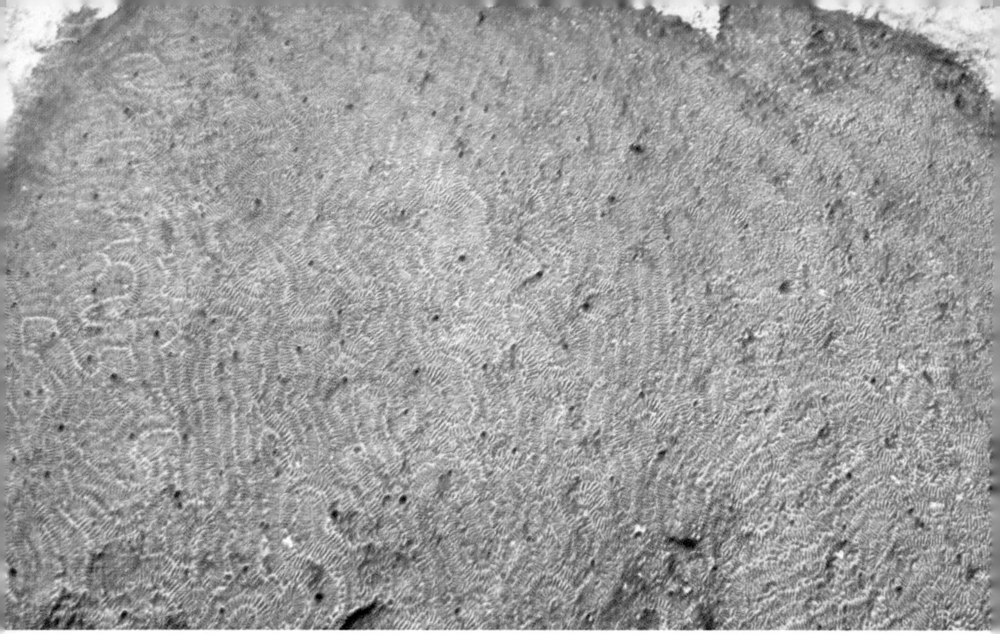

An unidentified species of *Cliona* which has completely taken over a coral head, probably *Diploria strigosa.* The septa of the coral are still visible, but the oscula of the sponge are apparent on the surface. Puerto Rico, Mona Island, Playa Sardinera, 15 m depth.

The boring sponge *Cliona delitrix* has nearly completely covered this coral head. The areas where the red *C. delitrix* and its associated zoanthid *Parazoanthus parasiticus* do not occur appear occupied by another boring sponge. Puerto Rico, Desecheo Island, 12 m depth.

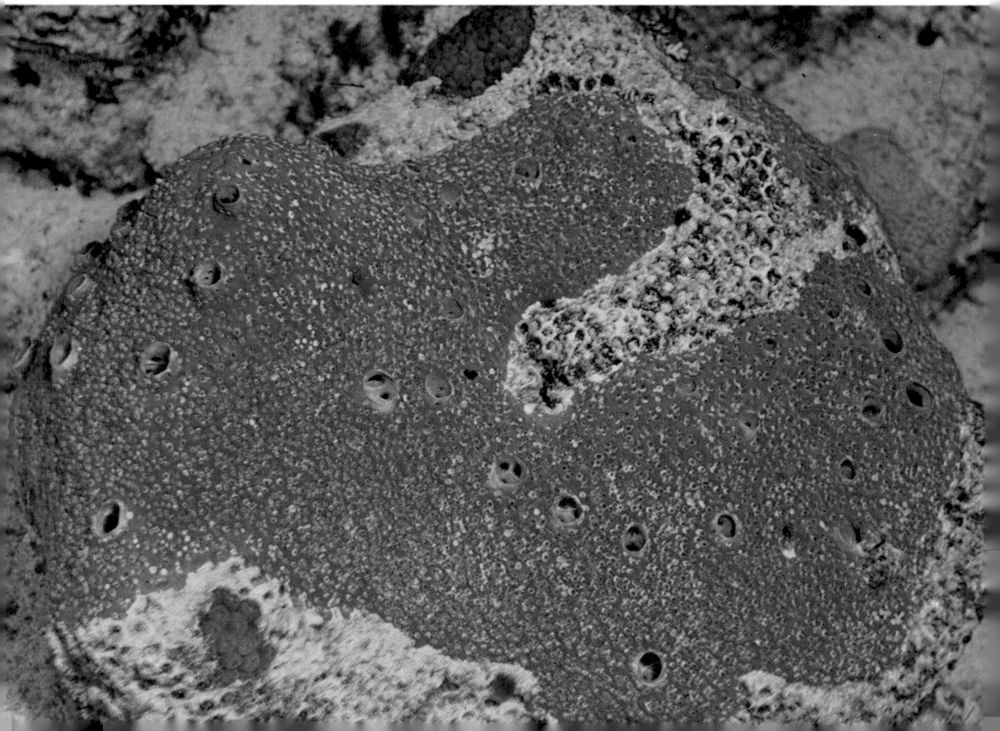

Tethya crypta on a rocky shelf with a considerable amount of sediment on its outer surface. Bahamas, Bimini, Turtle Rocks, 6 m depth.

Spherical orange sponge which is probably a member of the genus *Tethya*. Jamaica, Discovery Bay, 15 m depth.

to encourage observation and photography of the same without disturbing the natural environment. Reef areas, particularly those in resort or urban areas, are experiencing increased numbers of visitors and recreational use. The idea of taking portions of the reef home is totally inconsistent with a conservative attitude. For this reason it was decided that when possible all organisms in this book would be photographed *in situ* in an undisturbed condition. For example, all stony corals are illustrated as living specimens, many for the first time anywhere, rather than the bleached, dead skeletons figured in more classical publications. Obviously if it takes a dead skeleton to identify a stony coral, then a lot of living corals will be killed by interested observers simply to figure out what species they represent. The illustration of living specimens should eliminate this need in the course of routine identification.

The same is true for the gorgonians, black corals, molluscs and other groups which are today's curios sitting on the mantle and tomorrow's trash relegated to the closet. Such creatures are better left in their home, the sea, where they may thrive and reproduce, allowing others to enjoy them and ensuring that the reef will always be there. If you need to collect the forms and colors of the reef, do so in photographs.

Introduction to the Species Accounts

The descriptive accounts of individual organisms are arranged by phylum, with the animals preceding the plants. Within each phylum, species are considered by order and family. In many cases some information is included regarding the biology of these categories above the species level, but this book is in no way designed to serve as a textbook of marine invertebrate zoology or botany.

Each species account begins with the two-part scientific name of the organism, consisting of the genus and species, both in italics. The name(s) of the author(s) of the original scientific description of the species follows the scientific name. If the author's name appears in parentheses, the species was originally described in another genus and changed to the genus it is presently recognized under by a subsequent author. In some cases it is not possible to determine which species or even which genus a photographed specimen represents. In such instances this is noted by the lack of a specific name or the entire scientific name and, since no particular species is determined, no author is cited. Often in biological literature the date of the original description will be included after the name of the author(s), but in the present work this was not done as most of the papers containing the original descriptions of the organisms are not included in the bibliography. Reference to the more specialized works cited for each species or group will provide the references to many original descriptions for those interested.

Finally in some cases a commonly accepted (either among marine biologists or aquarists) common name is included, when such exists, although no published list of common names exists. In the many cases where no common

The burrowing sponge *Siphodictyon coralliphagum* has only its yellow-orange oscules exposed while the major mass of the sponge occurs beneath the surface of the coral head or rock substrate inhabited. Bahamas, Bimini, Turtle Rocks, 10 m depth.

The black color of *Adocia carbonaria* is distinctive and it can occur on reef areas adjacent to corals. Bahamas, Grand Bahama Island, 12 m depth.

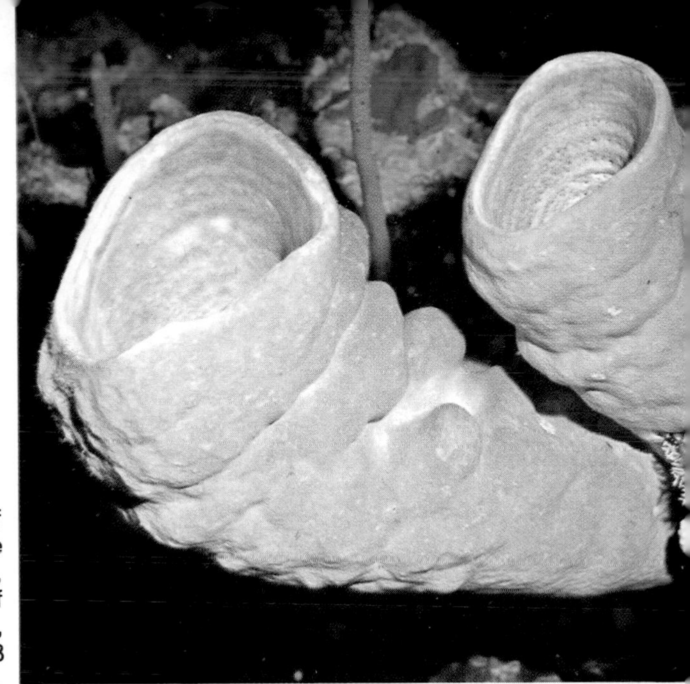

Two large tubes of *Agelus* sp. A, the trumpet sponge, on a vertical reef wall. Jamaica, Spring Gardens, 18 m depth.

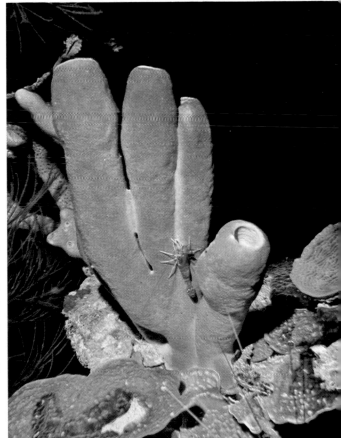

Tubes of the sponge *Agelus* sp. B photographed on the outer reef face at night with the shrimp *Rynchocinetes rigens* on its outer surface. Jamaica, Spring Gardens, 22 m depth.

name already exists, it has been decided not to attempt to introduce new common names. This task is better left to those preparing inclusive or monographic works for specific groups to avoid introducing inappropriate or confusing common names.

Depths, lengths and dimensions are given in the metric system. Depths are given in meters (m), with 1 m equalling approximately 3.3 feet. Other measures are usually in centimeters (1 cm = 0.01 m) or millimeters (1 mm = 0.001 m) and equal 2.54 cm per inch and 25.4 mm per inch respectively.

Rounded head of *Montastrea annularis* which has scrape marks on its surface produced by parrotfishes. While parrotfishes feed mostly on microalgae on rock surfaces, they occasionally sample the surface of stony corals.

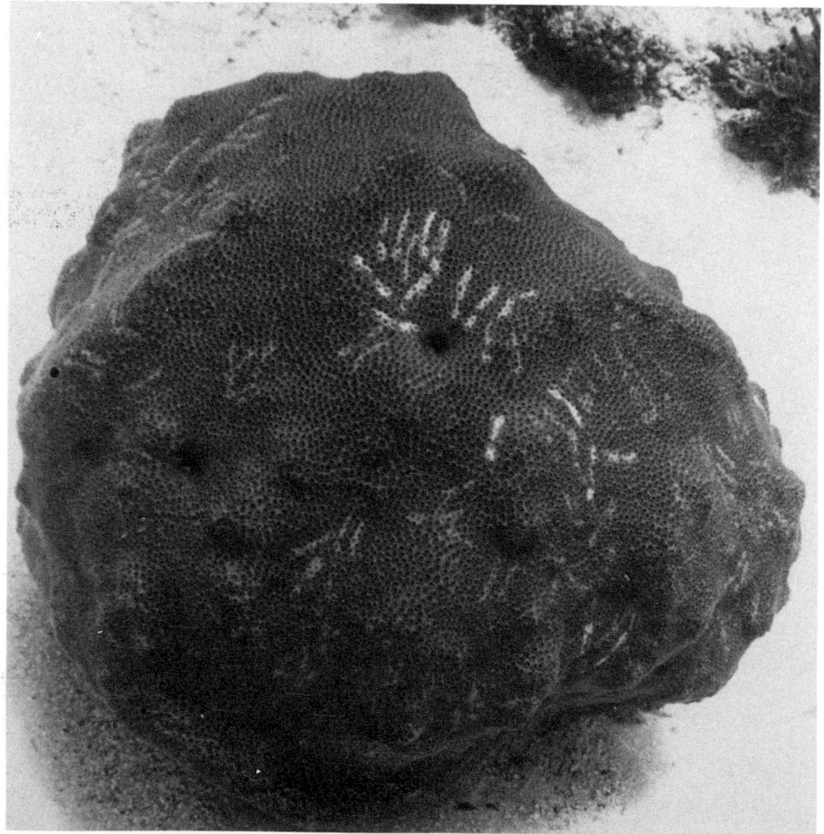

Phylum Protozoa

CLASS SARCODINA
ORDER FORAMINIFERA

Foraminiferans are single-celled protozoans having slender, often interconnected pseudopodia (protrusions of the cellular material) and being enclosed, except for the pseudopodia, in a simple or chambered shell (the test) composed largely of calcium carbonate. They are planktonic (pelagic) or bottom-dwelling (benthic); the benthic forms may or may not be attached to the bottom. Most species are small and, while extremely important to geologists due to their contribution to carbonate sediments and rocks, few are easily observed in the field. One species, discussed next, is an exception and is influential in the coral reef environment.

Homotrema rubrum (Lamarck)

The large red test of this foraminiferan, reaching up to 6 mm in diameter, is quite distinctive and may be found on the rocky undersurfaces of coral plates. The test takes on at least five morphological types, from encrusting or branching to the most abundant form, globose. A bare rock substrate or one only thinly covered with algae is needed for *H. rubrum* to begin growth; rock covered with boring sponges or encrusting bryozoans cannot be utilized. Single-celled animals and plants are utilized as food as evidenced by radiolarians and diatoms found in the digestive system of *H. rubrum.*

While occurring in the tropical regions of the Atlantic, Indian and Pacific Oceans, *H. rubrum* is so abundant on the reefs of Bermuda that the tests of dead animals, along with gorgonian spicules mixed with white coral and algal material, produce the famous pink beaches of that island. The test after death bleaches slowly from red to a lighter pink. The bathymetric range of *H. rubrum* includes all depths of suitable reefs from 3 to at least 20 m.

References: Emiliani, 1951; Rooney, 1970.

The brilliant orange sponge *Agelus* sp. C with a clump of the coral *Tubastrea aurea* growing on its surface. The occurence of the sponge and coral is unusual. Puerto Rico, Monito Island, 24 m depth.

Closeup view of the ostia and elongate oscules of *Agelus* sp. C. Puerto Rico, Monito Island, 20 m depth.

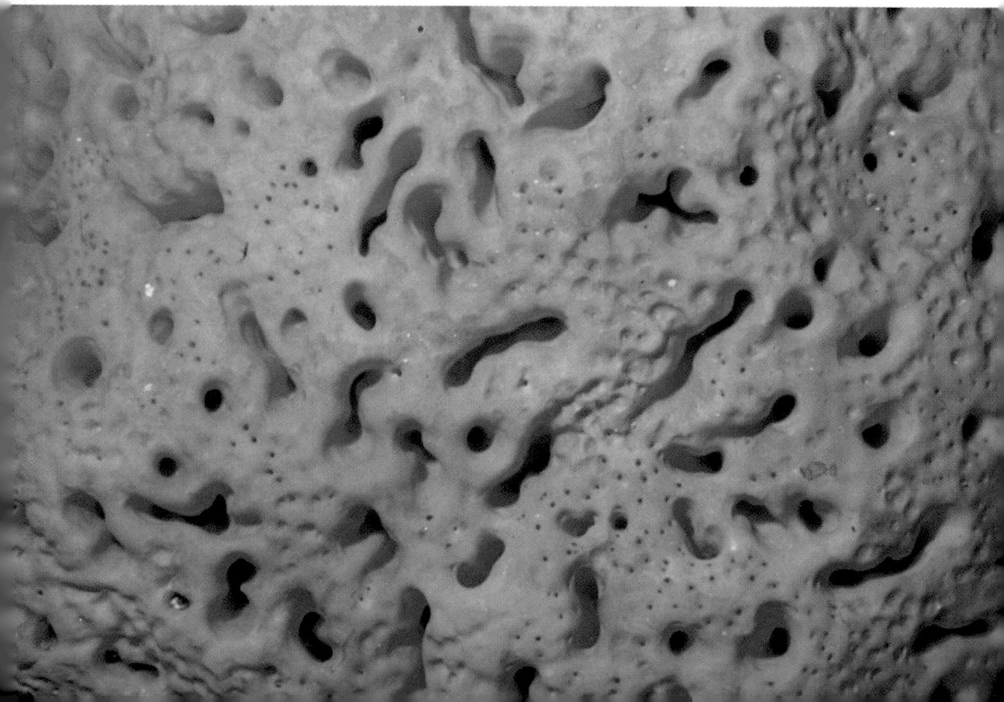

A pale specimen of the elephant ear sponge, *Agelus flabelliformis,* growing out from the seaward reef face. Jamaica, Spring Gardens, 27 m depth.

Tubular sponge, possibly *Agelus schmidti,* on a steeply sloping reef growing on the dead basal branches of the coral *Acropora cervicornis.* Puerto Rico, La Parguera, 22 m depth.

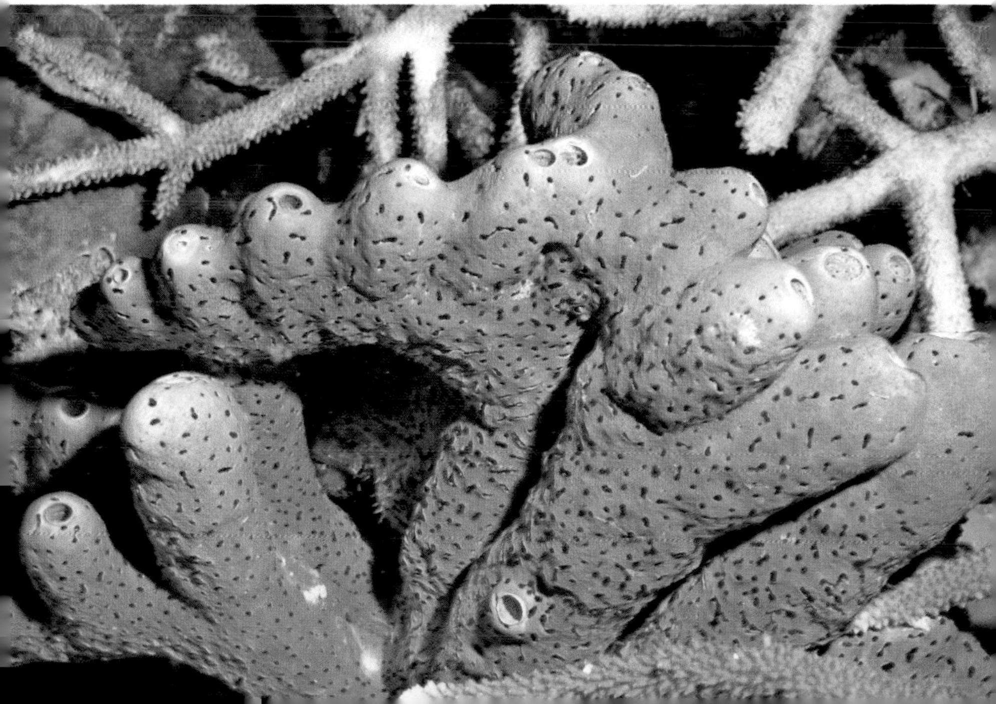

1) Large specimens of *Agelus* at 40 m depth at Discovery Bay, Jamaica.
2) Three-tubed specimen of *Verongia fistularis* at 15 m depth at Acklins Island, Bahamas.
3) *Verongia fistularis* at 10 m depth, Turtle Rocks, Bimini, Bahamas.
4) The shrimp *Lysmata wurdemanni* within the osculum of a *Verongia* sponge.

Phylum Porifera: Sponges

Sponges are multi-cellular animals typically attached to hard substrates and possessing various specialized cells but lacking organization of such cells into organs and tissues. Although their basic body plan is simple, some species attain surprising size (hundreds of pounds in weight out of water). All possess water current systems whereby water passes through channels in the body where food (bacteria, small planktonic organisms, larger organic aggregates) is filtered out. The currents are produced by the beating of the flagellae of collar cells (choanocytes) and enter through numerous openings called ostia (singular-ostium) and exit by one or more oscula (osculum).

Sponges are classified into four classes, three of which occur on Atlantic reefs, on the basis of the chemical composition and shape of their spicules (crystalline elements) and organic fibers comprising the skeleton. Differentiation of species is based on shape, color, resistance to tearing, smell, form and size of the spicules, surface texture, color when dried or in preservative, effect on human epidermis and a variety of other characters. They are perhaps the most difficult group of major reef animals to identify with certainty due to the variable nature of many species, the chaotic nature of the basic literature on the group and the scant attention that any, except the most common shallow water species, have received.

It is believed that sponges have been abundant in reef habitats since the Paleozoic (at least 200 million years). While they are ancient in their origin, sponges are considered a "dead-end group" and have not given rise, due to the basic limitations of their body plan, to any other present day group of organisms. This is not to underestimate their biological importance, however. In some areas of the reef the

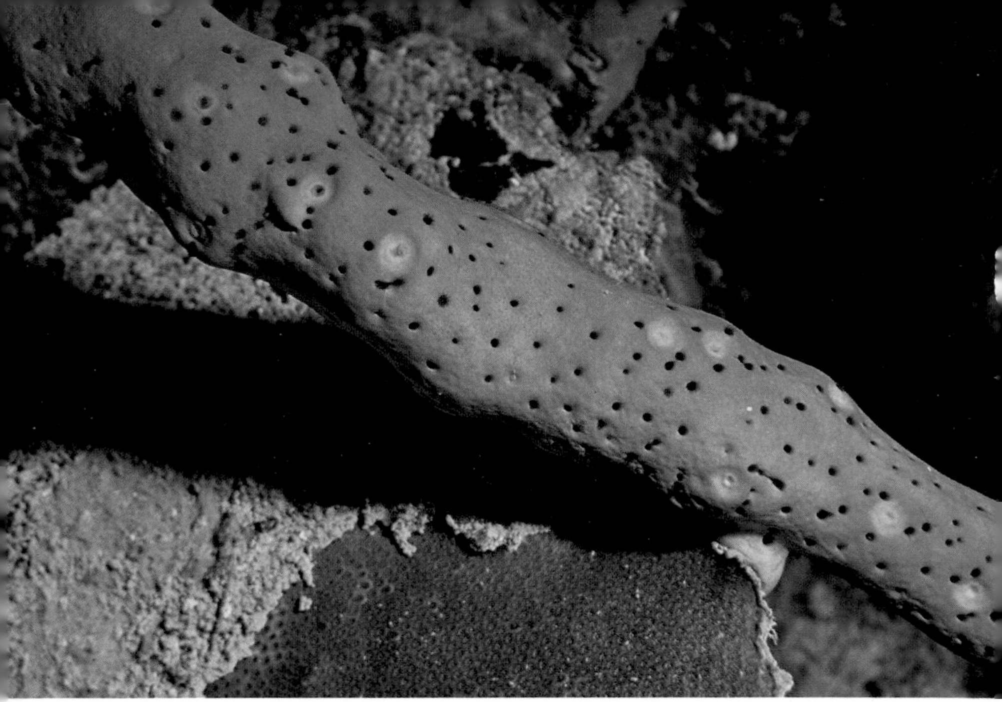

Cylindrical branch of *Agelus screptrum* which has a few small zoanthids growing on its outer surface. Jamaica, Rio Bueno, 20 m depth.

Branches of *Agelus screptrum* among corals and the filamentous black coral *Stichopathes lutkeni*. Jamaica, Spring Gardens, 24 m depth.

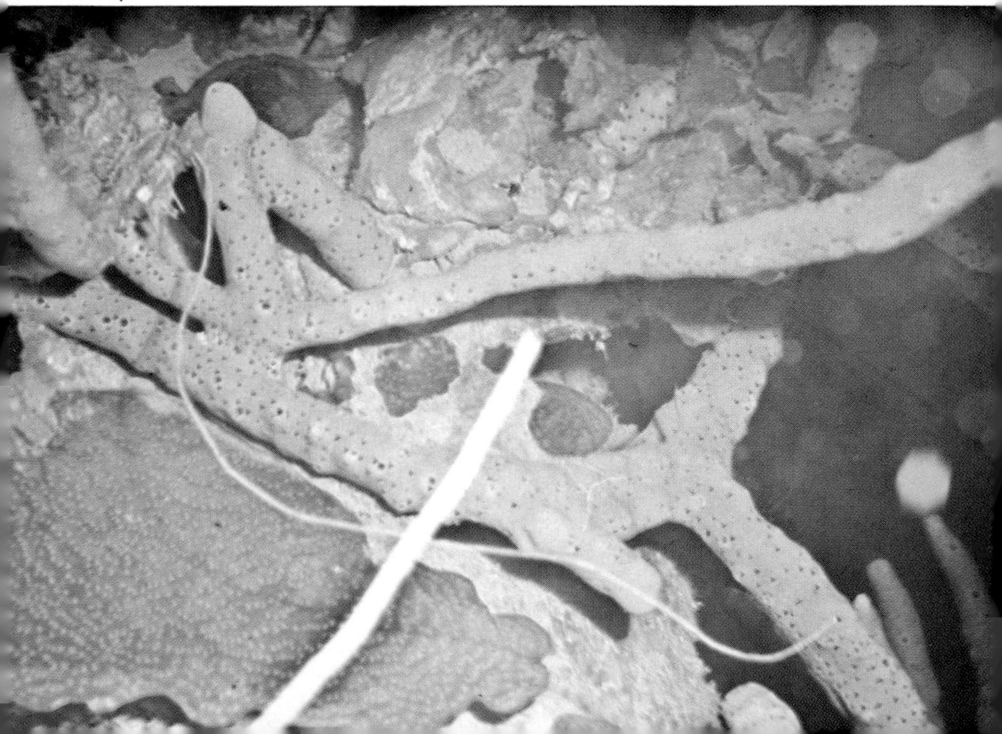

The conulose tubes of the sponge *Callyspongia vaginalis.* Puerto Rico, Desecheo Island, 12 m depth.

Closeup of the osculum of *Callyspongia plicifera*. Jamaica, Discovery Bay, 15 m depth.

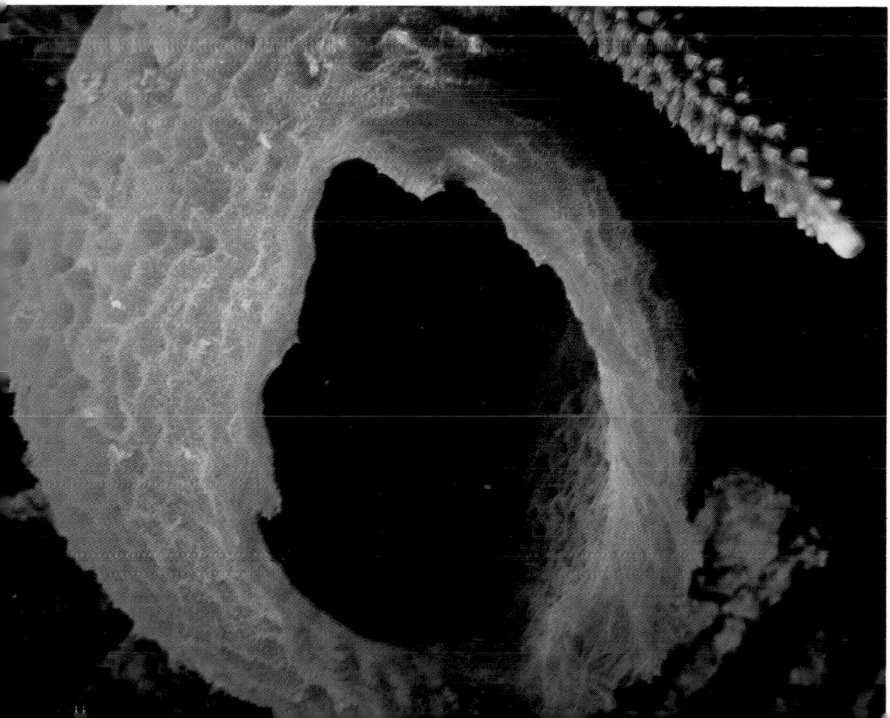

biomass of sponges present can exceed that of any other group, including reef-building corals, and the boring sponges are one of the major factors in the destruction of reef framework.

The class Hyalospongiae (Hexactinellida) does not occur on coral reefs, being restricted in tropical waters to depths below 200 m, and is not discussed further.

CLASS CALCISPONGIAE: CALCAREOUS SPONGES

The skeletons of Calcispongiae are composed of calcareous spicules. These spicules consist mostly of calcium carbonate, principally in the form of the mineral calcite. These sponges are generally small, inconspicuous and drably colored. They occur in marine waters, usually less than 200 m deep, in tropic and temperate areas. Two species are fairly common on Caribbean reefs.

Leucosolenia canariensis (Miklucho Maclay)

The bright lemon yellow, lacy masses of *Leucosolenia canariensis* can reach a size of over 10 cm in diameter. This sponge possesses an ascon type structure, the simplest found in sponges, and the reticulated network of asconoid tubes is soft and fragile. The minute ostia are distributed along the reticulations, and the osculum is the open end of a single or numerous coalesced ascon tubes. Small specimens are tuft-like in shape, while larger examples often have a series of gentle folds in the surface. The oscula in larger specimens are often located in rows on the apices of these folds.

Leucosolenia canariensis occurs on rock substrates of reefs, usually under a ledge produced by an overhanging plate-like coral. It is found adjacent to coralline algae, bryozoans and other organisms which constitute a community characteristic of the ledge environment. It can also grow around from beneath the undersurface of a coral to a point where it is openly exposed.

While originally described from the Canary Islands, the species is known widely in the Caribbean, from Dry Tortugas (Florida) and Bermuda. Its exact depth distribution is poorly known, but it reaches to at least 25 m at Mona Island.

References: de Laubenfels 1936a, 1950.

Small and large tube of *Verongia lacunosa* photographed at the Salt River submarine canyon, St. Croix, Virgin Islands at 30 m depth. A colony of *Diploria labyrinthiformis* is immediately below the sponge.

Callyspongia plicifera on the fore reef at 15 m depth, Discovery Bay, Jamaica. The sponge in the foreground is probably *Ircinia fasciculata* and the corals *Montastrea annularis* and *Acropora cervicornis* are visible among others.

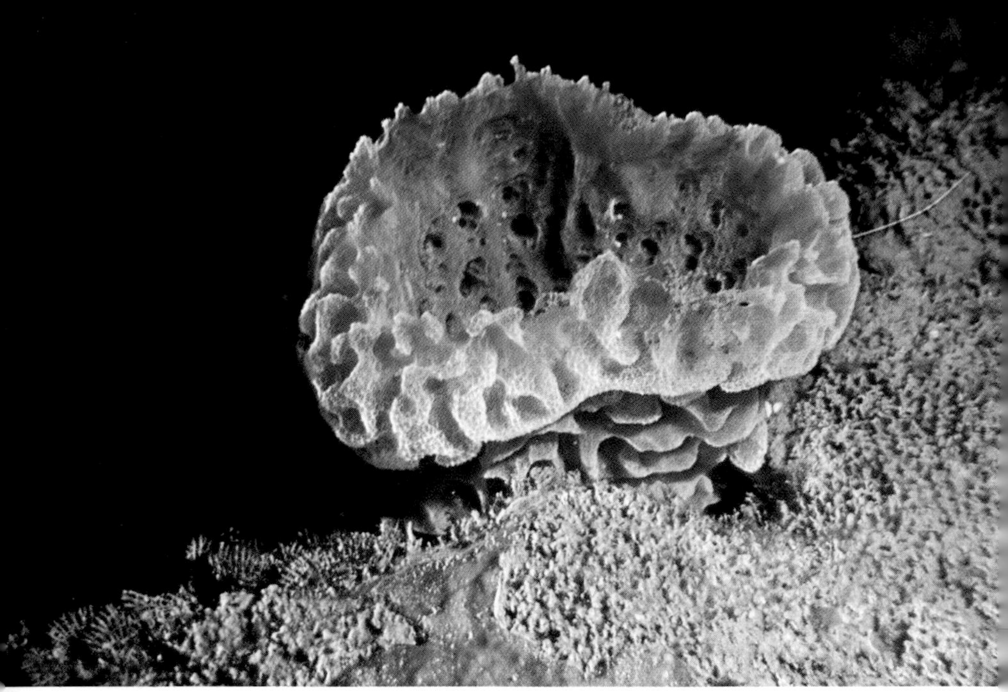

The fluorescent sponge *Callyspongia plicifera.* Puerto Rico, Rincon, 18 m depth.

The pinkish branches of the sponge *Thalysias junipera* with the symbiotic zoanthid *Parazoanthus swiftii* protruding from their surface. Bahamas, Eleuthera Island, Bamboo Point, 15 m depth.

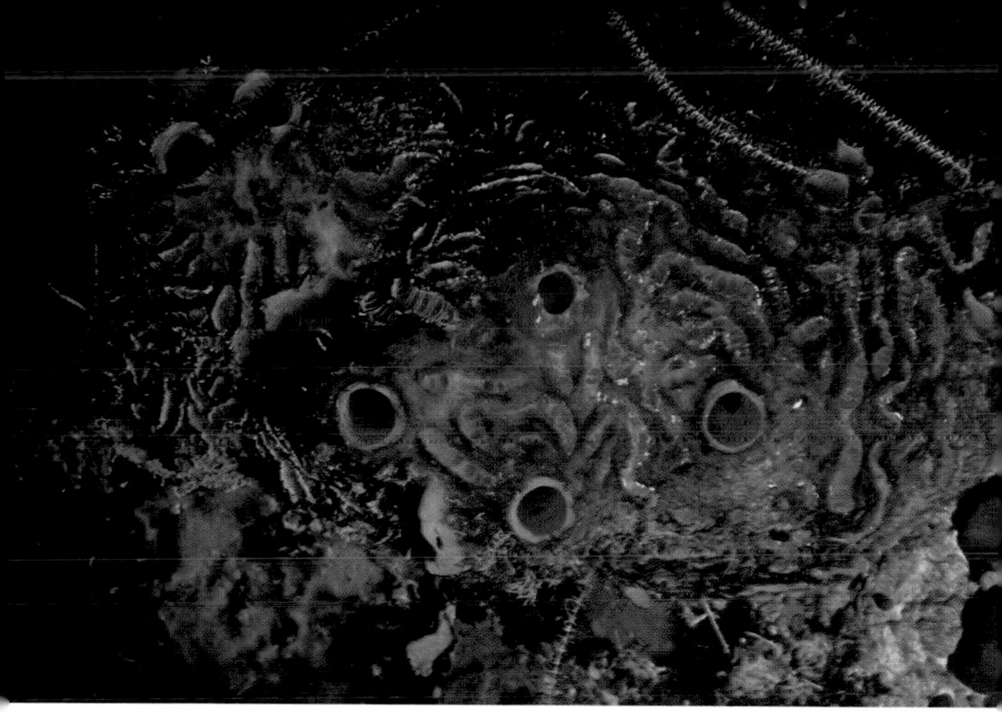

The sponge *Didiscus* sp. has its ostia in furrows on its surface. Much of the surface of this normally orange sponge is darkened by green algae in this individual. Jamaica, Rio Bueno, 20 m depth.

The sponge *Hemectyon ferox* grows horizontally over areas of substrate on the reef. Jamaica, Discovery Bay, 12 m depth.

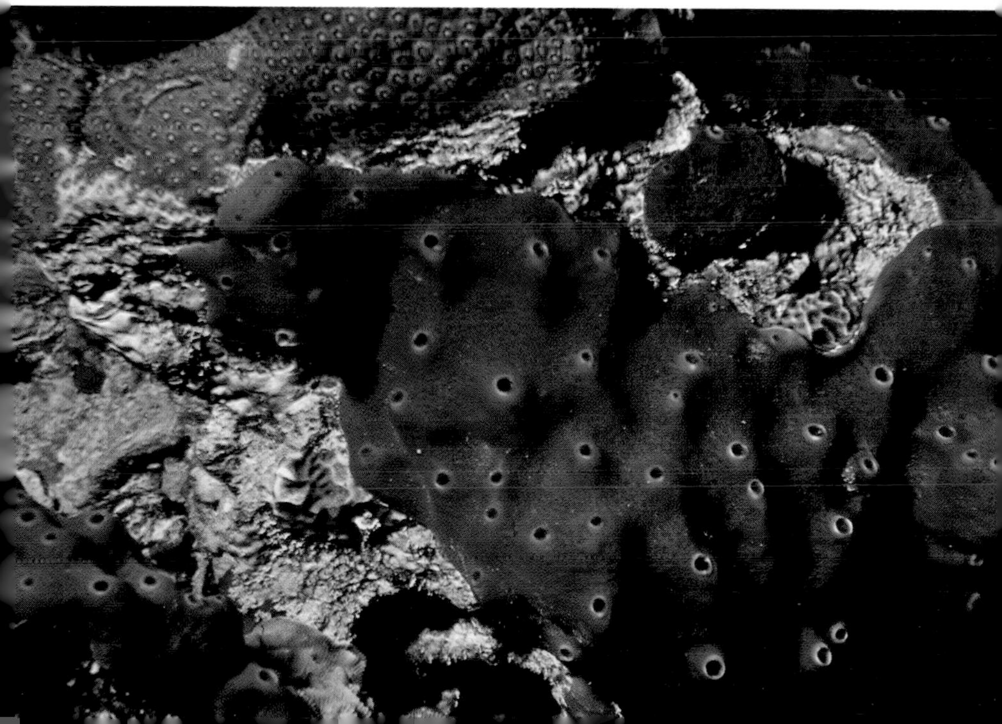

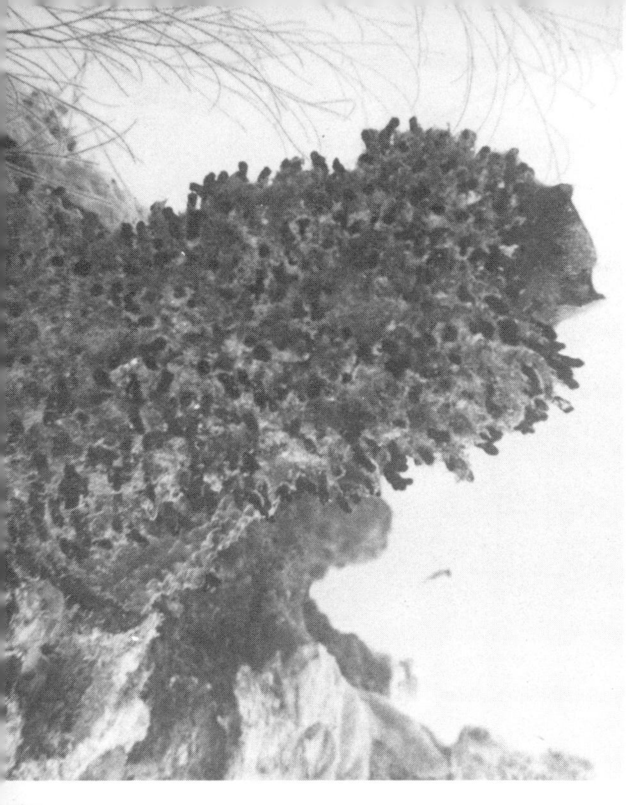

Left: An unidentified sponge which commonly occurs in drop-off areas at depth of 30 m and below. It has a large central osculum and a number of digitate processes on its surface. Photograhed at 40 m depth at Spring Gardens, Jamaica.

Below: Unidentified sponge, possibly a member of *Verongia,* on the fore reef at Discovery Bay, Jamaica.

Scypha barbadensis (Schuffner)

These small white urn or pitcher-shaped sponges are often found in groups at a minimum depth of 20 to 60 m under ledges that are overhanging, usually along vertical reef walls. They have thin, rigid walls and are quite fragile.

The distribution of the species of *Scypha* is poorly known, but *S. barbadensis* is definitely West Indian. Another species, *S. ciliata*, occurs in shallow water in Bermuda and Europe and a third, *S. linga*, is found on the American Atlantic coast.

Reference: de Laubenfels 1950.

CLASS DEMOSPONGIAE: DEMOSPONGES

The demosponges are the largest class of sponges, both in number of species and range of distribution. They range from intertidal to abyssal depths in the ocean and one family (Spongillidae) occurs in fresh waters, the only sponges to do so. They are encrusting to massive, ranging from nearly microscopic to over 2 m in diameter. Nearly all of the sponges encountered on coral reefs are members of this class.

Their skeleton consists of one- to four-rayed spicules (siliceous), "horny" proteinaceous fibers (spongin) or a combination of the two. Several genera lack a skeleton completely but are included in the class on the basis of other characters. A few members also have some calcareous spicules in addition to their siliceous ones.

SUBCLASS TETRACTINELLIDA

If spicules are present (some species lack them), siliceous tetraxons will occur. Tetraxons are four-rayed spicules, the rays extending out from a central point but not in a single plane. This subclass lacks spongin fibers.

ORDER CARNOSA

Chondrilla nucula Schmidt
Chicken-liver sponge

The smooth, shiny surface of this walnut-brown to yellow-brown sponge and its rounded, often hemispherical

The sponge *Mycale laevis* at one of the indentations that forms when it grows on the undersurface of the coral *Montastrea annularis.* One osculum of the sponge is visible in the center of the photograph. Jamaica, Discovery Bay, 30 m depth.

The sponge *Mycale laevis* living "free" without any associated coral colony. Puerto Rico, La Parguera, Mario Reef, 2 m depth.

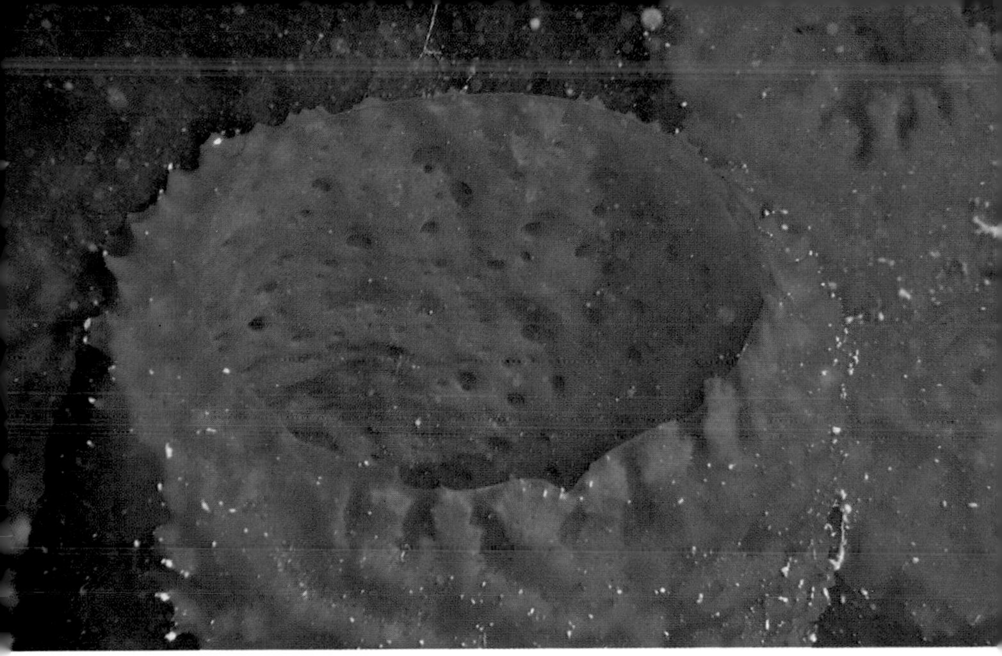

The sponge *Mycale* sp. appears black in deep water due to the selective filtration of sunlight by water, but when photographed using an electronic flash its true colors are revealed. Jamaica, Discovery Bay, 15 m depth.

The sponge *Ulosa hispida* occurs as encrustations with a conulose surface on rocky substrates. Puerto Rico, La Parguera, San Cristobal Reef, 9 m depth.

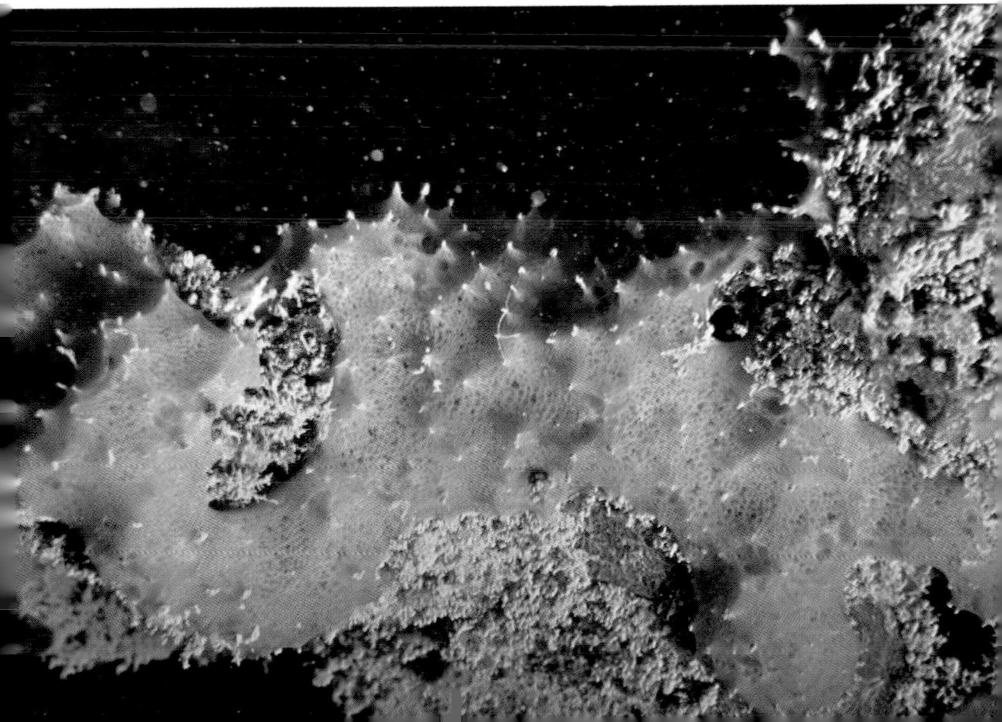

shape are distinctive characters. Its surface texture, not unlike fresh chicken liver, is partially the result of the almost total lack of a skeleton. It may form thick, irregular but always rounded encrustations on coral reefs or rocky areas. Large areas of the coral *Porites furcata* and entire colonies of *Siderastrea siderea* and *Diploria clivosa* are sometimes overgrown, probably by way of initially overgrowing dead portions of the corals. *Chondrilla nucula* also occurs on upper branches of *Porites porites* nipped off by the feeding of parrotfishes.

It is found in shallow waters of reef areas. *Chondrilla nucula* occurs throughout the West Indies, southern Florida and the Bahamas and is abundant in Bermuda. It also occurs in the Mediterranean and off western Africa.

References: Hechtel 1965, de Laubenfels 1950, Glynn 1973c.

SUBCLASS MONAXONIDA

Spongin fibers may or may not be present in members of this subclass. They all have siliceous monaxon spicules of fairly large size. Monaxons consist of a single straight or curved spine which has the ends ornamented in a variety of manners (pointed, knobbed, rounded).

ORDER HADROMERIDA

Family Clionidae

Cliona spp.

The genus *Cliona* is the largest group of the burrowing or excavating sponges. They occur in a wide variety of calcareous substrates (coral heads, alive and dead; branching corals; rocks; shells) and are quite destructive through weakening of support for the corals. Unlike the genus *Siphonodictyon*, the species of *Cliona* attack the non-living basal portion of coral colonies, not the living coral surface itself.

Burrowing is believed accomplished by secretion of minute amounts of acid by the cells adjacent to the calcareous substrate, undercutting fragments which are eventually passed out of the sponge by the excurrent canals. Although the damage is often not visible on the surface, large areas of the substrate can be riddled with chambers containing the sponge. In many species of *Cliona* and other burrowing sponges only the small oscules and ostial papillae are visible on the surface, but when cracked open a coral head may have nearly its entire inner volume taken over by the sponge. Other species may burrow, but also encrust on the outer surfaces of the coral. The destructive action of burrowing sponges is one of the factors limiting reef growth and is important in determining the morphology and species composition of the reef.

Many species of *Cliona* are brightly colored (blue, red, orange) and some, particularly those living in exposed habitats, may possess zooxanthellae, the unicellular algae occurring in stony and soft corals. Although the exact number is not known, there are at least ten species of *Cliona* at one locality (Jamaica), and one is considered further.

References: Pang 1973a, 1973b.

Cliona delitrix Pang

This is one of the burrowing sponges which also encrusts on the surface of its host. The sponge is red to red orange and covers up to 1 m in diameter of the substrate. Its oscules are large and prominent, being scattered over the surface. *Cliona delitrix* occurs in massive coral heads, both living and dead, and rarely, if ever, in branching corals. The sponge evidently kills the coral immediately around it and expands its growth outward. Smaller heads may be completely overgrown, while large ones will have the entire central portion occupied.

A white zoanthid, *Parazoanthus parasiticus,* dots the surface of most specimens of *C. delitrix,* but its relationship to the sponge is not known. Recorded from Jamaica and Puerto Rico at depths below 10 m.

Reference: Pang 1973a.

Family Tethyidae

Tethya crypta (de Laubenfels)

This black, inconspicuous sponge is globular or hemispherical in shape and the outer surface is often coated with

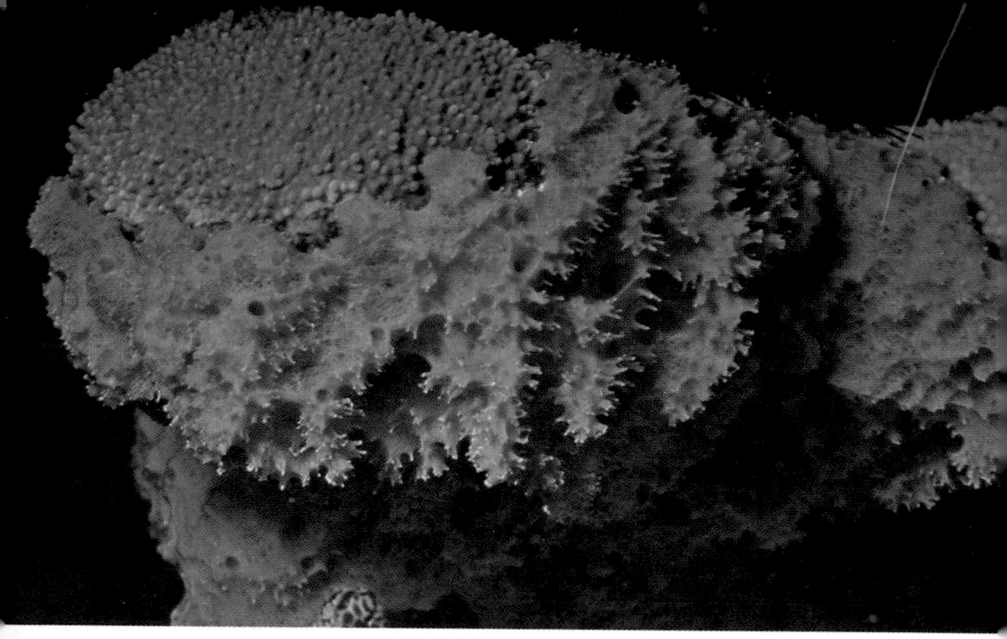

Ulosa hispida growing on the skeleton of a stony coral. Whether the sponge has killed the coral in the area it occupies or is simply covering an area where the coral was already dead cannot be determined from the photograph. Jamaica, Discovery Bay, back reef, 5 m depth.

The digitate red branches of *Haliclona rubens* are abundant on most reefs. Here they arise from the basal mass adjacent to the coral *Siderastrea radians.* Bahamas, Bimini, Turtle Rocks, 9 m depth.

The purple interconnecting branches of the sponge *Haliclona hogarthi* are quite slender and flexible. Jamaica, Rio Bueno, 15 m depth.

The sponge *Dasychalina cyathina* has a zoanthid associated with it in most instances. This individual has some areas of white sediment clinging to its outer surface. Jamaica, Discovery Bay, 15 m depth.

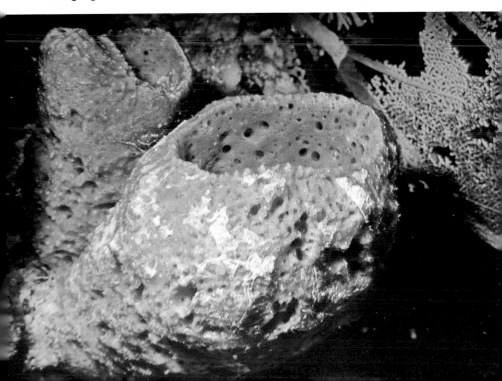

sediment or various epiphytic growths. As many as six separate oscula are found on the outer surface, and the shape of a particular individual can be varied somewhat since this species does not possess a rigid skeleton.

It occurs in back reef areas or on limestone shelves in sheltered areas. As newly settled individuals, *Tethya crypta* is open to predation by sea urchins, but once beyond a critical size they may live to an age of at least twenty years or more.

It is known from Jamaica, Puerto Rico and Bimini, Bahamas at 1 to 8 m depth.

References: Reiswig 1971a, 1971b, 1973.

Tethya spp.

The genus *Tethya* has many taxonomic problems. Most species are orange or red, spherical to ovoid in shape. The illustrated specimen has the oscules in a slight depression on the upper surface and algae commonly growing on the outer surface of the sponge.

ORDER POECILOSCLERIDA

Family Adociidae

Siphodictyon coralliphagum Rutzler

This burrowing sponge is quite different from the species of *Cliona* in that rather than attacking from the nonliving, basal portions of corals, its larvae evidently settle on the living coral surface, kill a portion of the coral to expose the substrate and then burrow in directly from that point. In large specimens a series of yellow-orange oscular tubes and chimneys protrude at intervals above the surface of the coral head, while the major portion of the sponge fills spherical or ovoid cavities beneath the surface. The oscular tubes are normally oriented vertically and the sponges occur on exposed coral heads in full light. The species of coral inhabited include *Diploria strigosa, Stephanocoenia michellini, Siderastrea siderea* and *Porites asteroides.*

Siphodictyon occurs in tropical reef environments only, and *S. coralliphagum* is known from several localities in the

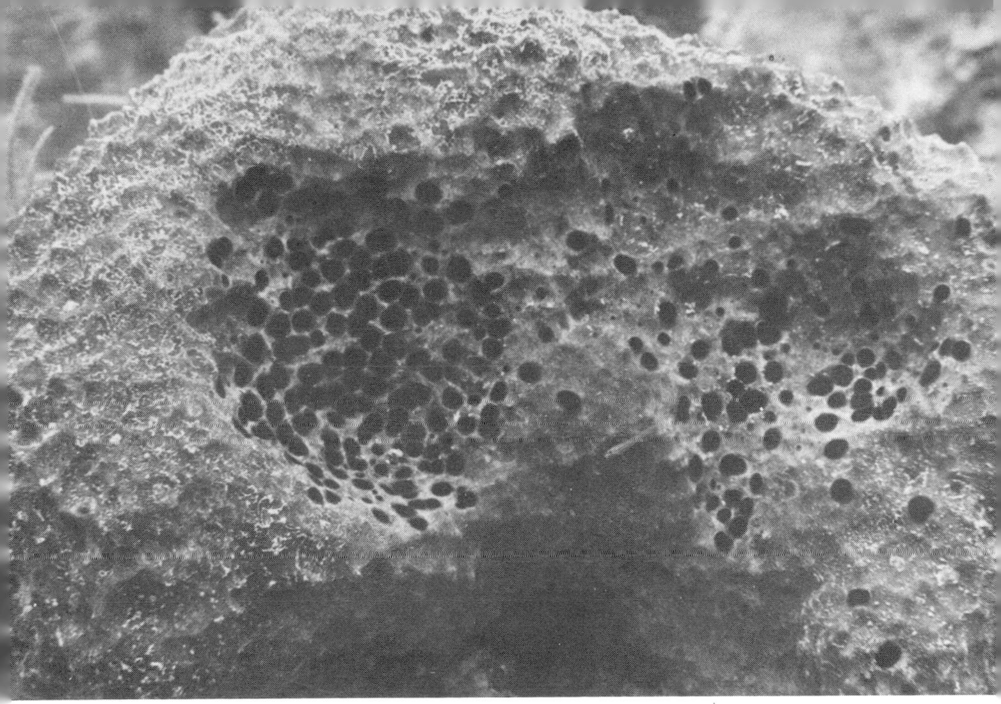

The sponge *Ircinia strobilina* with the closely grouped oscules clearly visible.

Multiple-tubed specimen of *Verongia lacunosa* occurring among plate-like colonies of *Montastrea annularis* at a depth of 35 m, Discovery Bay, Jamaica.

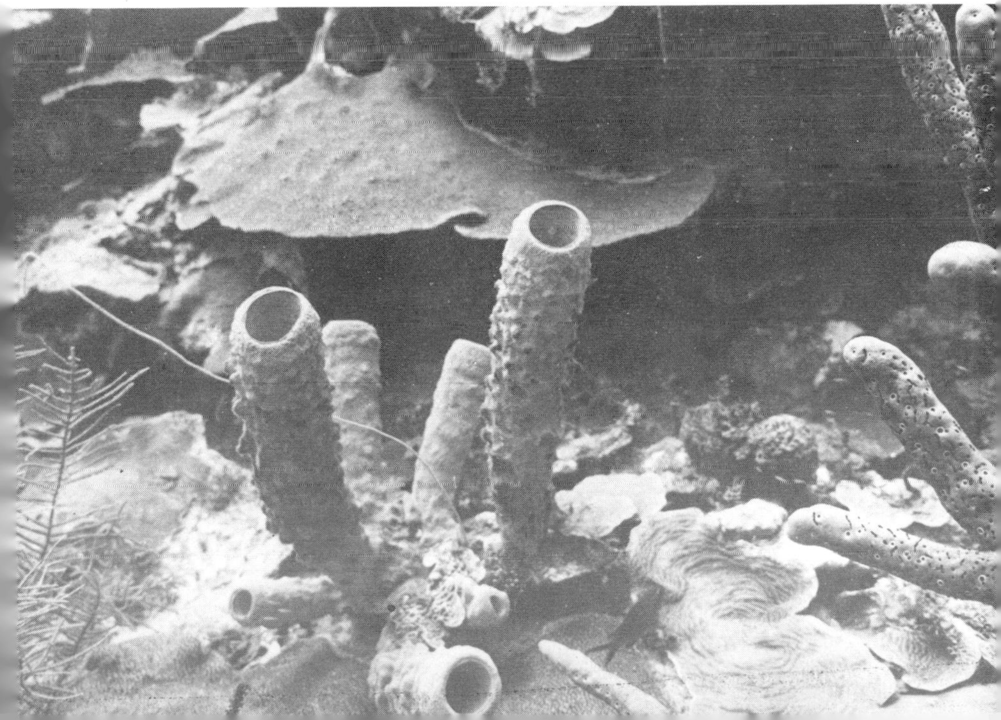

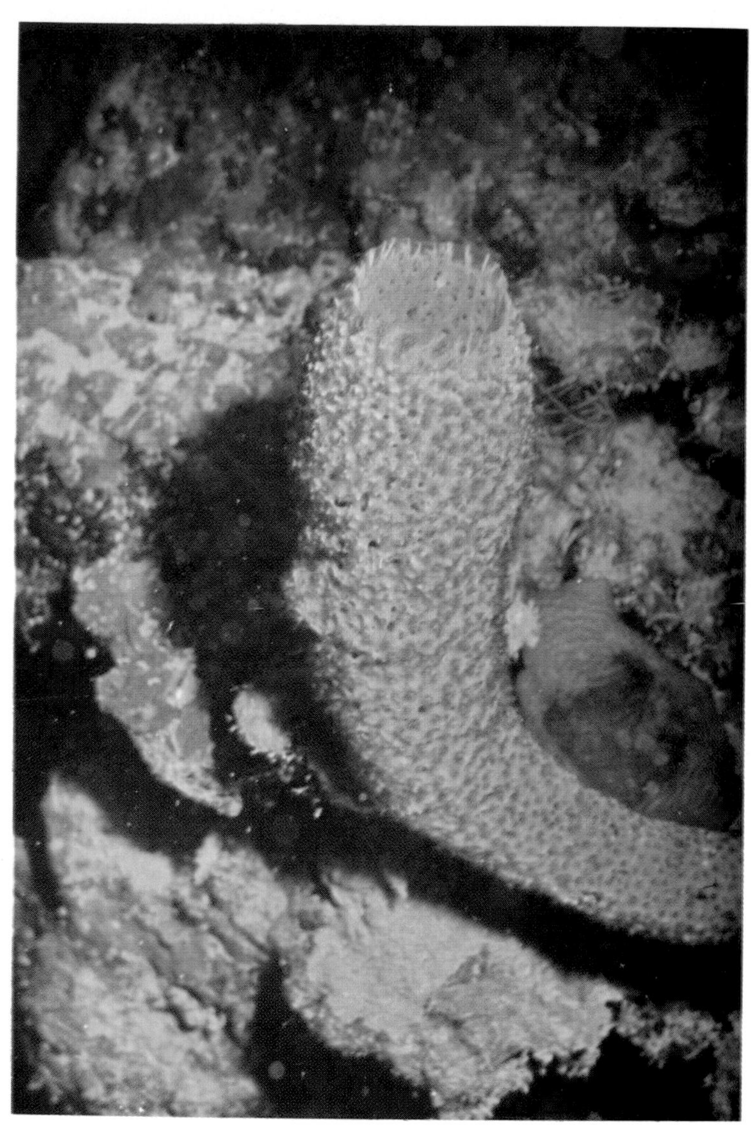

Vasiform individual of *Dasychalina cyathina* with associated zoanthids. Jamaica, Spring Gardens, 21 m depth.

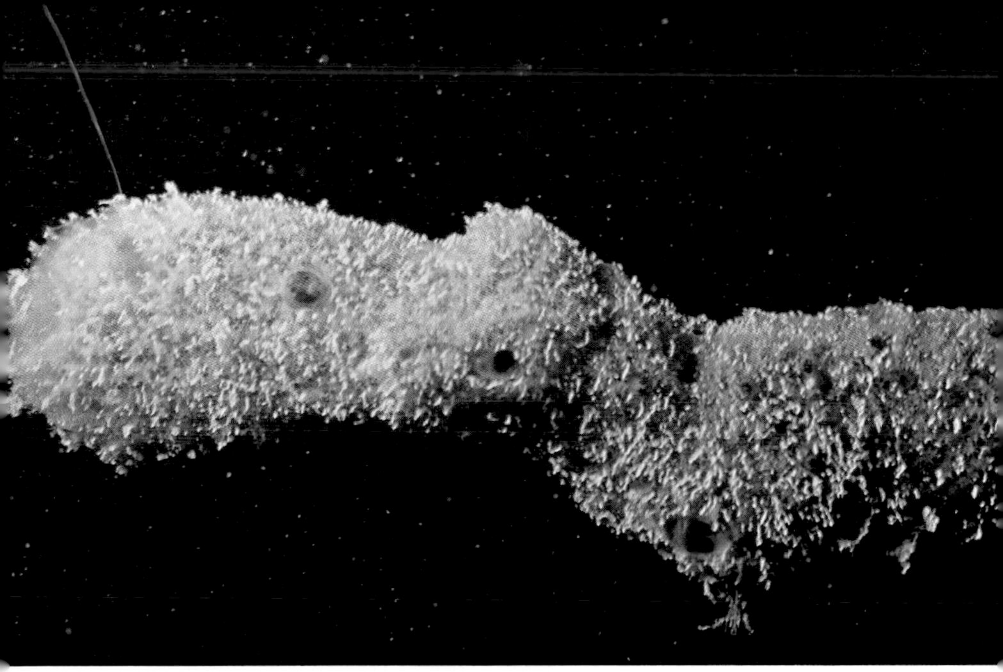

A branch of the sponge *Gelloides ramosa.* Puerto Rico, La Parguera, Laurel Reef, 12 m depth.

The sponge *Gelloides ramosa* with brittle stars, *Ophiothrix suensoni,* clinging to it at night. The zoanthid *Parazoanthus parasiticus* is also visible on the outer surface of the sponge. Bahamas, Crooked Island, Landrail Point, 15 m depth.

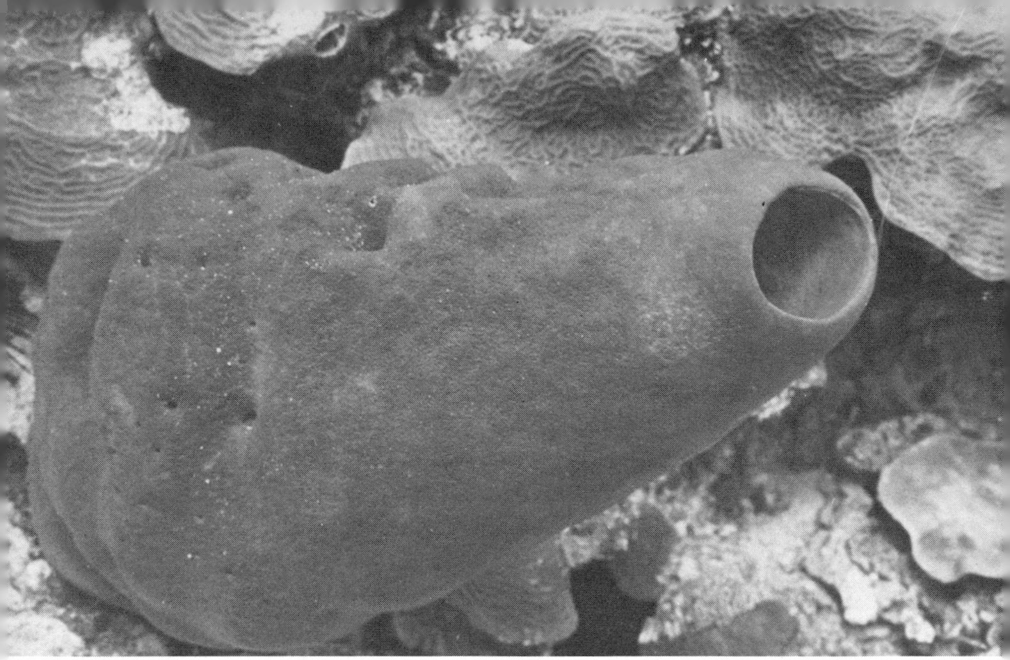

Oddly shaped specimen of *Agelus* sp. photographed at 20 m depth, Discovery Bay, Jamaica.

Three tubes of sponges of the genus *Agelus* growing with the coral *Madracis decactis* at 25 m depth, Discovery Bay, Jamaica.

Caribbean. At least two other species occur, *S. cachacrouense* Rutzler, which has grayish brown "pillows" with large oscules on the coral surface, and *S. brevitubulatum* Pang, with bright yellow cylindrical oscules. *Siphonodictyon coralliphagum* is known from 1.5 to 57 m in depth but is not commonly found above 10 m.

References: Rutzler 1971, Pang 1973a.

Adocia carbonaria (Lamarck)

This sponge may occur as a thick incrustation, low-lying branches or lobes. It is consistently black in color, and this alone serves to separate it from most other Caribbean sponges. The oscules are fairly small, scattered over the smooth outer surface and often have their edges raised slightly. The intake ostia also cover this surface, but due to their small size are not noticed unless the sponge is examined closely.

Adocia carbonaria occurs in both turtle grass (*Thalassia*) beds and reef areas. It occurs from a few meters to at least 15 m in depth. The species is found throughout the Caribbean, the Bahamas and possibly at Dry Tortugas, Florida.

References: Hechtel 1965, 1969.

Agelus spp.

The genus *Agelus* is a difficult group, in spite of the size of its members, in which to make a definitive identification. There are a number of undescribed species, and those already described have never been adequately reviewed. Identification of species should be considered tentative in the following discussion.

Agelus sp. A
Trumpet sponge

Probably undescribed, this large conical sponge occurs on deep reefs at depths of around 18 to 45 m in the West Indies. It often grows from behind a coral plate or ledge as do some other large deep-reef sponges. The walls are thick and smooth-sided with large rounded or knobby projections

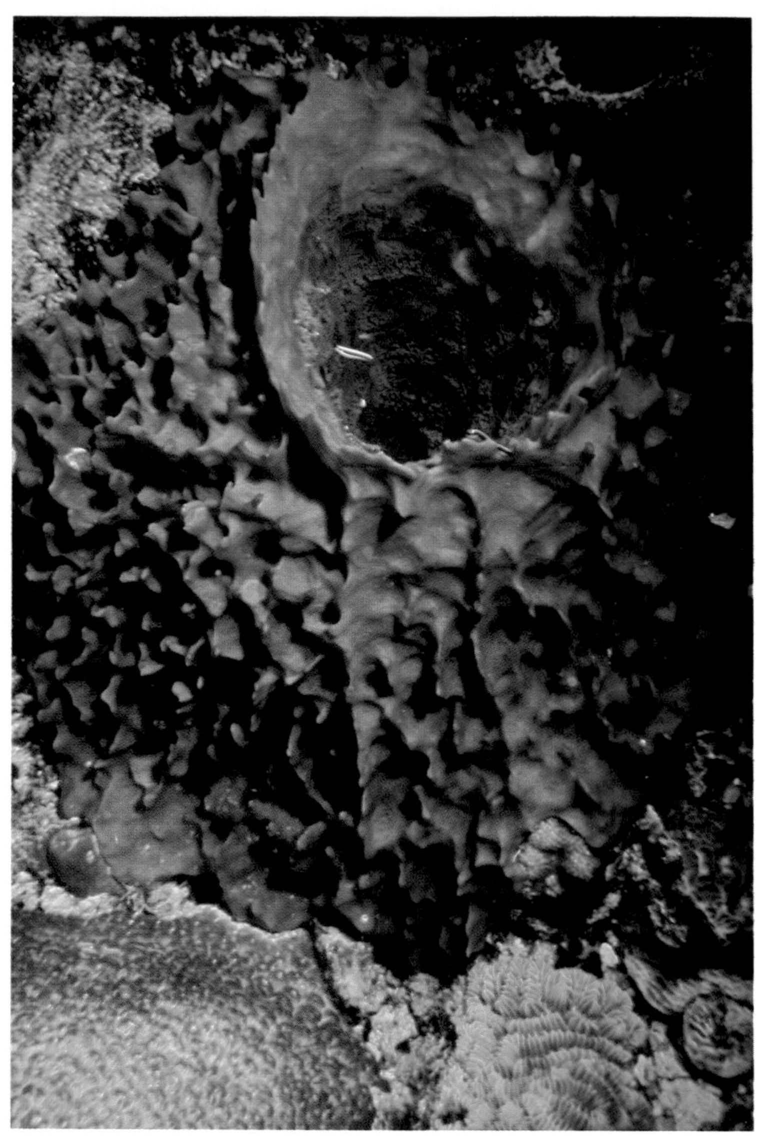

Xestospongia muta, the barrel sponge, is one of the largest of Caribbean sponges. It is almost stony in consistency and common in many reef areas. Puerto Rico, Desecheo Island, 8 m depth.

View towards the base of the cone-like cavity in the outer end of *Xesto-spongia muta* showing the lighter coloration of this area as compared to the outer surface of the sponge. Jamaica, Discovery Bay, 20 m depth.

The sponge *Xestospongia* sp. probably represents an undescribed species, but it is fairly common in occurrence on some reefs. Bahamas, Grand Bahama Island, 12 m depth.

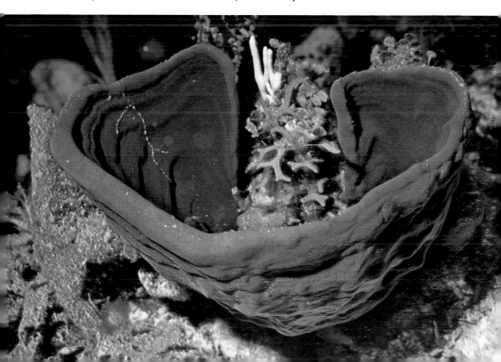

in some individuals; zoanthids rarely, if ever, occur on its outer surface. The wall at the rim of the atrium may be considerably thinner than in the mid-section of the sponge. The diameter of the sponge and the central lumen increase with increasing length, resulting in the conical shape of this species. The inner surface of the lumen is often, but not always, lighter in color than the external surface. Two adjacent tubes of the sponge may join along their outer surface with only a thin wall existing between their atria.

Agelus sp. B

A second, probably undescribed, species of *Agelus* has single or multiple candle-like tubular elements which are fairly consistent in diameter over their length. The outer surface is generally smooth-walled and the inner surface of the atrium slightly lighter in color. Orange-tan in exterior color, this species does not seem to associate with zoanthids.

It has been observed spawning, releasing clouds of smoke-like sperm and golden-yellow, sticky strands of eggs, in late July. The eggs persist for at least 24 hours in the vicinity of the spawning individuals. Various fishes and shrimps have been observed associated with this sponge also.

It occurs in the West Indies and the Bahamas in deep reef environments around 18 to 60 m.

Agelus sp. C

A brilliant red-orange species of *Agelus* occurs often on Caribbean reefs. This sponge can be massive and take a variety of forms. Some are urn- or barrel-shaped and up to nearly 1 m in height. Others are fan-like or encrusting. The sponge is flaccid and the surface has elongate and circular depressions with much smaller pits on the portions of the surface between the depressions.

Algae and the coral *Tubastrea aurea* have been observed growing on the surface of this sponge.

Known from Jamaica and Puerto Rico (Mona and Desecheo Islands) at 15 to 30 m.

Agelus flabelliformis (Carter)
Elephant ear sponge

While it may be only a few centimeters thick, this flattened sponge can reach huge proportions, approaching 1 by 2 m. It may be brown, brown with white edging, or nearly white in color and attached to the substrate near one edge. They are quite fragile due to their size, thin structure and weak attachments. While normally a deep reef inhabitant (to at least 100 m in depth), it has been observed as shallow as 6 m in a dark, still cave at Desecheo Island.

References: Hartman 1973, Hechtel 1969.

?*Agelus schmidti* Wilson

This species can be described as a series of vertical or nearly vertical tubes with apical oscules, often connected, arising from a nearly horizontal basal or encrusting element. The entire mass of the sponge may comprise dozens of vertical elements and occupy a sizable area. There are circular to elongate depressions (5-25 mm wide) on the outer surface occcupied by a zoanthid, *Parazoanthus* sp., that is dark reddish brown in color. Some masses tend to be lobate, but the oscula are still oriented vertically. The outer surface is orange-tan or reddish tan in color. It is found in deep reef clear water areas at 15 to 45 m. It evidently occurs widely in the Caribbean.

Reference: Hechtel 1969.

Agelus screptrum (Lamarck)

This species consists of a series of cylindrical branches, often a meter in length between junctions, which spread horizontally or vertically over the bottom. Oscules are scattered over the surface, occasionally but not always tending to occur in rows on one side without respect to the vertical. It is orange-tan in color and often possesses zoanthids on the outer surface.

References: Hartman 1973, Hechtel 1969.

The sponge *Neofibularia nolitangere* should not be touched as it can produce a serious dermatitis. Sponge-dwelling neon gobies, such as the white form of *Gobiosoma horsti* seen here, are often found with this sponge. Jamaica, Discovery Bay, 25 m depth.

The atria of the sponge *Neofibularia nolitangere* have the small white parasitic polychaete worms, *Syllis spongicola,* clearly visible in their walls. While these worms are present in other species of sponges, they are most easily observed in *N. nolitangere.* Jamaica, Discovery Bay, 15 m depth.

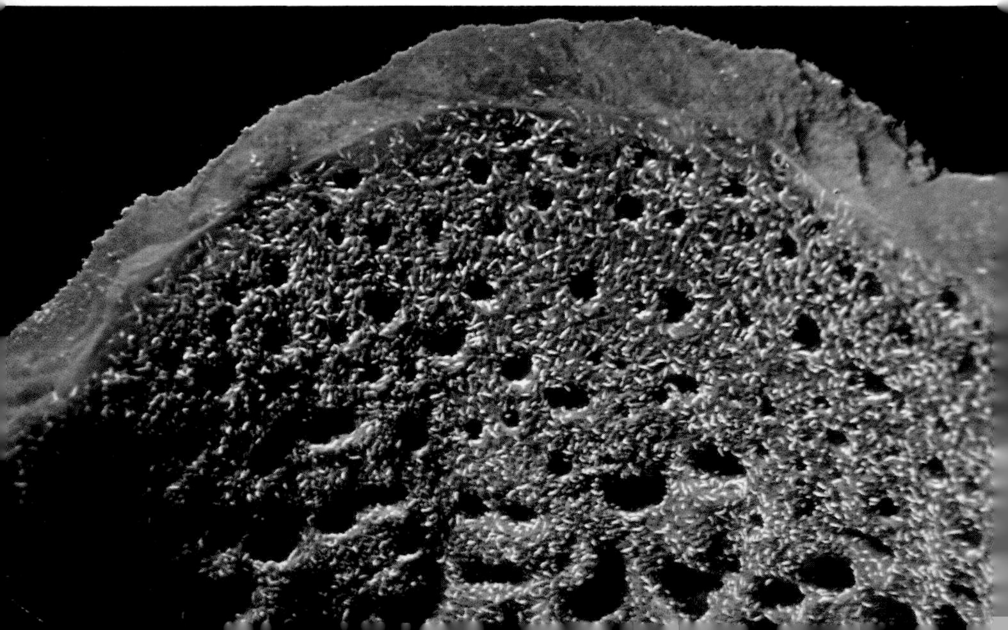

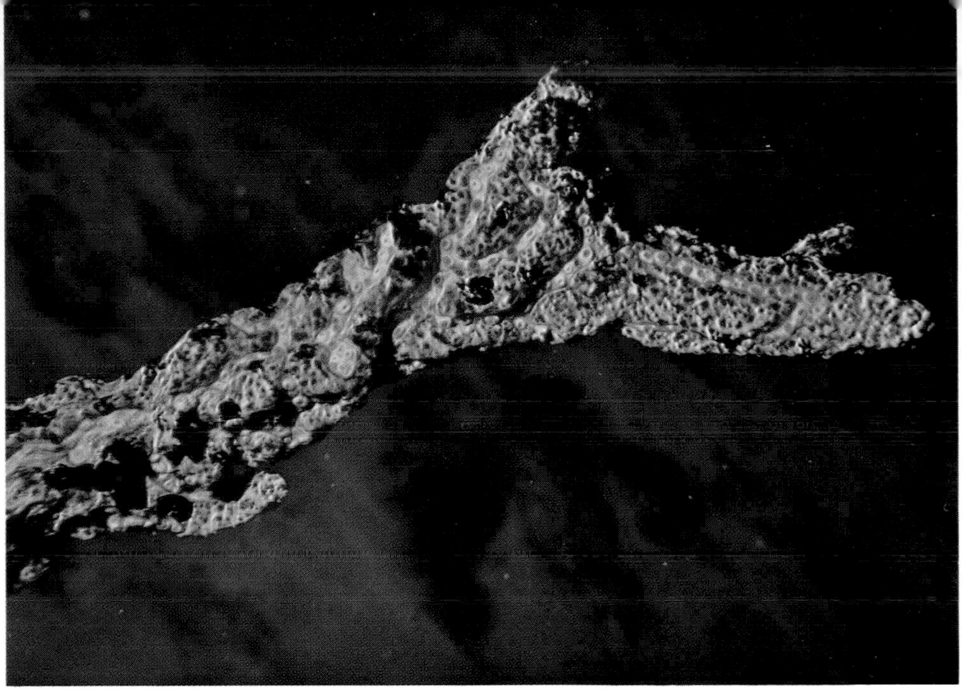

The sponge *Iotrochota birotulata* is normally found with the orange zoanthid *Parazoanthus swiftii* which occurs as winding chains of polyps on the sponge. Jamaica, Discovery Bay, 10 m depth.

The stinker sponge, *Ircinia fasciculata*, has its oscules distributed widely over the sponge rather than in one localized area. Bahamas, Acklins Island, Datum Bay, 20 m depth.

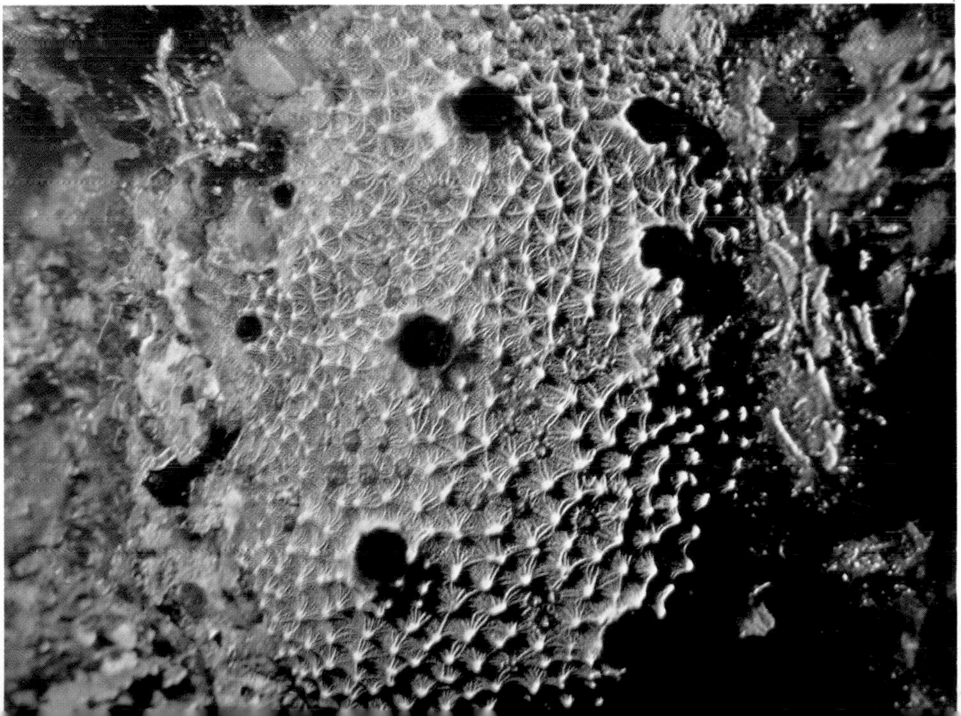

Family Callyspongidae

Callyspongia vaginalis (Lamarck)

This common, shallow water reef sponge may consist of a single thin-walled tube to bunches of as many as 20 to 30 tubes. In those consisting of multiple tubes, they may be arranged in clusters, rows or combinations of both. Each tube may be 8 cm or more in diameter and up to 1 m high. Even the largest tubes are still thin-walled compared to other tube-shaped sponges. The color of *C. vaginalis* is quite variable, often purple, gray, lavender or violet. The outer surface may be either smooth or peaked.

This may be the most common sponge of any size in some reef communities, and it also occurs in rocky habitats. The skeleton is quite stiff, enduring for some time after death. The zoanthid coelenterate *Parazoanthus parasiticus* occurs with *C. vaginalis* in many localities, with the number of zoanthid polyps per sponge quite variable.

The species occurs from water as shallow as 2 m to at least 20 m. It is widespread in the Caribbean, the Bahamas, Bermuda and southern Florida.

References: Hechtel 1965, de Laubenfels 1950.

Callyspongia plicifera (Lamarck)

A vasiform species which occurs as a single element or a cluster of elements. Some specimens tend to be cylindrical, but all have elaborately pitted and sculptured outer surfaces. The color is generally a dull purple, but the sponge often fluoresces a light blue, particularly around the apical opening and various portions of the outer surface. Individual sponges reach over 40 cm in height.

It is confined to reef communities on solid substrates but can occur in fairly shallow water (4 m). Its maximum depth is at least 23 m. It is known from the Caribbean, the Bahamas and southern Florida, but not from Bermuda.

References: de Laubenfels 1936a, Hechtel 1965.

Family Microcionidae

Thalysias junipera (Duchassaing and Michelotti)

This sponge is red, sometimes with a pinkish cast, and

has masses of interconnected branches. The ends of the branches are knob-like, and the sponge as a whole has been described as "lumpy." The masses reach 40 cm high at times, and the orange zoanthid *Parazoanthus swiftii* may occur on it, making this a particularly colorful combination.

Thalysias junipera occurs from southern Florida, where it has been termed common, through the West Indies at depths generally of 6 to 30 m.

Reference: de Laubenfels 1936a.

Didiscus sp.

This undescribed species of *Didiscus* has the ostia in grooves meandering over the sponge surface and the oscules scattered on raised portions of the surface. Much of the surface is covered with algae and sediment, giving it a dark appearance, but the grooves of the ostia are free of these growths and are consistently red-orange. The area around the oscular openings and the tissue inside are also red-orange in color.

This sponge is generally found on vertical walls in deep reef environments, where it can be fairly common. Known from Jamaica, Honduras and Mona Island, Puerto Rico. Its known depth range is 18 to 40 m.

Family Raspailiidae

Hemectyon ferox Duchassaing and Michelotti

This sponge is a red-brown encrusting species with the conically raised oscules scattered on the surface. *Hemectyon ferox* covers sizable areas of substrate and may be capable of growing over some corals. Contact should be avoided with this sponge as it causes erythema and swelling of affected skin.

Spawning has been observed in late August, with the parent sponge bearing mucous-like strands of eggs. The species is common on reefs in the Caribbean, particularly in the 13 to 17 m range. Its maximum known depth distribution is 6 to 54 m.

Reference: Halstead 1965.

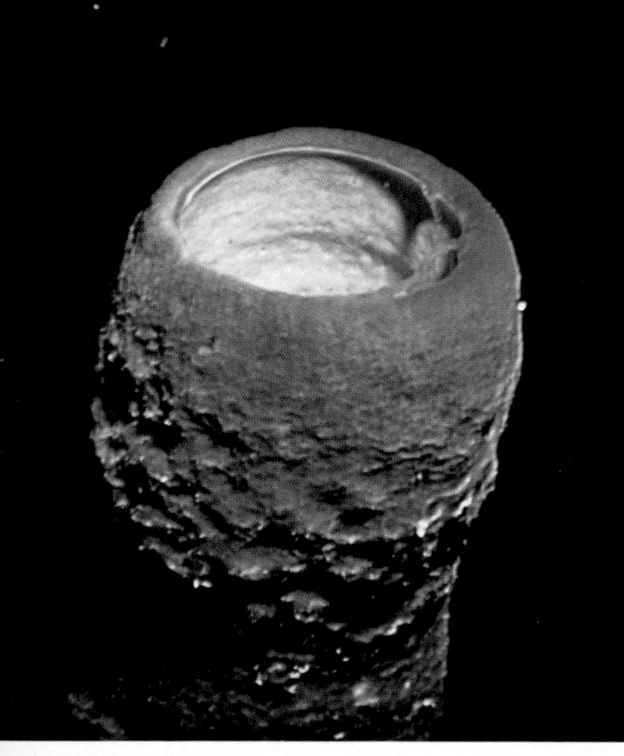

The tubular sponge *Verongia lacunosa* is limited to a deep reef environment. Bahamas, Eleuthera Island, 24 m depth.

The sponge *Ircinia strobilina* has its oscules in groups on the upper surfaces of the sponge. Its outer surface is conulose with radiating patterns at each conule. Jamaica, Discovery Bay, 15 m depth.

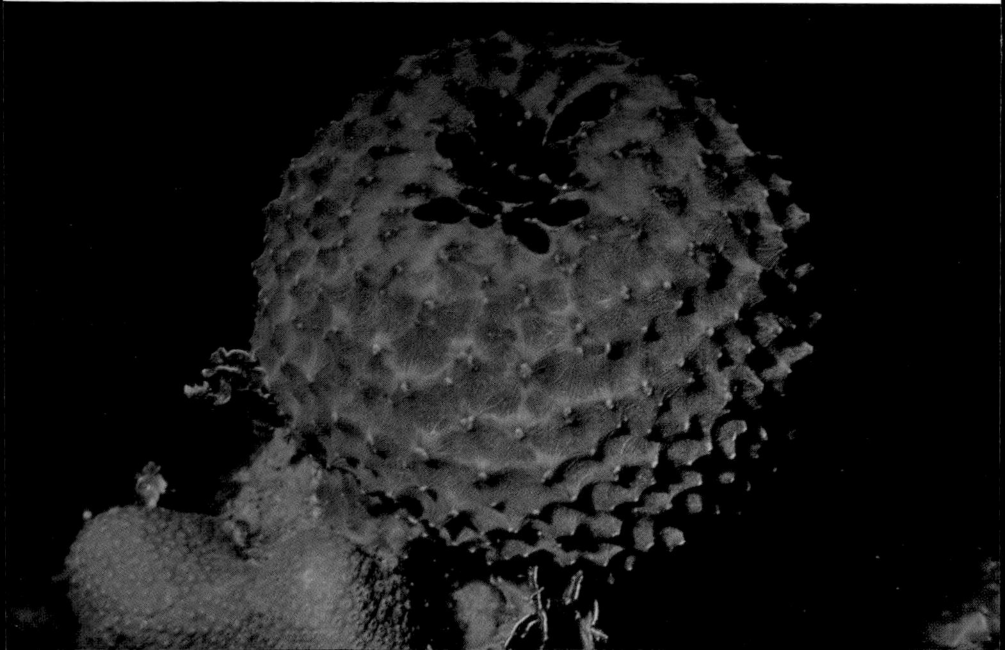

Verongia archeri has the longest and thinnest tube wall of any of the Caribbean tube sponges. These tubes can reach 2 m in length and the sponge is restricted to the deep-reef environment. Jamacia, Discovery Bay, 30 m depth.

Shown is a group of large demosponges which includes *Verongia gigantea* (yellowish, barrel-like sponge) and other species of *Verongia* and *Agelus*. Jamaica, Discovery Bay, 40 m depth..

Family Mycalidae

Mycale laevis (Carter)

This sponge has a relationship with corals which is evidently unique among Caribbean sponges. It occurs on the undersurfaces of coral plates, particularly *Montastrea cavernosa, M. annularis, Mycetophyllia lamarckiana, Porites asteroides* and *Agaricia agaricites*, and has the ability to alter the shape of the growing coral plate. The sponge, usually a pale to vivid orange, causes the edge of the coral plate to be indented at fairly regular intervals by folds from which the oscules of the sponge open outward. Often the undersurface of entire groups of coral plates is occupied by this sponge. It can even occur between the fingers of coral such as *Madracis* sp.

Mycale laevis is also occasionally free-living as an encrusting to massive form, but although the entire range of the species is 1 to 80 m, it is generally found associated with coral below 25 m. The relationship between the coral and this sponge is not well known, but may be beneficial for both organisms. The sponge may protect the undersurface of the coral from attack by destructive burrowing sponges, and the coral plate provides an excellent habitat for the sponge. Certain coral-sponge colonies have been observed over a period of years, and the relationships observed were stable.

The species is known from the West Indies, the Bahamas and Florida.

Reference: Goreau and Hartman 1966.

Mycale sp.
Strawberry sponge

The outer surface of this red to orange-red sponge, consisting of one to seven (rarely) tubular elements with single large terminal oscula, is strongly conulose to spiny and the walls thin (1-3 cm). While each tubular element may reach 50 cm in height, the basal attachment to the substrate is not increased in size as the sponge grows in length and somewhat in diameter.

The brilliant color of *Mycale* sp. is not apparent at the depths it inhabits (below 13 m) due to the filtration of nearly

Large individual of *Mycale* sp. occurring adjacent to an *Ircinia* sponge at 20 m depth, Discovery Bay, Jamaica.

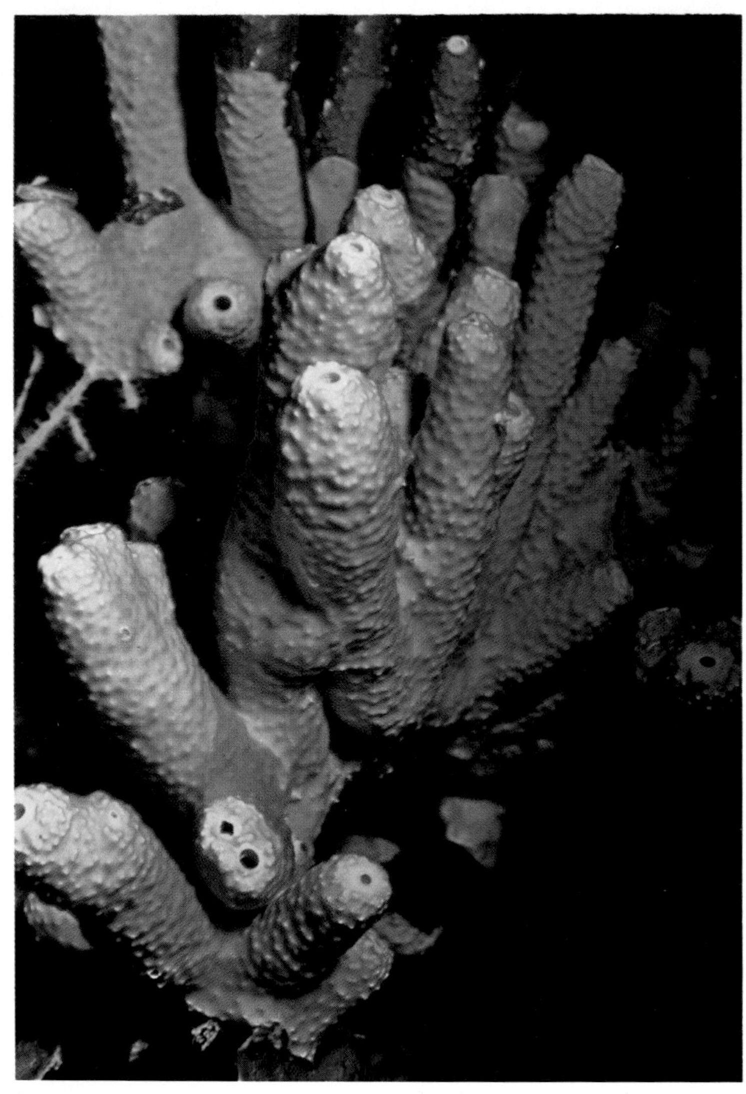

Verongia longissima has branches rather than tubular growths like the other species of *Verongia* illustrated. Jamaica, Discovery Bay, 30 m depth.

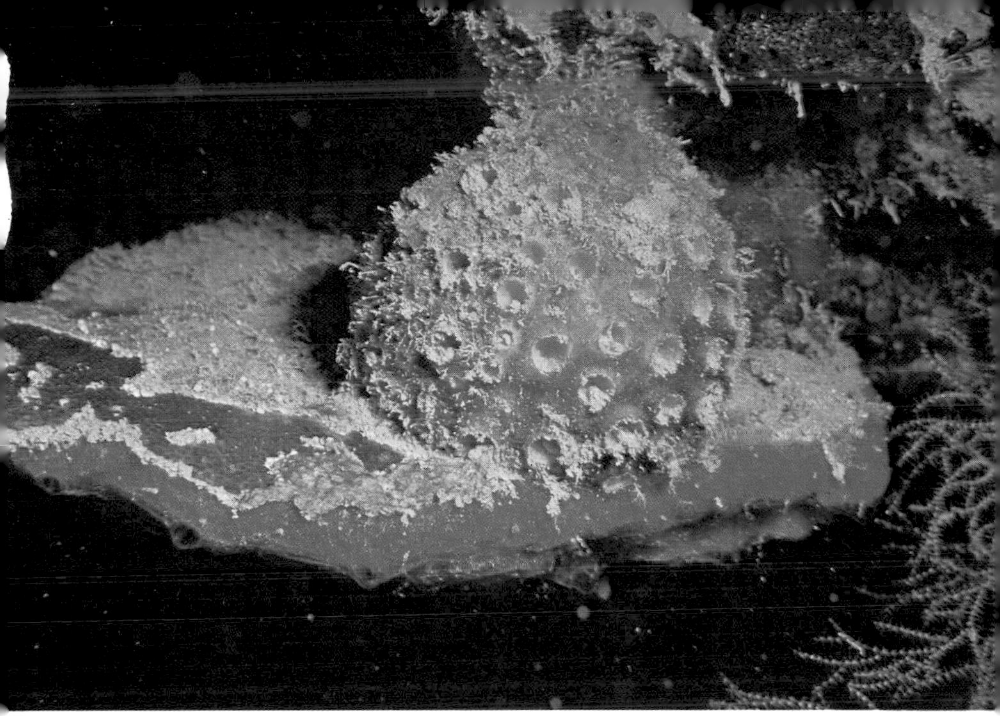

Unidentified demosponge. Jamaica, Spring Gardens, 27 m depth.

Unidentified demosponge. Puerto Rico, La Parguera, Laurel Reef, 12 m depth.

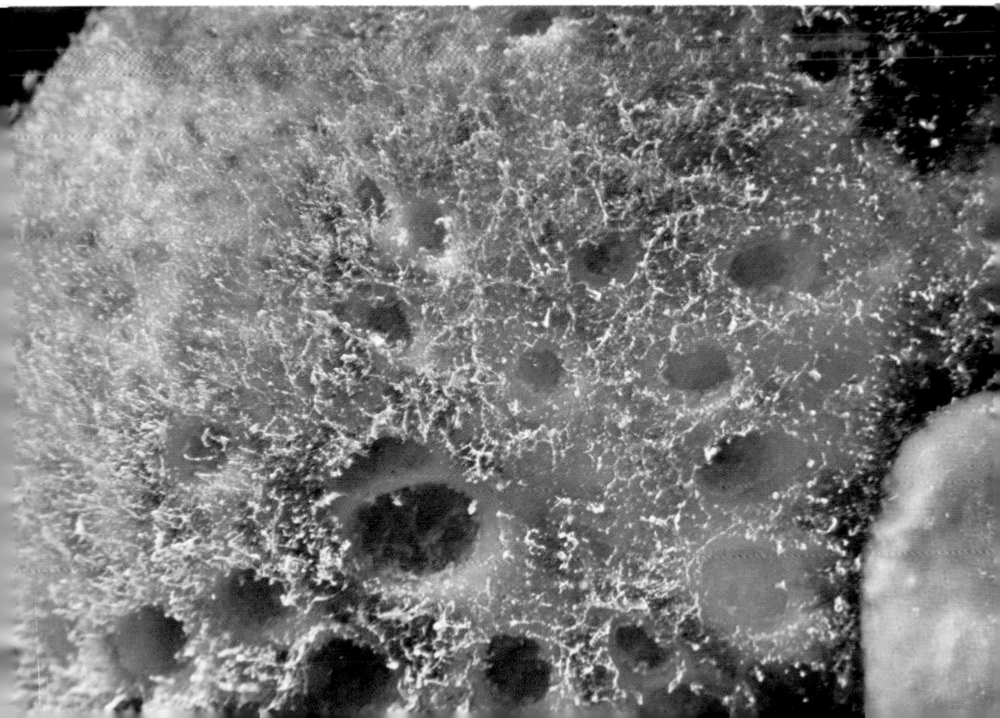

Unidentified tube sponge growing nearly surrounded by the sponge *Ircinia strobilina* which is growing on top of a colony of the coral *Siderastrea radians*, depth 12 m, Discovery Bay, Jamaica.

Steeply sloping deep-reef environment at 40 m depth, Discovery Bay, Jamaica. Two large *Verongia* sponges occur among large numbers of plate-like *Montastrea annularis.*

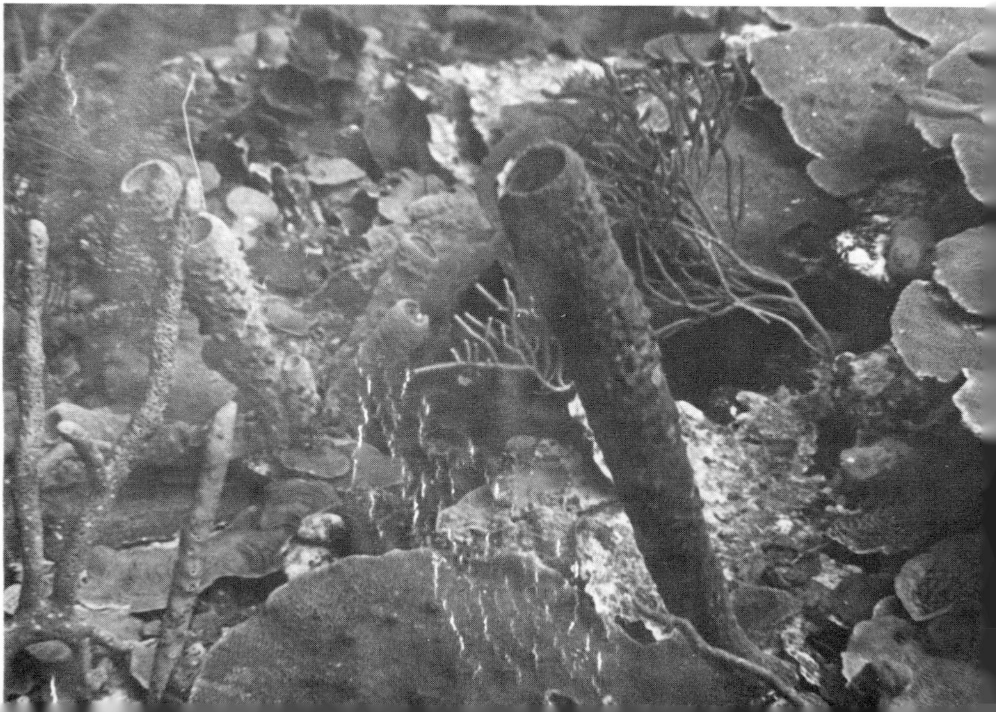

all the red wavelengths at that depth. The sponge, instead, appears black or very dark purple unless a light or photographic flash is used to reveal the red coloration.

Excellent work on the general biology of this species exists. It is characterized by a short life expectancy, rapid growth and rapid replacement. It occurs in clear water areas on outer reefs, often steeply sloping, and attaches to a variety of flexible substrates, such as the stalks of dead gorgonians or loose coral rubble. The flexible substrate reduces the tendency of storm-induced waves and surges to tear the sponge loose from its substrate. Even so, damage from storms is believed to be the greatest cause of mortality. The species can inhabit solid substrates only in "quiet" situations such as can be found in deep-reef areas. It occurs often in transitional areas between sandy areas and solid reef, and young specimens, growing on rubble in open sand channels, have been observed to grow faster than others occurring on the reef proper.

Both sexes may occur in one sponge, and spawning occurs from May to October in Jamaica. Growth is considerably greater in non-spawning periods, indicating that a sizable amount of energy must be expended on the production of gametes.

The species is known from Jamaica, Grand Cayman Island, Puerto Rico, the Bahamas and the Dry Tortugas, Florida. Its depth range is from 13 to 62 m.

References: Reiswig 1971a, 1971b, 1973.

Ulosa hispida Hechtel

This dull orange sponge occurs as thick to thin encrustations on rock surfaces or attached to organisms with a hard skeleton. Its surface is conulose (raised into tiny peaks) and it has a slimy texture. It occurs from mangrove roots to reef areas and seems broad in its environmental requirements.

It is known from several locations in the Caribbean Sea at shallow to moderate depths.

Reference: Hechtel 1965.

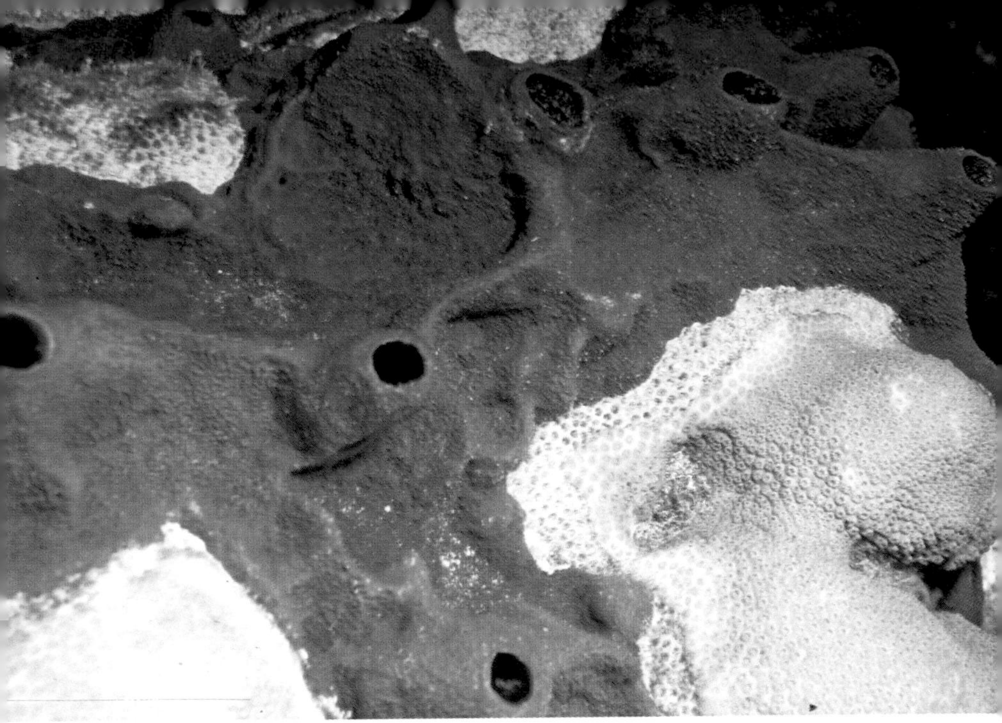

Unidentified demosponge. Jamaica, Discovery Bay, 27 m depth.

Unidentified demosponge. Puerto Rico, Mona Island, 15 m depth.

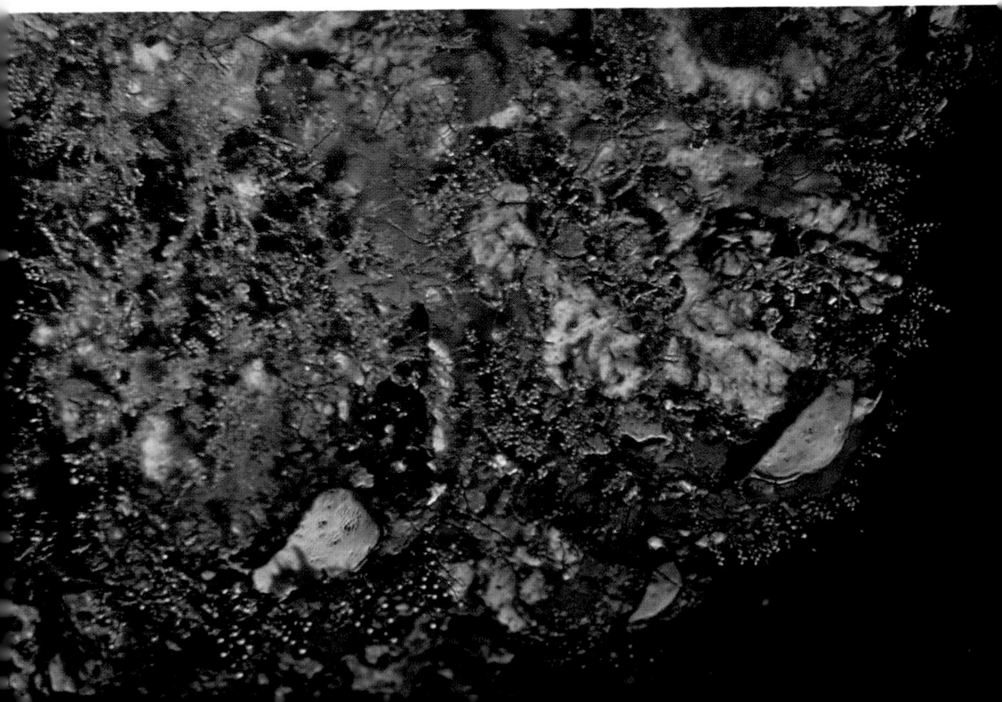

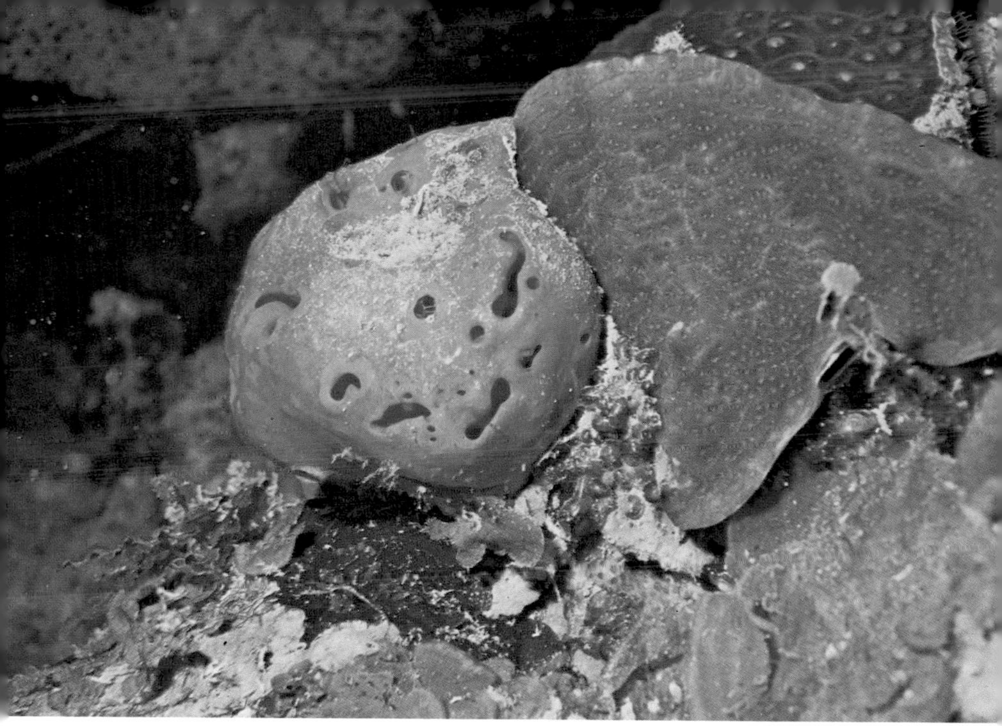

Unidentified demosponge. Jamaica, Discovery Bay, 24 m depth.

Unidentified demosponge. Bahamas, Bimini, Turtle Rocks, 10 m depth.

ORDER HAPLOSCLERIDA

Family Haliclonidae

Haliclona rubens (Pallas)

The dull red branches of *Haliclona rubens* can be found in abundance on coral outcrops. These branches, which arise from a basal encrusting mass, may reach 40 cm in length and are 1-4 cm in diameter. The oscules are small, scattered over the surface of the branches. Rarely the species is massive, usually as a fist-sized hemispherical shape. The color has been reported to vary somewhat, from dark chocolate brown, carmine red or gray-brown to nearly black.

The abundance of this species in a given locality can vary widely over time. It is, however, typical of reef communities and occurs at depths of 1 to 20 m. It occurs throughout the Caribbean, the Bahamas and southern Florida, but there are no definite records from Bermuda.

References: de Laubenfels 1936a, Hechtel 1965.

Haliclona hogarthi Hechtel

This sponge consists of slender (less than 1 cm diameter) purple interconnecting branches. It is soft and compressible, with the surface being without conules. *Haliclona hogarthi* occurs from mangrove areas to reefs at depths to 30 m and is known from southern Florida and the Caribbean.

Reference: Hechtel 1965.

Dasychalina cyathina de Laubenfels

This purple vasiform sponge is surprisingly poorly known considering its relative abundance in some areas. The vases reach at least 30 cm in height and are thin-walled in most instances. The outer surface of the sponge is rugged, and often zoanthid polyps are present. There is an erect stiff fringe around the apical opening.

It is known from several locations in the Caribbean, the Bahamas and southern Florida. Crinoids (Echinodermata) have been observed within the vase-shaped lumen with their arms protruding out the apical opening. The depth range of this species is at least 8 to 23 m.

References: de Laubenfels 1936a, 1953.

Ircinia strobilina at 20 m depth, Discovery Bay, Jamaica with two branches of *Haliclona rubens* growing adjacent to it. The close grouping of the oscules in *I. strobilina* and its conulose surface are clearly visible.

Scattered corals on the edge of a sediment chute at 50 m depth at Discovery Bay, Jamaica. The sponge *Haliclona rubens* is visible in the center of the photograph.

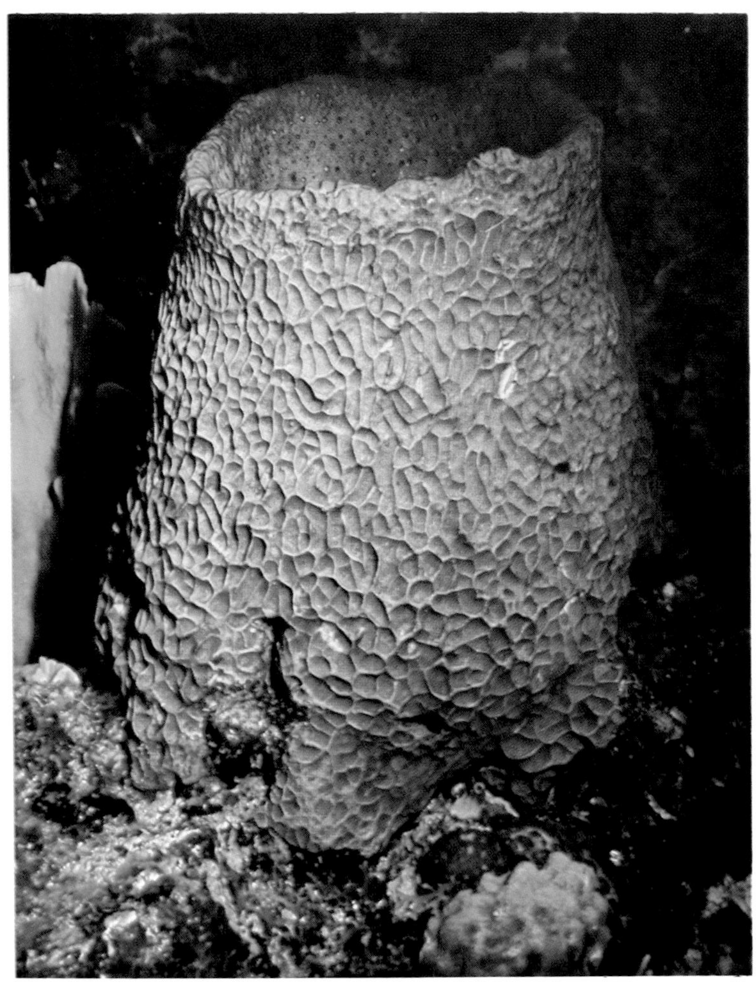

Unidentified demosponge. Puerto Rico, Desecheo Island, 20 m depth.

Unidentified demosponge. Jamaica, Rio Bueno, 15 m depth.

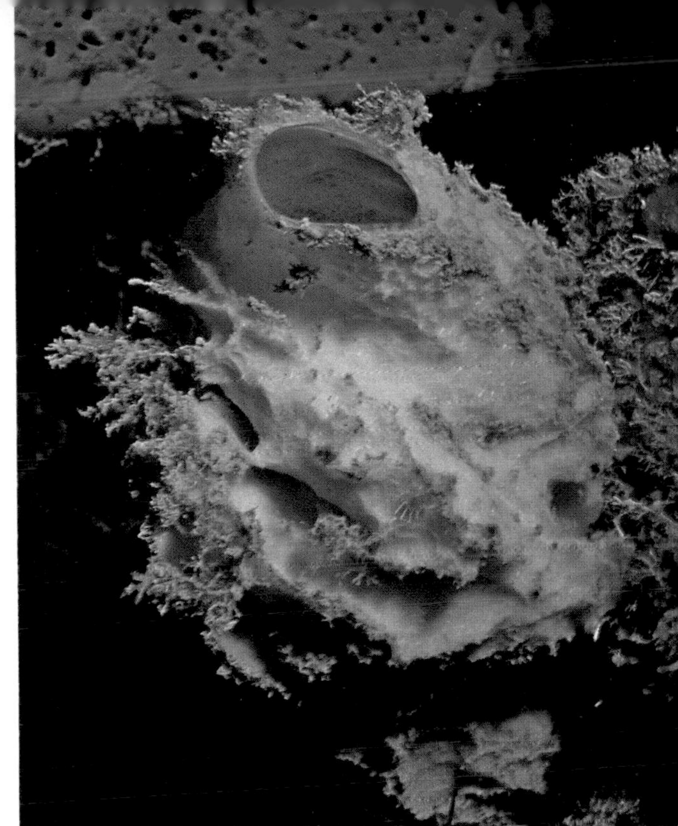

Unidentified demosponge. Jamaica, Rio Bueno, 15 m depth.

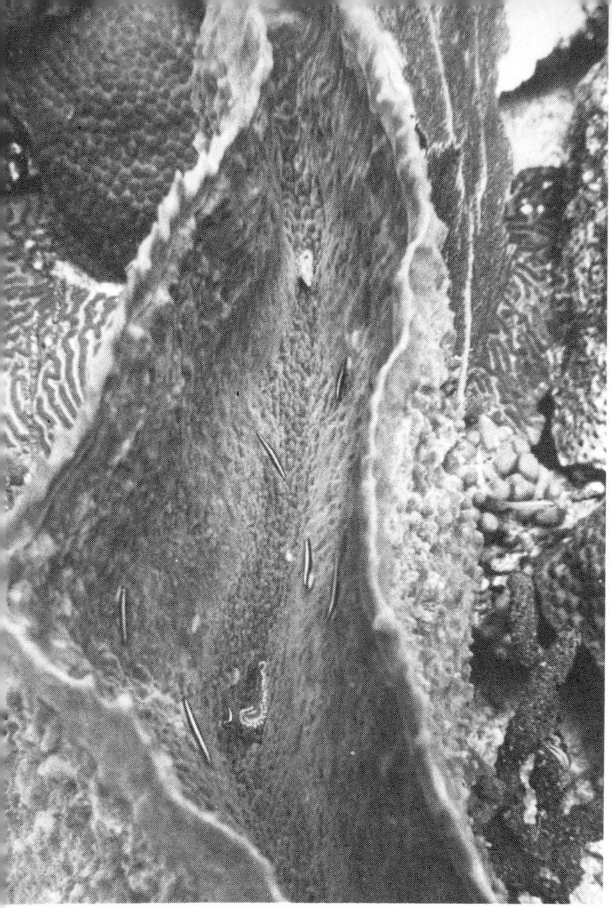

Left: Vertical view into the opening of the barrel sponge, *Xestospongia muta,* at 15 m depth off the west coast of Barbados. This species in this area is elongate in cross section, rather than the typical round shape, possibly in response to current direction. Small gobies, *Gobiosoma prochilos,* occur on the inner surface of the sponge.

Below: Large specimen of the toxic sponge *Neofibularia nolitangere* at a depth of 24 m at Discovery Bay, Jamaica.

Family Desmacidonidae

Gelloides ramosa (Carter)

This sponge is generally branching with diameters of 1 to 2 cm, but also may be flattened along surfaces. Its surface has numerous conules raised 2 to 5 mm. It may have the zoanthid *Parazoanthus parasiticus* associated with it, but this is not invariably so.

It is known from the Caribbean at moderate depths.

Reference: Hechtel 1965.

Xestospongia muta (Schmidt)
Barrel sponge

This sponge is normally the largest species in volume that would be encountered on western Atlantic reefs. It can take a limited variety of forms, but the most typical is a barrel shape with the outer end having a cone-like cavity. This form may reach 1.5 m in height with nearly as great a diameter. This sponge reaches such a size that divers can sit in the cone-like osculum. The weight of such sponges out of water would be several hundred pounds and, since most larger sponges are relatively slow-growing, the age of large *X. muta* is probably considerable. In some locations, perhaps due to local current conditions, the shape is distorted longitudinally, resulting in an elongate rather than circular cross section and a narrow osculum running the elongate dimension of the sponge.

No matter what form the sponge may take, its consistency (hard, almost stony) and outer surface texture (extremely rugged, often buttressed) remain constant. While normally dark brown at diving depths, it becomes lighter in color at greater depths and is white or white suffused with pink at its greatest depths around 120 m. It is one of the few shallow-water reef sponges, being found in as little as 8 m depth, that reaches beyond the lower limit of coral growth. It occurs on flat to vertical surfaces and has even been observed growing on nearly completely overhanging rock surfaces at Isla Monito.

Various fishes of the family Gobiidae (gobies) may occur with it. *Risor ruber*, an unusual tusked goby, has been ob-

Unidentified demosponge. Jamaica, Discovery Bay, 54 m depth.

Unidentified demosponge. Jamaica, Discovery Bay, 15 m depth.

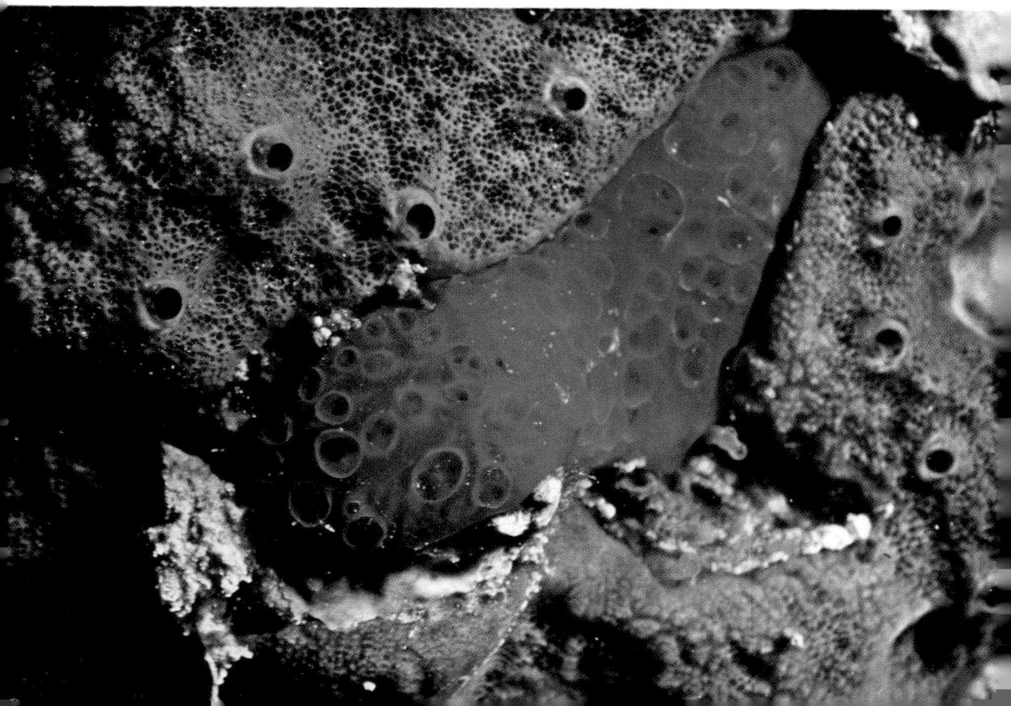

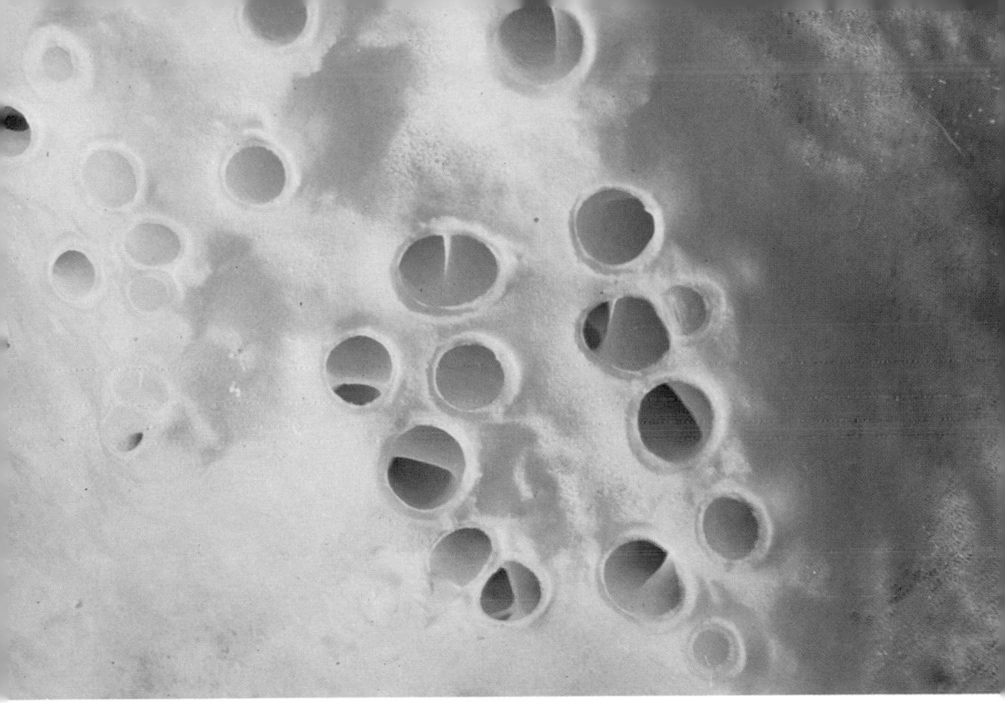

Unidentified demosponge. Jamaica, Rio Bueno, 20 m depth.

Unidentified demosponge. Little Cayman Island, 18 m depth.

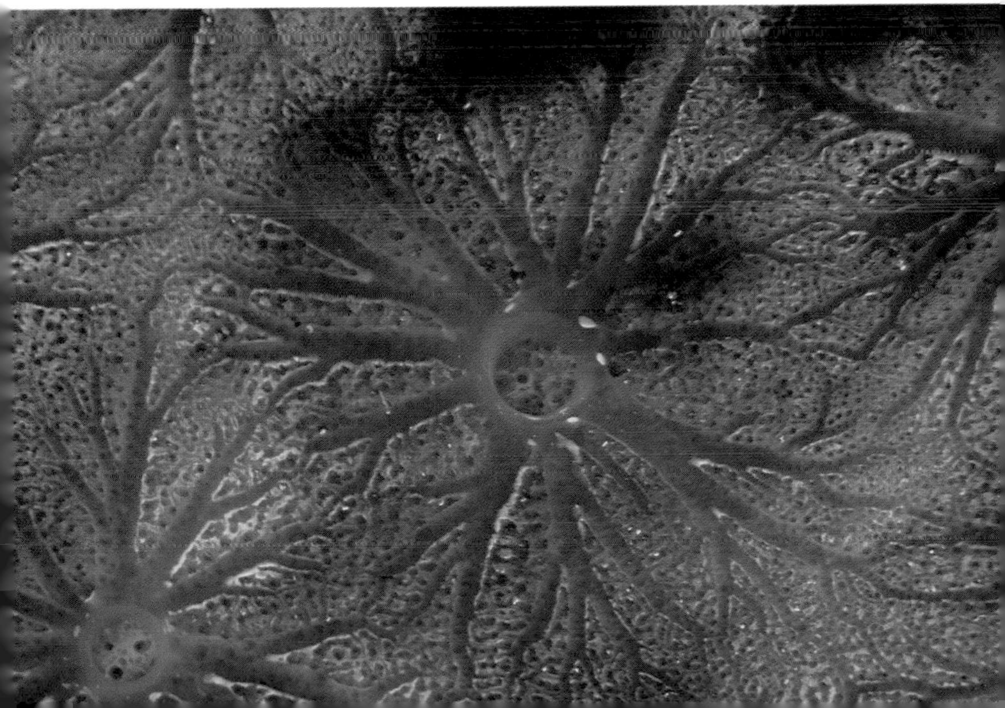

served within the cone-like osculum at night, and a variety of invertebrates can occur on its surfaces at night.

It occurs throughout the Caribbean, the Bahamas and Florida reefs, and may be present in Bermuda.

References: de Laubenfels 1953, Hartman 1973.

Xestospongia sp.

This dark brown normally cup-shaped sponge is a species of the genus *Xestospongia* which may well be undescribed. It is hard and stiff in consistency and the inner and outer surfaces are fairly smooth but not slick to the touch. Occasionally one portion of the side of a cup-shaped sponge will not be present. An unidentified zoanthid can also occur with this sponge.

Known from the northern Bahamas at around 12 m. This species has been observed releasing sperm ("smoking") in early April.

Neofibularia nolitangere (Duchassaing and Michelotti)

This dark brown massive sponge can be found in a variety of forms. Often it is large, over 1 m across, with several large cloacal openings with smaller oscules on the inner walls. The sponge around each cloacal opening may be conical or cylindrical, but the walls are invariably thick. The inner surface of the cloacal cavity may appear to be buttressed in some large individuals.

Neofibularia nolitangere can also occur as a series of rounded, connected mounds with the cloacal openings more restricted. In all growth forms the outer texture, although generally smooth, is not shiny and the walls of the cloacal cavity are much rougher than the outer surface. Usually small white particles can be seen protruding from the cloacal walls. These are the parasitic polychaete worm *Syllis spongicola*, not part of the sponge.

This sponge should not be touched as it produces a dermatitis reaction in humans which includes smarting sensations and numbness of the affected skin. Individuals can become sensitized to contact with this sponge and have a more serious reaction.

The species occurs on reasonably level bottoms, often among corals, in depths from 3 to at least 46 m. It is common on West Indian reefs but has been taken as far north as North Carolina.

The polychaete worm *Syllis spongicola* is parasitic on this and several other species of western Atlantic sponges. Various fishes, such as the gobies *Gobiosoma horsti* and *G. chancei*, live associated with this sponge and others and feed largely on *S. spongicola*.

This sponge has been observed spawning (releasing sperm) in Jamaica in late October.

References: Hartman 1967, Reiswig 1970.

Iotrochota birotulata (Higgin)

This species forms aggregations of branches, occasionally anastomosing, either erect or sprawling. The branches, 1 to 4 cm in diameter, may be 50 cm in length. They are purplish to black but often have layers of emerald green cells over much of their surface. These green cells have dendritic patterns and are the outer layer of subdermal canals which lead to the inconspicuous oscules. Quite often the orange zoanthid *Parazoanthus swiftii* occurs as winding chains of polyps on the branches of *I. birotulata*, adding to the colors of the sponge. The species emits a purple exudate when squeezed and is one of the favorite prey items of some of the sponge-eating fishes in the West Indies.

The species is common in the West Indies, the Bahamas and southern Florida, but is not recorded from Bermuda. It is known from a depth of 2 to at least 15 m.

References: de Laubenfels 1936a, 1953, West 1971.

SUBCLASS KERATOSA: HORNY SPONGES

The skeleton of keratose sponges is composed of spongin fibers without siliceous spicules. The fibers usually form a network which often has hard materials (rocks, etc.) from the environment incorporated into it. The surface of these sponges is often leathery and may be smooth or covered with elevations, often pointed, termed conules. The commercially

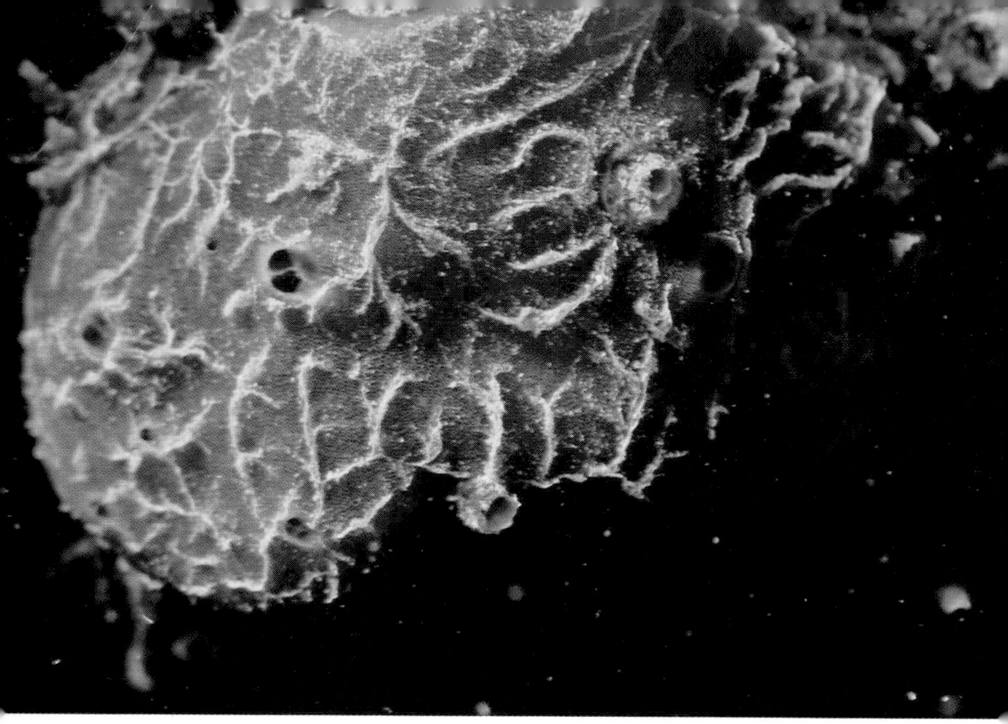

Unidentified demosponge. Puerto Rico, La Parguera, Mario Reef, 3 m depth.

Unidentified demosponge. Jamaica, Discovery Bay, 15 m depth.

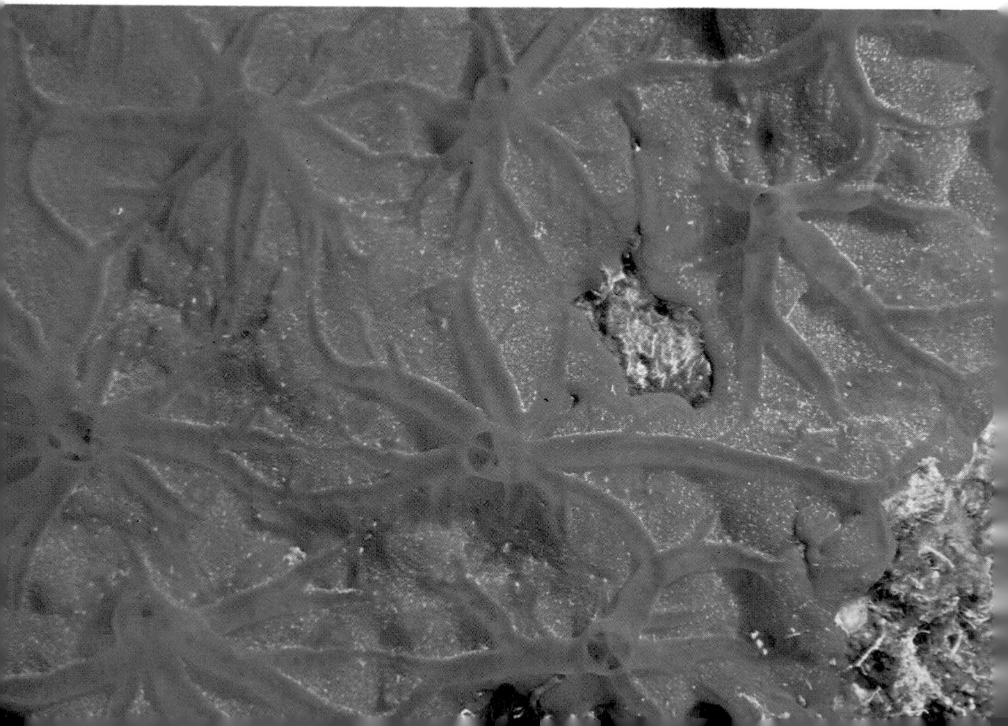

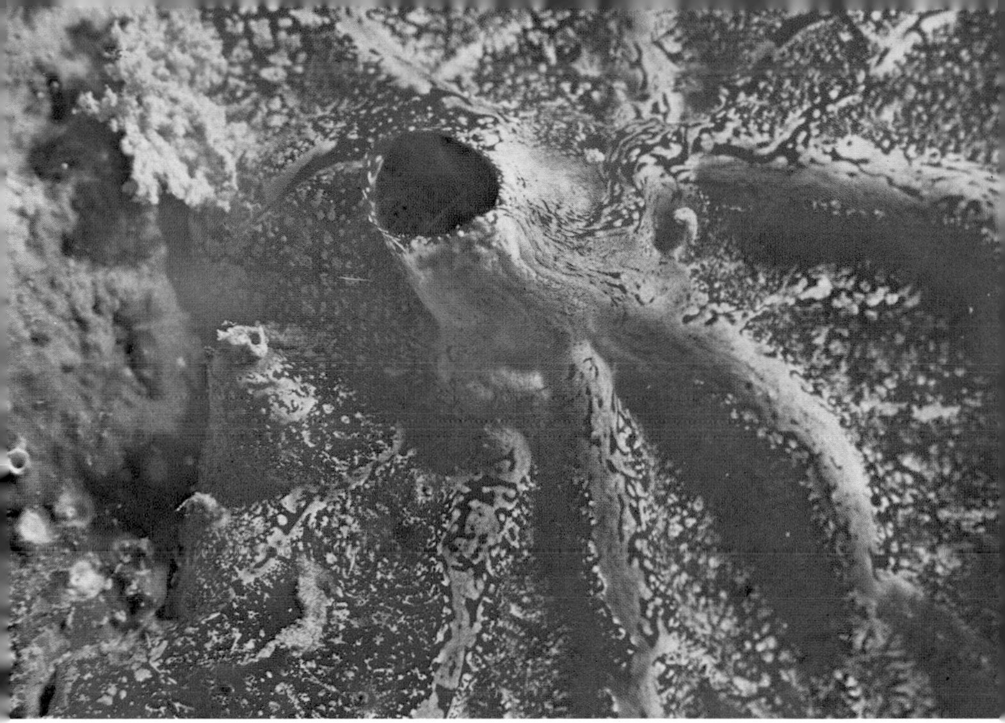

Unidentified demosponge. Puerto Rico, La Parguera, Laurel Reef, 12 m depth.

Unidentified demosponge. Jamaica, Rio Bueno, 15 m depth.

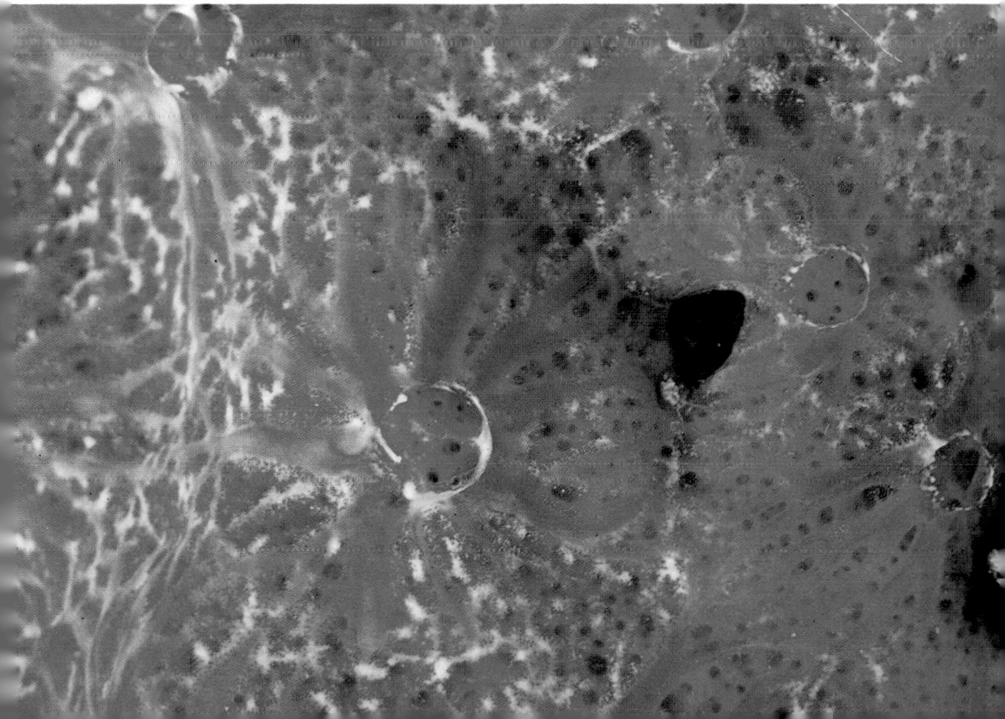

valuable sponges, although not normally occurring on reefs, are members of this subclass.

Ircinia fasciculata (Pallas)
Stinker sponge

This sponge can occur in a variety of forms, including a single or fused series of lobes, branched, subspherical and rarely cup-shaped. The oscules never occur in groups (as in the following species) and are dark brown or black with the tissue within them dark. Lobes over 15 cm in diameter are common and the conulose surface texture is characteristic. The conules have networks of easily visible lines and ridges radiating out and connecting them. The outer surface is tough and difficult to tear, while the sponge is somewhat compressible.

The stinker sponge, so-called due to its fetid odor when removed from the water, is found in a wide range of shallow-water habitats. It occurs on reefs, often adjacent to corals, and can be one of the most abundant sponges. Plant cells, possibly symbiotic, have been reported from the surface layers of this sponge.

It is widespread in the western Atlantic from North Carolina and Bermuda throughout the West Indies. The species may occur worldwide in tropical waters although there is some doubt as to whether Pacific and Indian Ocean specimens are truly this species. It is known from depths of 1 to 20 m.

References: Hechtel 1965, de Laubenfels 1936a, 1948, 1950.

Ircinia strobilina (Lamarck)

This species is a massive sponge, often lobate or cake-shaped, reaching over 50 cm in diameter. The oscules are in groups near the center of the upper surfaces, their rims and inner surfaces black. The gray outer surface, like that of *I. fasciculata*, is conulose with radiating patterns at each conule and sometimes having connecting ridges. *Ircinia strobilina* is tough in consistency, being difficult to tear, and has an unpleasant odor. The species could be confused with the loggerhead sponge, *Speciospongia vesparia*, which can occur in reef areas and has grouped oscules on the upper surface. *Ircinia strobilina* is easily distinguished by its easy compressibility as compared to the incompressible, woody consistency of the loggerhead sponge. The latter species also possesses spicules which the keratose *I. strobilina* does not.

It is common throughout the West Indies and in some localities, such as Dry Tortugas (Florida), has been reported as the most abundant sponge. Also known from the Bahamas and Bermuda, and there are records, possibly representing a similar species, from the Mediterranean Sea and the Pacific Ocean. Its known depth range is between 0 and 17 m.

References: de Laubenfels, 1936a, 1950, Hechtel 1965.

Verongia fistularis (Pallas)
Yellow-green candle sponge

The cylindrical tubes of *Verongia fistularis* are as characteristic of the well developed Caribbean medium depth coral reef as are most species of corals. The color, unlike any other sponge occurring on the reef, has been described variously as mustard or greenish yellow. There may be several tubes united at the base in a single sponge, each with a cylindrical lumen and terminal cloacal opening (not a true osculum). These resemble hollow candles and range from small (a few cm) to nearly 50 cm in height with a diameter of over 8 cm. Small digitate processes are often found on the apical rim of these sponges.

The intense yellow-green color of *V. fistularis* at depths as great as 40 m may be due to fluorescence of material in the sponge surface since at that depth nearly all the yellow wavelengths of sunlight penetrating the surface have been absorbed by the water. Extracts of *V. fistularis* have been found to fluoresce at certain wavelengths of light (blue) which penetrate water well.

The individual tubes are soft, capable of being squeezed shut by a hand. When handled roughly or taken from the water the sponge turns a dark purple which also stains the hands of the person touching it. Although apparently harmless, this purple stain will persist for several days.

Various fishes, including the sponge cardinalfish,

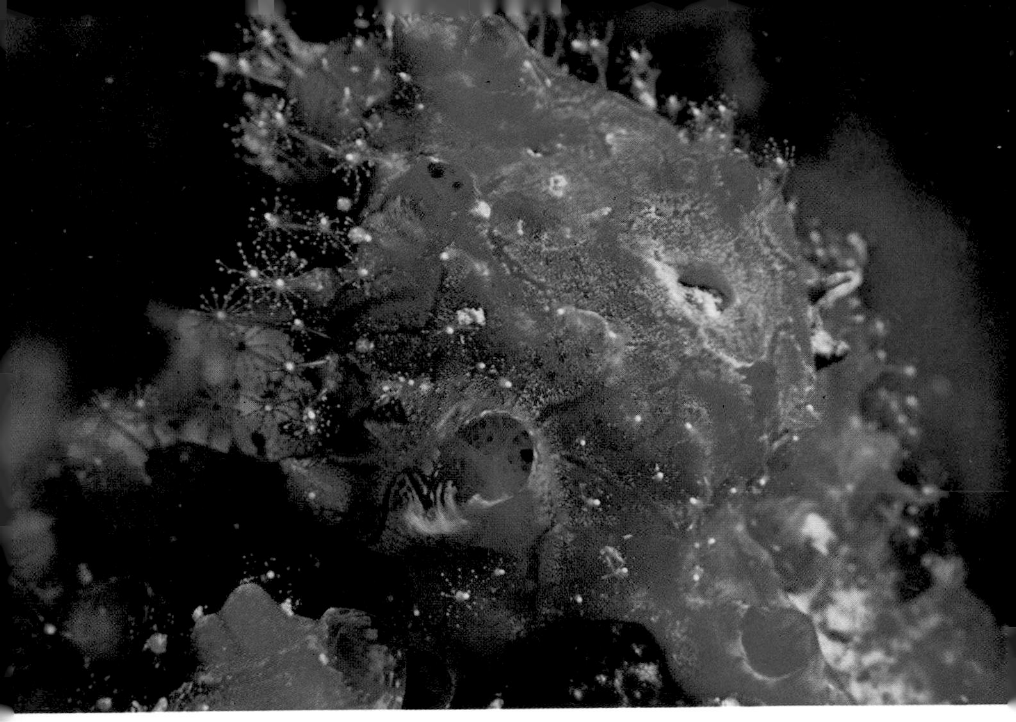

Unidentified demosponge with unidentified hydroid, Puerto Rico, La Parguera, Laurel Reef, 12 m depth.

Unidentified demosponge. Jamaica, Rio Bueno, 20 m depth.

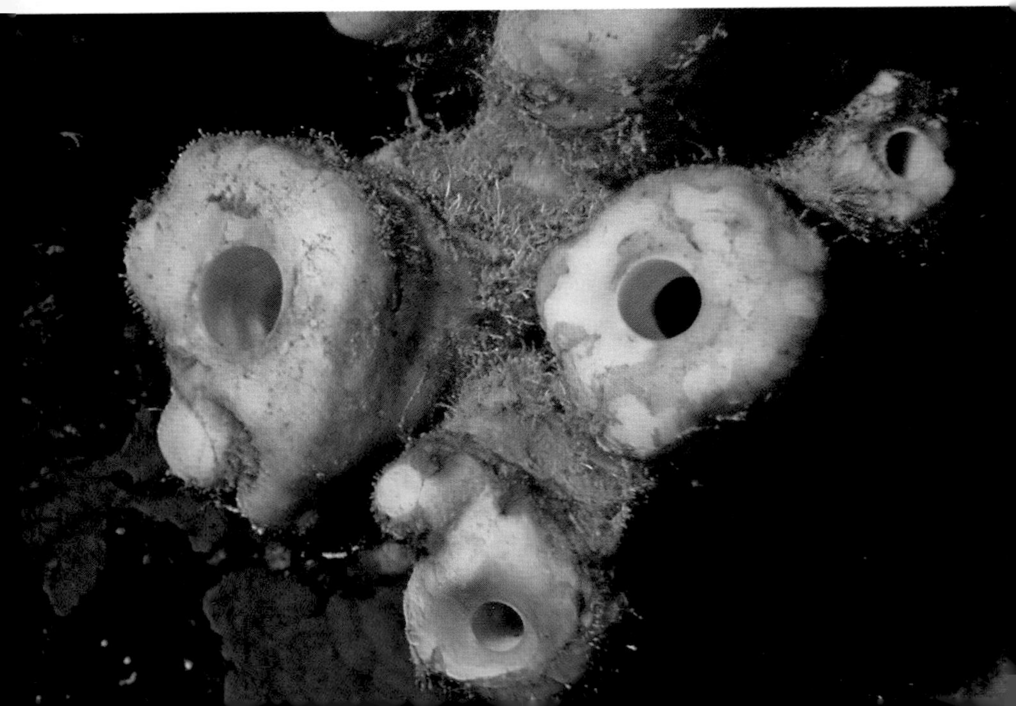

Unidentified demosponge. Jamaica, Discovery Bay, 54 m depth.

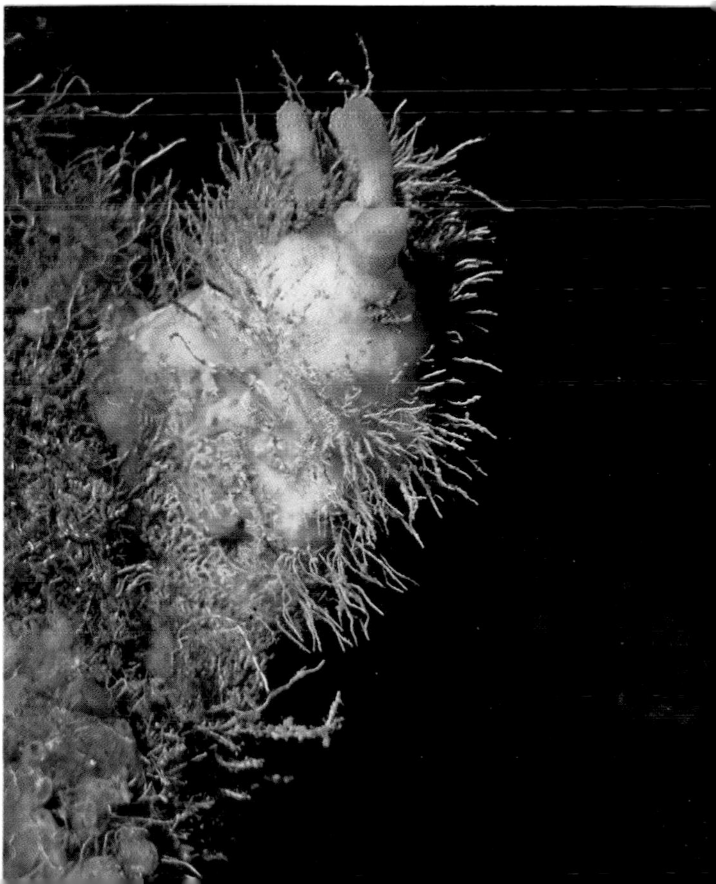

Unidentified demosponge. Puerto Rico, Desecheo Island, 6 m depth.

Phaeoptyx xenus, and species of *Gobiosoma* (*Elacatinus*) gobies, may be found within the lumen.

The sponge occurs between 5 and 40 m on reefs, usually facing open water. While fairly rare along the Florida coast, it is common at Dry Tortugas, Florida. It occurs throughout the Caribbean and the Bahamas and is recorded from Bermuda and the Brazilian coast.

References: de Laubenfels 1948, 1950, Hechtel 1965, Read, *et al.* 1968.

Verongia lacunosa (Lamarck)

A thick-walled, tubular species of *Verongia* which usually occurs as single tubes, although occasionally multiple-tube individuals are observed. The outer surface is strongly convoluted or covered with circular pits or elongate grooves. The sponge is various colors (greenish yellow, pinkish lavender, red-brown) and reaches over 1 m in height and 10 cm in diameter with a large apical cloacal opening.

This strictly deep reef species is known from several localities in the West Indies and is common on many Bahamian reefs. It has been recorded attached to the coral *Porites furcata* but normally occurs near, but unassociated with, corals. It occurs at depths of 20 m to at least 50 m.

Reference: de Laubenfels 1948.

Verongia gigantea (Hyatt)

This large urn- or tube-shaped sponge is easily identified by several characteristics. It always consists of a single element, has a large apical osculum and may be yellow-green, yellow or olive-green in color. The outer surface is "lumpy" and there are smaller reticulations on the "lumps." The inner surface of the central atrium is smooth and bright yellow, but is peppered with the openings of the exhalant canals. Individuals reach a volume of 120 liters (about 40 gallons) and nearly 1 m in height. The basic proportions of individuals do not appear to vary greatly with growth, the total diameter nearly half or more of the height. The walls of *V. gigantea* are thick, reaching as much as 10 cm. Other species of

Right: Unidentified tubular sponge, possibly of the genus *Verongia.*

Below: The massive sponge *Verongia gigantea* is yellowish or olive in color and occurs only on deep fore reef environments. This large specimen was photographed at 40 m depth at Discovery Bay, Jamaica.

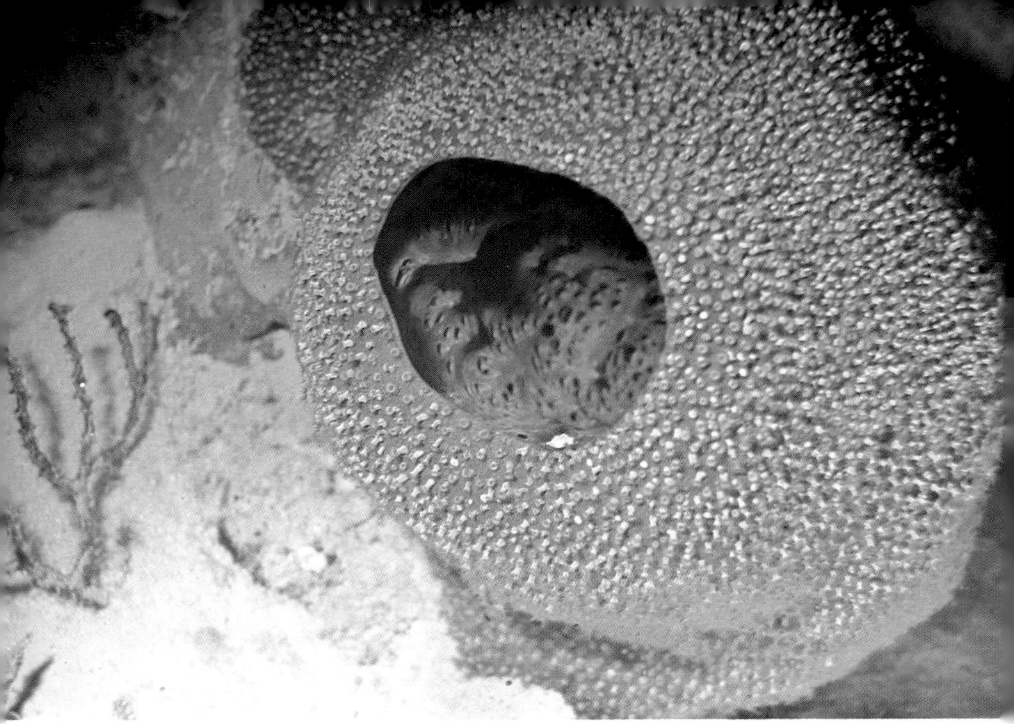

Unidentified demosponge. Jamaica, Discovery Bay, 54 depth.

Unidentified demosponge. Bahamas, Grand Bahama Island, 15 m depth.

The sclerosponge *Ceratoporella nicholsoni* is the largest of Atlantic sclerosponges and occurs in deep-reef caves. At greater depths they may be found out in the open or beneath overhanging ledges. Jamaica, Spring Gardens, 20 m depth.

The sclerosponge *Stromatospongia vermicola* is always found growing with tubes of serpulid worms in dark, cavernicolous locations. Puerto Rico, Desecheo Island, 10 m depth.

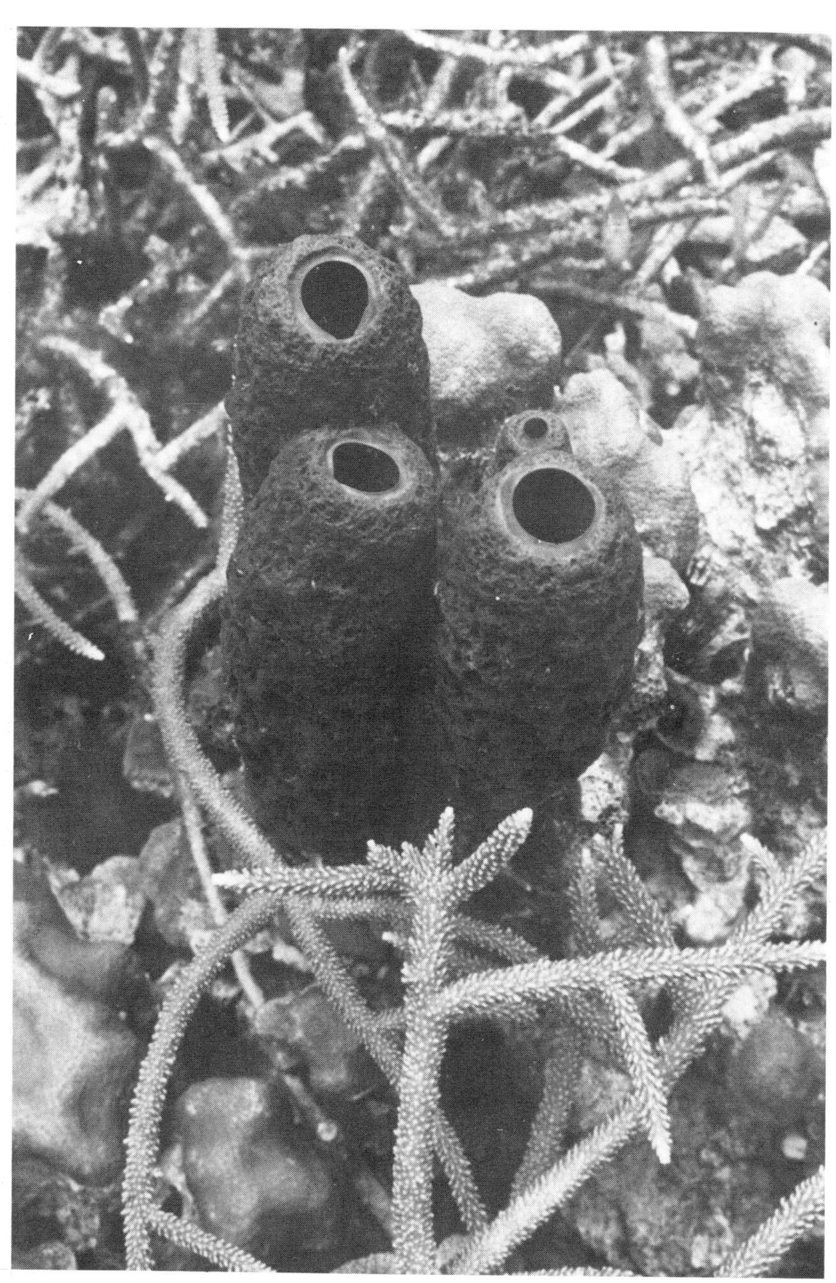

Specimen of *Verongia fistularis* nestled among the corals *Montastrea annularis* and *Acropora cervicornis* on the fore reef at 15 m depth, Discovery Bay, Jamaica.

Verongia (*V. fistularis, V. lacunosa, V. archeri*) are normally much more slender than *V. gigantea.*

Fortunately some superb work exists on the biology of *V. gigantea.* The water pumping activity of the species, while normally high, undergoes periodic cessations of slightly less than one hour every few days for unknown reasons. There is no synchronization of these cessations among a population on a reef, and they are not related to environmental conditions. Synchronous long term cessations (of a few days) have been noted at other times for *V. gigantea*, but are believed to be due to massive sperm release (so much that it clouds the water overlying the reef) by other species of sponges.

Verongia gigantea is restricted to solid exposed reef areas with clear water generally between 25 and 52 m. Populations are seldom large in terms of individuals per unit area, but the large size of individual sponges can result in a high standing crop. They are slow growing, with a long life expectancy (at least 50-100 years) and slow replacement. The young begin growth beneath flattened coral plates, growing out from underneath and expanding their holdfast as they increase in size. The slow growth of *V. gigantea* may be due to the presence of a parasitic polychaete worm, *Syllis spongicola*, in densities of 50-100 worms per milliliter of sponge! Symbiotic bacteria also occur in *V. gigantea.*

Each sponge produces only gametes of a single sex (dioecious) and they probably spawn during a limited period in the winter. The geographic range is poorly known, the species being definitely recorded from only Jamaica and the Bahamas.

References: Reiswig 1971a, 1971b, 1973.

Verongia archeri (Higgin)

This is a deep reef species which is relatively poorly known. It consists of a single to several tubes which can be as much as 2 m in length but very slender. The walls are thin (only a few cm) and the soft tube is easily squeezed shut with the hand. The exterior is reddish brown, often having a variety of filamentous growths occurring on it, with the central

The small mammilate masses of *Stromospongia norae* are found only on the walls and ceilings of caves. Jamaica, Pear Tree Bottom, 20 m depth.

Goreauiella auriculata has a thin dish-like skeleton with the excurrent canals of the sponge clearly visible in undisturbed individuals. Jamaica, Pear Tree Bottom, 20 m depth.

The hydroid *Halocordyle disticha* can occur on a variety of substrates as long as there is some current flow. Puerto Rico, Aguadilla, 6 m depth.

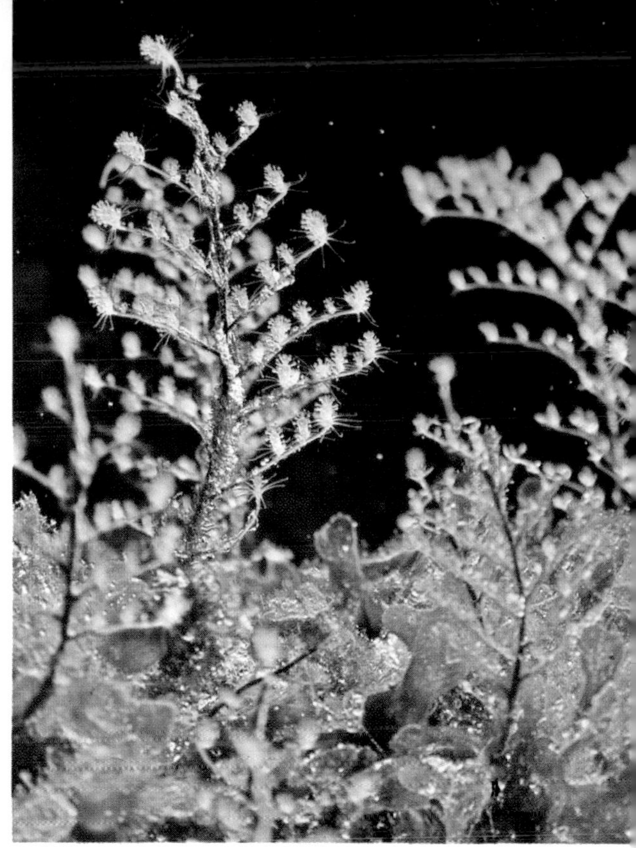

The sclerosponge *Goreauiella auriculata* can be extremely abundant on the roofs of dark caverns in outer reef areas of moderate depths. Jamaica, Pear Tree Bottom, 30 m depth.

atrium cream colored.

Tubes may grow vertically, horizontally or slightly downward. If several tubes comprise a single sponge, the tubes radiate from nearly a single point. The water pumping pattern is similar to that of *V. gigantea*, with a consistent high level and periodic cessations of short duration.

Massive sperm releases, appearing as smoke issuing from the large apical osculum, have been noted for *V. archeri* during February in Jamaica. Various fishes including gobies (*Gobiosoma*) and cardinalfish (*Phaeoptyx xenus*) may occur in the lumen.

Geographic distribution is poorly known, with definite records from Jamaica, the Bahamas and Yucatan. This sponge is strictly a deep reef inhabitant and is seldom found above 30 m depth.

References: Reiswig 1970, 1971a.

Verongia longissima (Carter)

This species is different in its general appearance from the preceding species of *Verongia*. It is ramose (branching) with the digitate-like branches only a few cm in diameter. The branches often anastomose (reconnect) and there is little tendency for the sponge to be massive. Fine conules are found on the surface of the sponge and small apical oscules occur on some branches.

Color varies from gray to yellow, although most accounts describe it as drab or colorless.

Verongia longissima is abundant in the West Indies and is known from Dry Tortugas, Florida. Its depth range is known to be from 15 to 25 m.

References: de Laubenfels 1936a, 1948.

UNIDENTIFIED DEMOSPONGIAE

A number of unidentified Demospongiae have been figured. Many of these are deep reef species and may well be undescribed. Others are inconspicuous or members of the encrusting sponge fauna. These encrusting species are particularly colorful and are perhaps the most poorly known group of Caribbean reef sponges. Many species are similar in appearance and their forms variable to conform to this environment.

CLASS SCLEROSPONGIAE: SCLEROSPONGES

One of the most exciting discoveries in coral reef biology during the 1960's was the occurrence of sclerosponges in caverns and deeply shaded areas on West Indian reefs and their importance in the cave and deep reef community. Sclerosponges have a hard skeleton consisting of aragonite (calcium carbonate), silica spicules and organic fibers with the living portion of the sponge forming only a thin veneer on the previously deposited material. Some species reach as much as 1 m in diameter and are the only living organisms depositing large amounts of both silica and calcium carbonate. The strength of the skeleton and the firmness of its attachment to the substrate can be attested to by one incident where the hydrographic wire from a 23 m-long research ship was attached around a 1 m in diameter unidentified sclerosponge in the Bahamas. The winch was unable to tear lose this sponge and the ship was also unable to pull it loose by steaming with both engines!

Close resemblance of the living sclerosponges to the fossil stromatoporoids (variously considered as hydrozoan coelenterates or sponges) has been noted, as well as similarities to certain fossil chaetid "corals." The living sclerosponges have been suggested as a fourth class of Porifera and here are considered as such.

References: Hartman and Goreau 1970, Lang *et al.* 1975.

Ceratoporella nicholsoni (Hickson)

This is the largest, most abundant and most distinctive of the western Atlantic sclerosponges. The orange or peach colored masses reach 1 m in diameter and 50 cm in thickness Mammillate processes bearing excurrent canals and oscula occur on the finely-pitted living surface with the canals forming visible star-shaped patterns. The incurrent pores are small (about 0.1 mm in diameter) and are located over the pits.

The feather-like hydroid *Sertularella speciosa* often occurs as groups. Jamaica, Discovery Bay, back reef, 6 m depth.

A closeup view of *Sertularella speciosa,* with the individual polyps visible on the branches of the hydroid. Puerto Rico, La Parguera, Laurel Reef, 12 m depth.

The sympodial hydroid *Thyroscyphus ramosus* has delicate branches which arise alternately from the main branches. Jamaica, Rio Bueno, 18 m depth.

In hydroids with a sympodial growth pattern, such as *Cnidoscyphus marginatus*, the primary axis of the branch is produced by a succession of polyps rather than a single apical polyp. Little Cayman Island, 18 m depth.

Like all sclerosponges, *C. nicholsoni* has the skeleton formed of three components. Siliceous spicules are embedded in an organic matrix which becomes included in the aragonite as the sponge grows. Many of the spicules dissolve or erode later, but *C. nicholsoni* shows a series of nearly concentric growth rings when sliced in half with a rock saw. The resulting skeleton is dense, about twice that of scleractinian (stony) corals and approaching that of solid limestone.

This species is cavernicolous (cave-dwelling) in water less than 40 m deep, but at depths of 70 m it may be found on exposed faces competing with deep-water stony corals for space. The sponge is found in zones of twilight, be it in cave or deep reef on steep, well drained of sediment rock faces, and is known to occur as deep as 200 m. The densest populations are found in caves, and the species is capable of constructing reef framework under conditions and at depths below those required for reef-building corals.

Ceratoporella nicholsoni is known from a variety of localities, including Jamaica, Cuba, Belize, the Bahamas, Mona Island and Desecheo Island. The lower depth limit of the species may be determined by the presence of a thermocline in some areas. Zoanthids have been found on some individuals of *C. nicholsoni*.

References: Hartman and Goreau 1966, 1970.

Stromatospongia vermicola Hartman

The most unusual Caribbean sclerosponge is the encrusting *Stromatospongia vermicola*. It has always been found growing associated with tubes of a serpulid worm forming a tangled, flattened mass of apricot to salmon-pink sponge and worm tubes. Little is known of this relationship or how it begins, but the sponge tissue often reaches to the insides of the serpulid tubes.

The species is found exclusively in deeply shaded locations, usually the rock walls of caves or on the underside of large corals in fairly deep water. The largest individuals (up to 40 cm diameter) are found below depths of 60-70 m, and those above 30 m are small. The known depth range extends from 10 to 95 m and, while known from only a few localities, the species should be widespread in the West Indies.

The species is encrusting; a specimen 20 cm in diameter may be only a few centimeters thick. Evidence exists that the larvae may be "incubated" by the parents.

References: Hartman 1969, Hartman and Goreau 1970.

Stromospongia norae Hartman

This sclerosponge forms small (up to 5 cm on a side by 4 cm high) nodular or mammillate masses. The living sponge is cream to ecru beige in color and has the excurrent canals in radiating patterns on the outer surface.

Stromospongia norae occurs only on the walls and ceilings of dark reef caves. In the proper habitat it can be quite abundant, with as many as 400 individuals having been recorded in a one meter square. It has always been found associated with a serpulid worm tube which it encrusts. The exact relationship between the sponge and the worm is unknown. The species is known only from Jamaica at 8 to 85 m but is generally found above 35 m.

References: Hartman 1969, Hartman and Goreau 1970.

Goreauiella auriculata Hartman

The small yellow dish-like or auriculate (ear-like) individuals of *Goreauiella auriculata* are extremely delicate. The edges of the thin sponge are turned upward or curled downward and the entire specimen is attached by a short stalk or peduncle to rock substrates. While specimens reach 16 cm in long dimension, the thickness may be only 2 or 3 mm.

Goreauiella auriculata is a true cavernicole, growing attached to the ceilings of the darkest reef caves with the living tissue surface directed downward. Because of its propensity for the darkest localities, it is almost essential that divers carry lights to observe these creatures.

The living surface of the sponge has multibranching patterns and occasionally commensal zoanthids which cause the sponge to form upright processes on its usually smooth surface.

The species is known from several localities, including

Cnidoscyphus marginatus can occur in dense stands such as shown here on a rocky face. Puerto Rico, Monito Island, 6 m depth.

The hydroid *Aglaophenia allmani* has feather-like branching and is capable of stinging humans. Puerto Rico, Rincon, 20 m depth.

The hydroid *Plumaria habereri* is a large species reaching 30 cm in height and is generally branched in a single plane. Puerto Rico, Mona Island, 15 m depth.

Aglaophenia allmani often has small spots of white on the final branches as is shown in this photograph. Caicos Islands, Providenciales, 20 m depth.

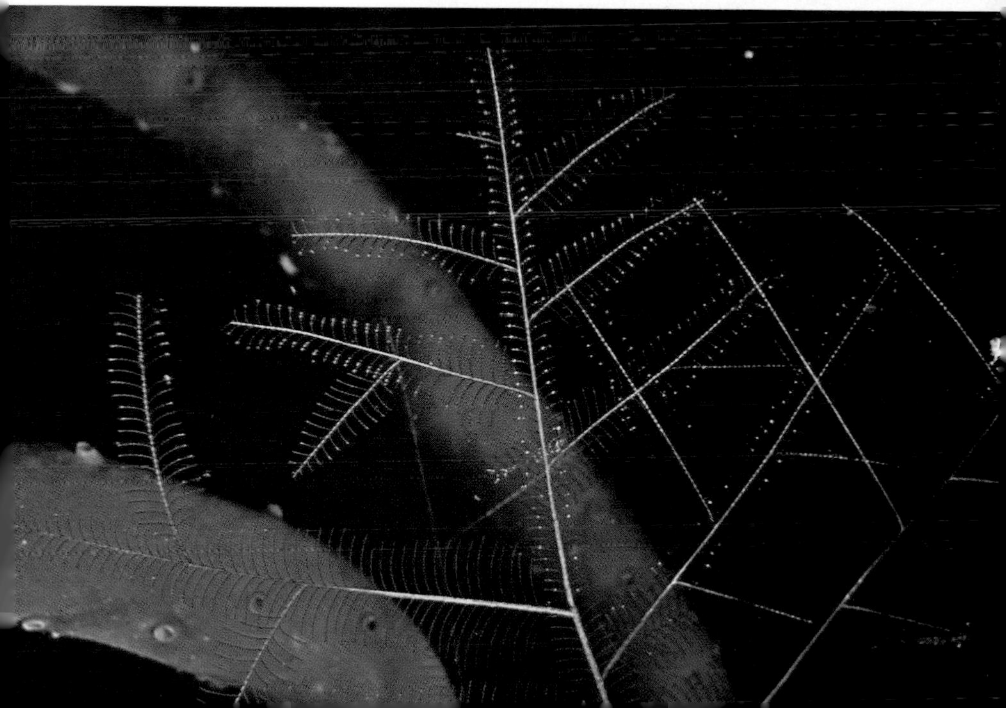

Jamaica, the Bahamas and Mona Island at depths between 6 and 70 m. While it occurs in many of the small but dark reef caves, it is associated with a number of species of ahermatypic (non-reef-building) corals, including *Desmophyllium* and *Caryophyllia*, in some of the larger reef caves.

References: Hartman 1969, Hartman and Goreau 1970.

A species of *Agelus* on the deep reef at Discovery Bay, Jamaica, depth 30 m. Because of systematic confusion in this group of sponges it is impossible to identify this specimen to species.

Phylum Coelenterata (Cnidaria)

The cells of Coelenterata are organized into tissues, and the body usually consists of three layers: an external epidermis, an internal gastrodermis and between these a mesoglea which may or may not be cellular. Coelenterates are radially symmetrical, at least superficially, and generally possess stinging cells, called nematocysts, which are of diagnostic value.

Nematocysts occur in many forms and are found variously in the epidermis and the gastrodermis. They are triggered by a variety of stimuli, injecting a venom into a prey organism, and once discharged cannot be used again. New ones are produced to replace those discharged.

Coelenterata have two basic morphological types, the polyp and the medusa. The polyp is attached to a substrate at the aboral end and has a mouth, and usually tentacles, on the free (or oral) end. The medusa is unattached, with a bell-shaped dome having tentacles on its margin and a mouth on the concave side. Many coelenterates have alternating polyp-medusa stages, while others have one stage reduced or even absent. The polyps may often reproduce asexually by budding and can occur either singly or as a colony. In a number of groups the polyps lay down an exoskeleton of calcium carbonate.

CLASS HYDROZOA

The hydrozoans usually have a polyp and a craspedote (possessing a velum) medusa stage; a few species may lack one stage. Those that are of any significance on reefs have the polypoid portion of the life cycle predominating over the free-swimming medusoid portion. The class is divided into

This feather-like hydroid, *Gynangium longicauda,* grows from a rhizome so that abundant stands can occur. Puerto Rico, Mona Island, 15 m depth.

The hydroid *Solanderia gracilis* looks more like a sea fan than a hydroid. It occurs on rocky substrates, often in areas swept by strong waves or currents. Puerto Rico, Desecheo Island, 4 m depth.

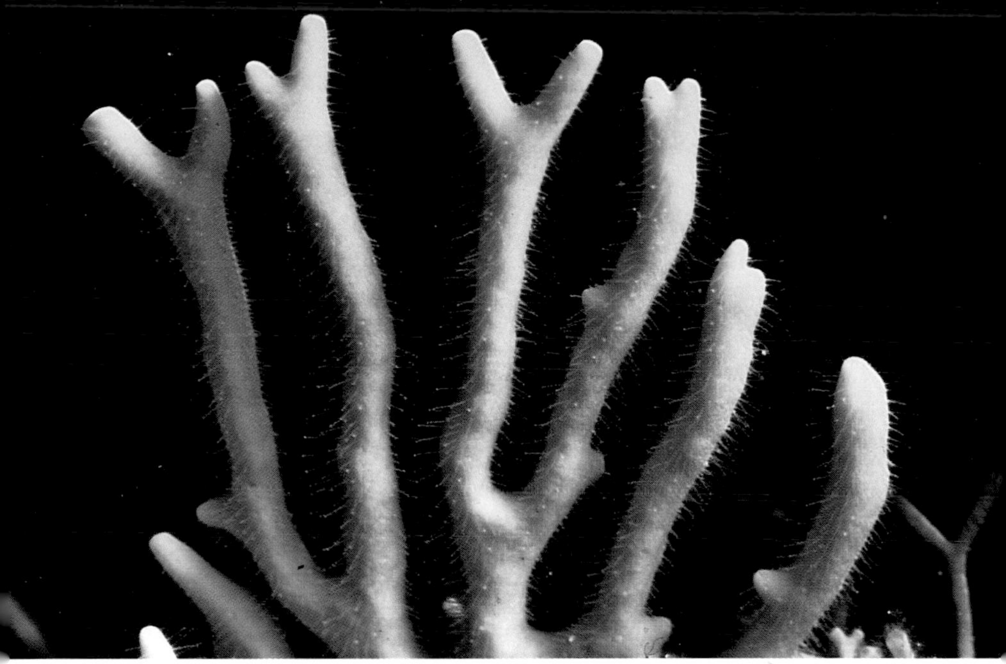

These digitate processes of *Millepora alcicornis* have the dactylozooids clearly visible protruding from the skeleton. Bahamas, Eleuthera Island, 10 m depth.

A flattened colony of *Millepora alcicornis* with some small brittle stars on it. Bahamas, Grand Bahama Island, 10 m depth.

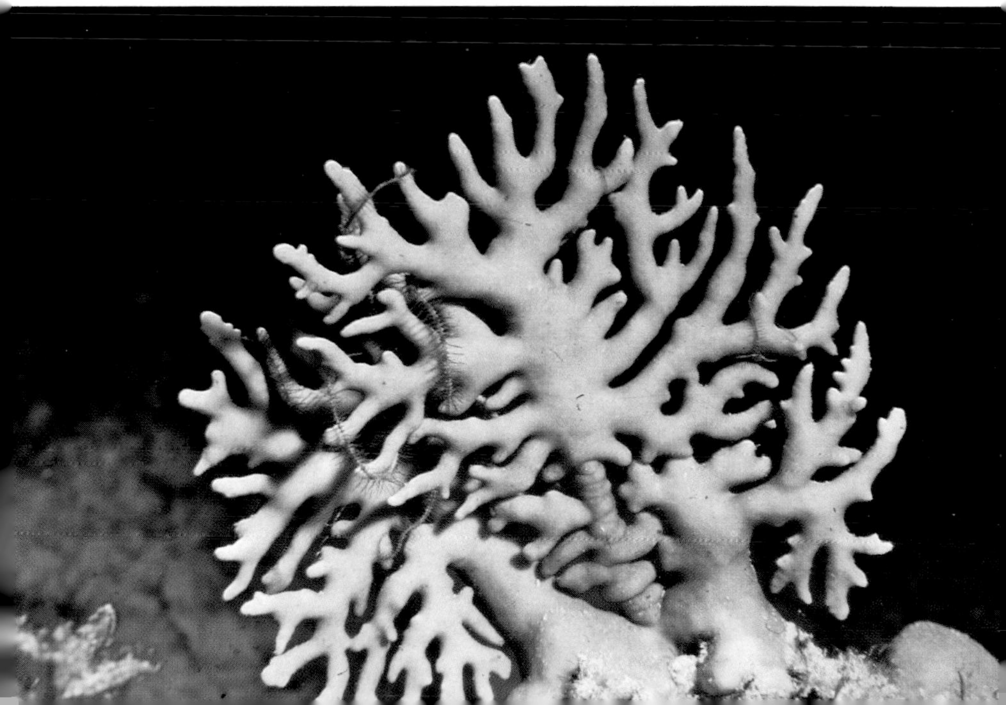

five orders, of which only three are of any significance on Atlantic reefs. Some orders (Milleporina and to a lesser extent Stylasterina) produce a calcium carbonate skeleton of varying significance in reef construction.

Most reef-dwelling hydrozoans are colonial, although solitary species do exist. More than one type of polyp may exist in a colony, with specializations for feeding, reproduction or defense. A coelenteron (digestive tube) is continuous throughout the colony.

The Hydrozoa often possess powerful stinging nematocysts. The order Milleporina derives its common name, fire corals, from the pain occurring when humans contact these colonies and cause the nematocysts to discharge. Various members of the order Hydroidea can produce a more instantaneous and painful response in humans on contact, but fortunately are not as abundant generally on reefs as millepores.

References: Hyman 1940, Bayer and Owre 1968.

ORDER HYDROIDEA: HYDROIDS

The hydroids, which include most species of the Hydrozoa, are solitary or colonial with the polypoid generation much more extensively developed than the medusoid generation. The solitary hydroids are not important on Caribbean reefs, but the colonial species can be conspicuous members of the reef community. The typical colonial hydroid is a small tree-like structure resembling a plant; it has a non-calcified skeleton which is somewhat flexible. A typical polyp consists of a base, stalk and hydranth (terminal mouth and tentacles). The hydroids are grouped into two suborders on the basis of having (Thecata) or lacking (Athecata) a hydrotheca, the chitinous cup-like structure surrounding much of the hydranth.

Many species of hydroids can occur on Caribbean reefs. They often occur in widely different environments and often have large, occasionally circumtropical, geographic ranges. While most species are not a nuisance, the larger members of genera such as *Plumaria* can sting painfully when contacted, a pain which is more intense and immediate than that of fire coral, *Millepora* spp. Fortunately the pain usually does not persist. Various hydroids are common in fouling communities and often are the sources of discomfort when contacting piles or ropes that have been submerged for some time.

ORDER ATHECATA

Family Halocordylidae

Halocordyle disticha (Goldfuss)

This species is a common athecate hydroid, in some locations being the most common species of its order. Colonies may be 10-12 cm high and have a terminal polyp on the primary stalk. They occur most often in areas with some current flow.

Halocordyle disticha is known circumtropically in shallow water. In the western Atlantic it is known from the Florida Keys to Venezuela.

Reference: Vervoort 1968.

ORDER THECATA

Family Sertulariidae

Sertularella speciosa Congdon

This large feather-like hydroid reaches about 20 cm in height and often occurs in clumps. It is found in areas of current flow at depths usually below 9 m.

Sertularella speciosa is known from southern Florida and Bermuda to the Caribbean.

Reference: Vervoort 1968.

Thyroscyphus ramosus Allman

This thecate sympodial hydroid is similar in appearance to *Cnidoscyphus marginatus*, but differs in the operculi which cover the opening of the hydrotheca when the polyp is retracted. It reaches at least 16 cm in height and can be partially covered with algae.

It is known from southern Florida to Colombia in the Caribbean, and there is one record from the Indian Ocean.

Reference: van Gemerden-Hoogeveen 1965.

The blade-like branches of *Millepora complanata* are a common sight on the fore reef areas. Puerto Rico, Desecheo Island, 6 m depth.

The low box-like structure of *Millepora squarrosa* in a wave-swept fore reef area. It is difficult to separate the three species of *Millepora* particularly since intermediate examples of each type can often be found. Puerto Rico, La Parguera, San Cristobal Reef, 5 m depth.

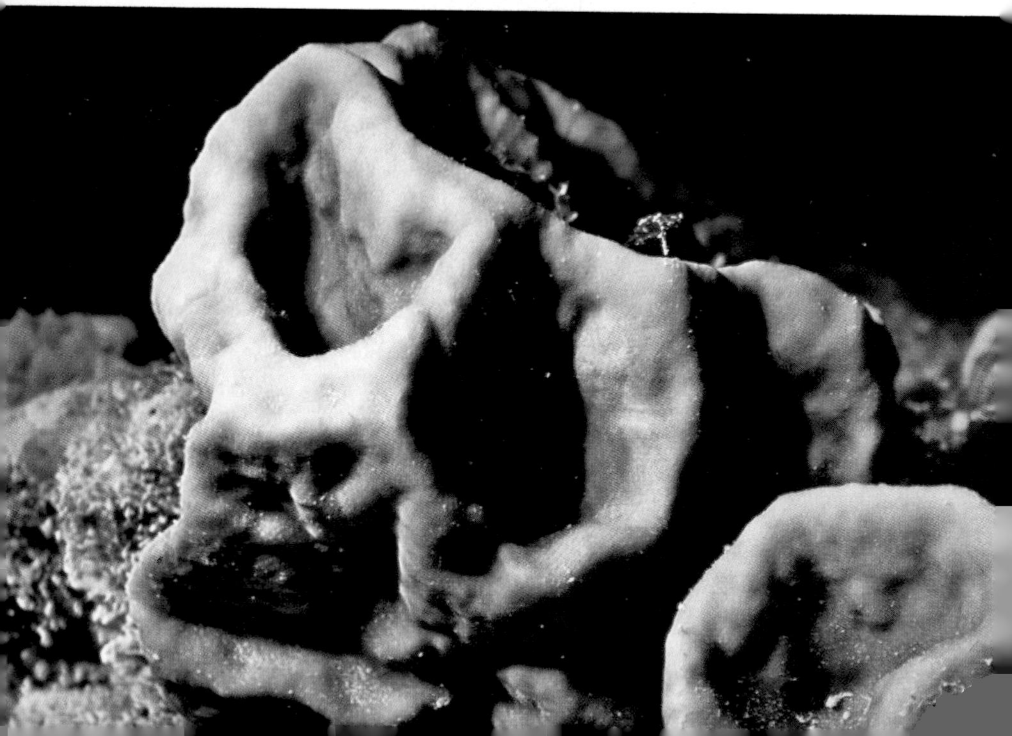

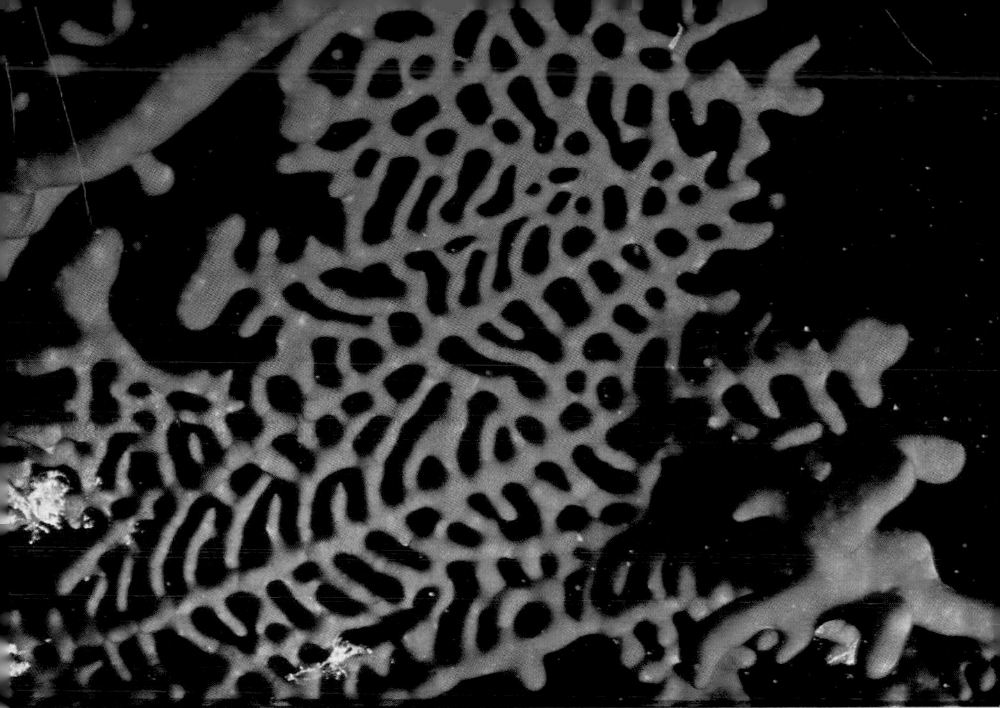

Millepora fire coral encrusting a sea fan skeleton. This type of *Millepora* is as potent a stinger as the more typically seen form, and its presence on a normally nonstinging substrate can produce an unpleasant surprise for the diver. Puerto Rico, La Parguera, 10 m depth.

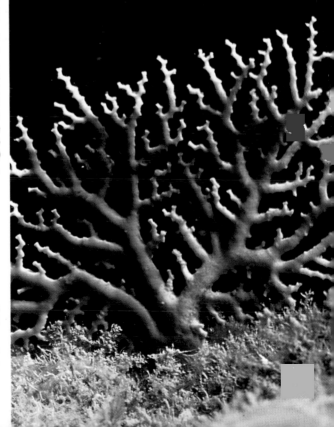

The delicate stylasterine coral *Stylaster roseus.* Jamaica, Rio Bueno, 18 m depth.

Cnidoscyphus marginatus (Allman)

This species is a common large hydroid of areas with some current flow. The polyps are quite large and occur in a sympodial growth pattern where the primary axis is produced by a succession of polyps rather than a single axial polyp. Branched and unbranched colonies occur and reach at least 10 cm in height. The branches are often covered with algae.

Cnidoscyphus marginatus is known from southern Florida and the Bahamas to Venezuela at depths from 1 to 45 m.

Reference: van Gemerden-Hoogeveen 1965.

Family Aglaopheniidae

Aglaophenia allmani Nutting

The form of the branching in this species is characteristic, although somewhat irregular. The polyps are small and the final branches often have small spots of white. In spite of their small size, the polyps produce a potent sting when contacted.

The species is known from the tropical western Atlantic only, with records from Florida, the Bahamas, the West Indies and Brazil. It occurs at moderate depths (about 15 m).

Reference: van Gemerden-Hoogeveen 1965.

Family Plumariidae

Plumaria habereri Stechow

This species is a large hydroid, reaching 30 cm in height, branched generally in a single plane. It branches irregularly at acute angles and the finest branches have pinnules arranged like a feather. The stem and larger branches are brownish black and the most distal portions yellowish.

The species occurs with the plane of the colony perpendicular to the current, often growing horizontally out from vertical rock surfaces. Specimens have been reported partially covered with zoanthids. This hydroid stings painfully on contact. *Plumaria habereri* is known from Indonesia, Japan and the Caribbean (Curacao, Puerto Rico) at depths between 10 and 50 m.

Reference: van Gemerden-Hoogeveen 1965.

Colony of the hydrozoan coral *Millepora complanata* located at the juncture of the *Acropora cervicornis* and *A. palmata* zones on the fore reef, Discovery Bay, Jamaica. The finely divided tips of the blades in this case resemble what could be called *M. alcicornis* and demonstrate why classification of these hydrozoans based on morphology is difficult.

Another extreme of blade development in *Millepora complanata*, this specimen representing the "typical" *complanata* morphology.

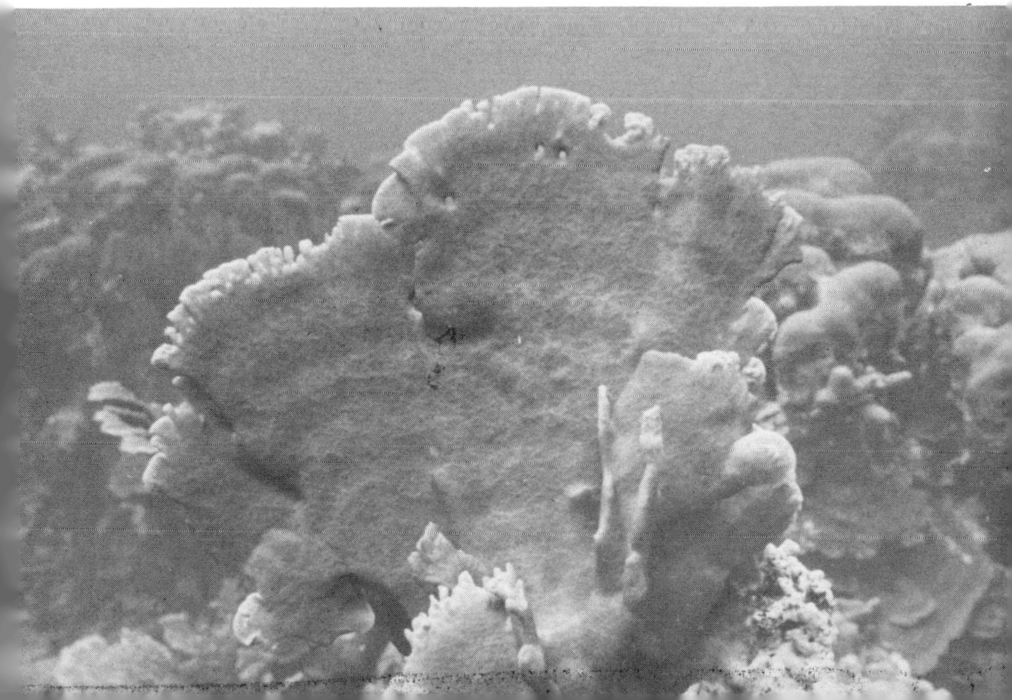

Occasionally all-white colonies of *Stylaster roseus* are found. Jamaica, Rio Bueno, 24 m depth.

The moon jelly, *Aurelia aurita*, is seen occasionally in great numbers over coral reefs. Bermuda, 3 m depth.

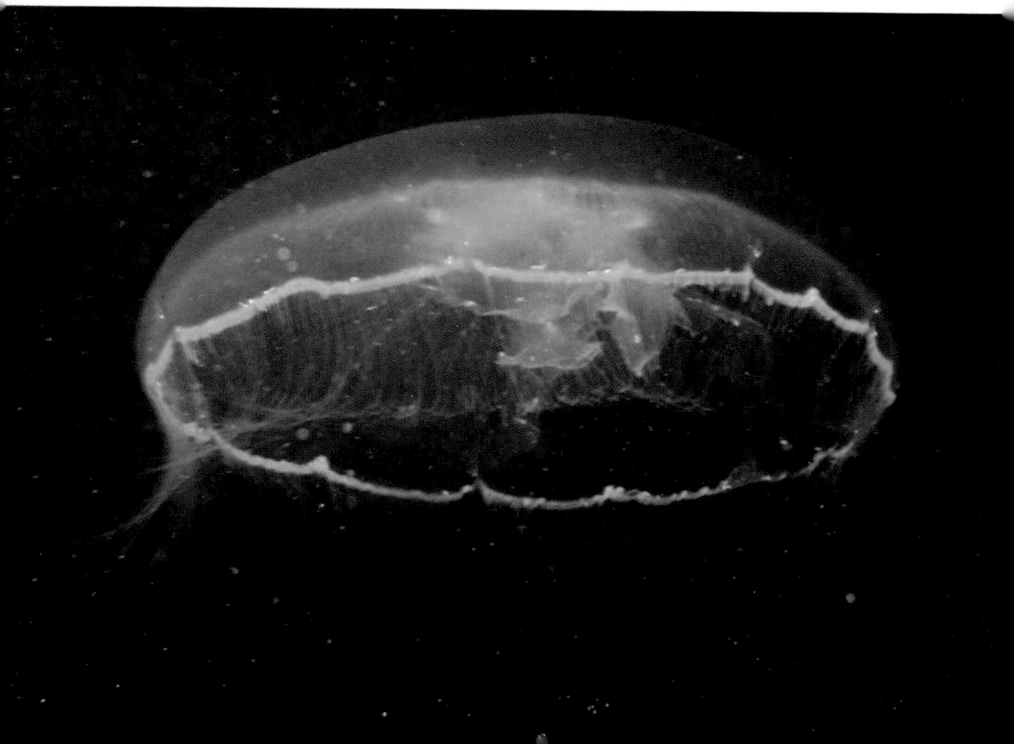

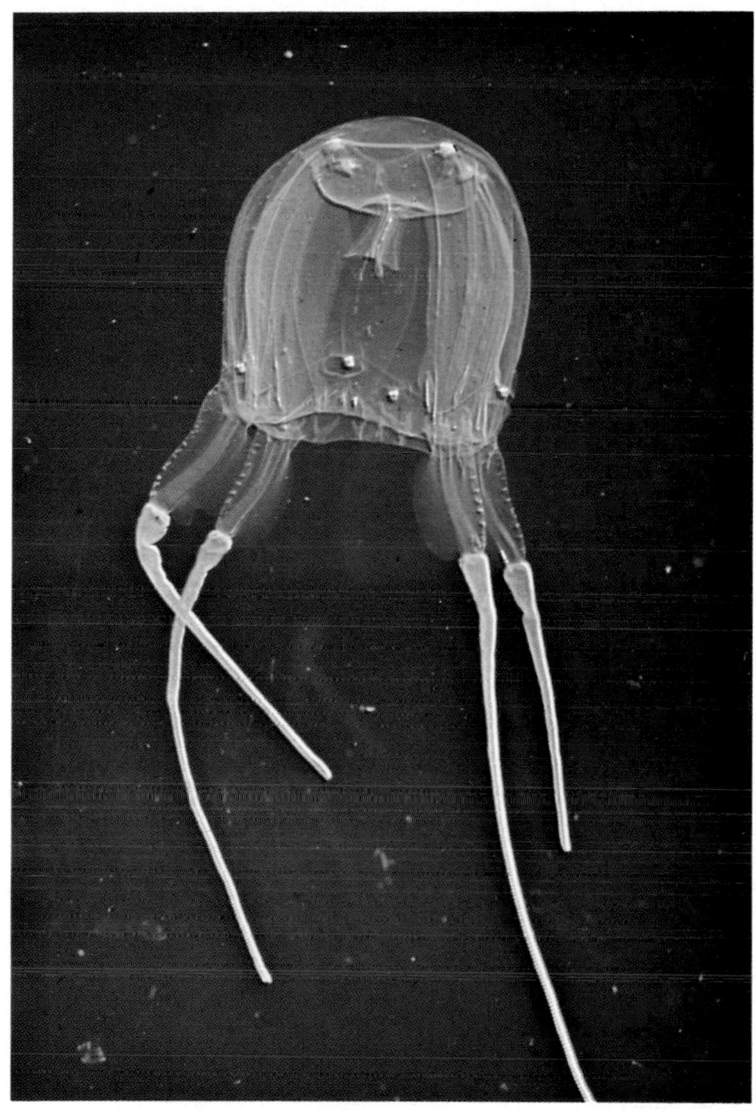

The cubomedusae *Carybdea marsupialis* is not particularly dangerous, but is quite similar in appearance to its more potent Caribbean relatives such as *Carybdea alata.* Sea wasps are usually seen only at night in any great numbers but their occurence is quite variable and difficult to predict. Puerto Rico, Aguadilla, 1 m depth. (Photograph by A. Charles Arneson.)

Millepora alcicornis thinly encrusting a now dead gorgonian. At times the only evidence observable at any distance that *Millepora* is encrusting something is its characteristic yellow brown color.

Digitate processes of *Millepora alcicornis.* The dactylozooids can be seen protruding from pores in the skeleton and resemble clear "hairs" when observed under water.

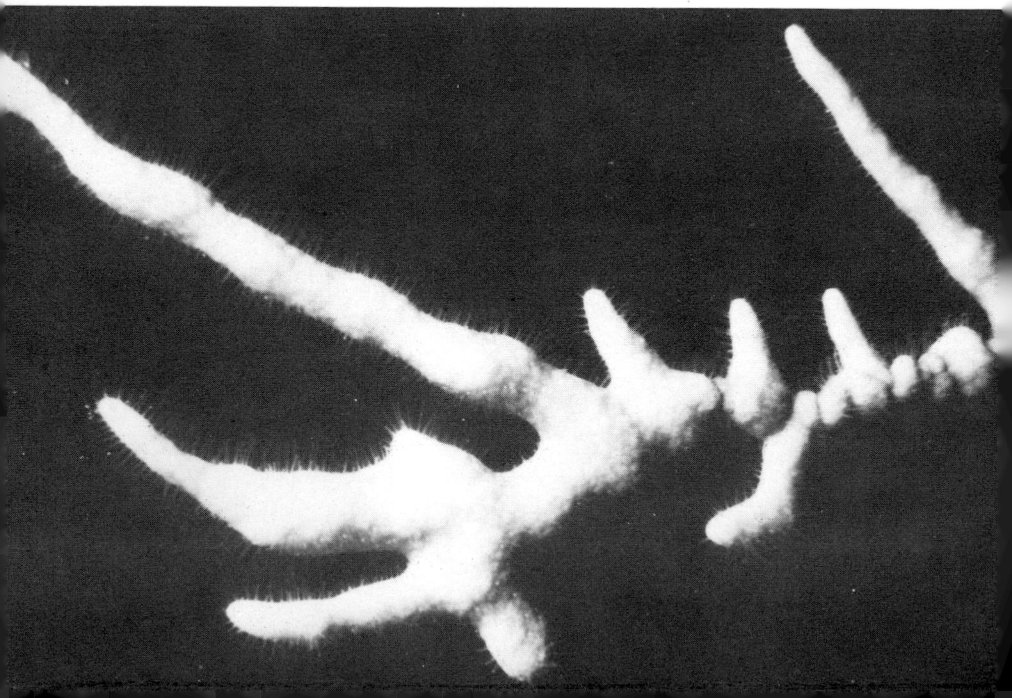

Gynangium longicauda (Nutting)

This tall, slender feather-like hydroid grows from a creeping rhizome, forming groups of colonies. It occurs in clear water areas with some current flow. It is known from several locations in the Caribbean at 12 to 60 m.

Reference: Vervoort 1968.

Solanderia gracilis Duchassaing and Michelin

This attractive hydroid looks more like a sea fan, gorgonian or black coral than a member of the Hydrozoa. It had previously been classified as a gorgonian and a "horny sponge" on the basis of its dried skeleton, but after examination of the polyps it was decided that this circumtropical genus belongs in the hydroids.

The colonies are single planar and somewhat fan-like, reaching at least 30 cm in height. All polyps on a single colony are either male or female and there is no apparent difference in the general form of the colonies between the sexes.

This species is known fairly widely in the Caribbean and occurs at depths as shallow as 1 m down to at least 35 m. It is most common in wave swept rocky areas with clear water, often growing off vertical faces. It can occur considerably deeper along vertical faces where there is some current. It is very common around Mona and Desecheo Islands.

Reference: Vervoort 1962.

ORDER MILLEPORINA: FIRE CORALS

The polyps of fire corals protrude through pores in a massive calcareous skeleton that the polyps produce. Two types of polyps exist in the colony, the gasterozooids which are feeding polyps and the dactylozooids or defensive polyps. The gastric cavities of all polyps are interconnected, however. Usually each gasterozooid, protruding through the larger pores in the skeletons surface, is surrounded by several dactylozooids arising from smaller pores.

Millepora spp.

Three described species of western Atlantic *Millepora*

The gorgonian *Telesto riisei* is often considered a fouling organism in the West Indies as it grows readily on pilings in many harbor areas. On reefs where it occurs the water is generally turbid. Jamaica, Discovery Bay, 3 m depth.

The white polyps of *Telesto riisei* are usually expanded during the day making identification of this species usually simple. Puerto Rico, La Parguera Laurel Reef, 12 m depth.

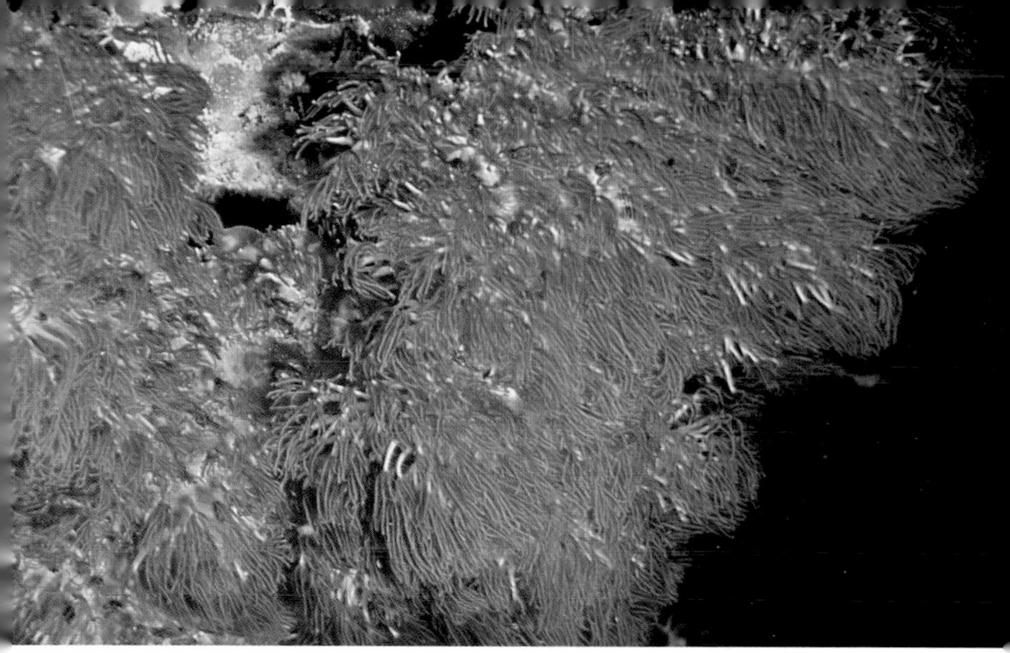

The gorgonian *Briareum abestinum* has extremely large polyps which are normally found expanded. It is often encrusting as is shown here on rocky substrates. Jamaica, Spring Gardens, 12 m depth.

The gorgonian *Iciligorgia schrammi* has its polyps only along the sides of the branches and the colony as a whole is only branched in a single plane. Bahamas, Cat Island, 30 m depth.

exist: *M. alcicornis* Linnaeus, *M. complanata* Lamarck and *M. squarrosa* Lamarck. They differ only in the morphology of the corallum (the massive skeleton) and are often considered ecological variants of a single species. *Millepora alcicornis* is a branched form, often finger-like, and can occur in a single plane or branched in all directions. *Millepora complanata* has blade-like branches extending vertically which are connected with one another only at their bases. The final form, *M. squarrosa*, has upright plates united to form a boxwork structure. Whether these forms truly deserve status as species is unimportant in the present consideration. The various forms will occur adjacent to one another on occasion.

The branched form, *M. alcicornis*, occurs somewhat deeper than the others, while *M. squarrosa* is found in heavy surf or in areas exposed to air in the troughs of waves. Under extreme wave conditions or when covering the remains of another organism, *Millepora* can be encrusting. Entire sea fans may be covered by *Millepora*, retaining the delicate meshwork of the original skeleton but coloring it the typical yellow-brown. Colonies of *Millepora* may also grow on the outer portion of the stalks of dead gorgonians. Barnacles and serpulid worm tubes may occur on the sides of the blade-like forms of *Millepora*.

The species are found from deep fore reef areas to back reefs. They can contribute significantly to reef structure in the areas immediately adjacent to the reef crest. The species are known from southern Florida and the Gulf of Mexico to Brazil.

References: Boschma 1948, Stearns and Riding 1973, Mattraw 1969, Roos 1971.

ORDER STYLASTERINA

The Stylasterina are a small tropical and subtropical group which resemble the Milleporina. They differ in that their dactylozooids (defensive polyps of the colony) lack tentacles and the gasterozooids (nutritive members) have a small upright spine in the bottom of the cup in the skeleton. They generally form branching calcareous growths, although one encrusting species is known from California.

Planula larvae are produced by sexual reproduction.

Family Stylasteridae

Stylaster roseus (Pallas)

The small fragile colonies of *Stylaster roseus*, the only western Atlantic species of the Stylasterina on reefs, reach 10 cm in height and are branched only in a single plane. The fan-like colonies commonly occur in caves or crevices, often growing on inverted surfaces and occasionally (as at Mona Island) on open vertical rock faces. They may be white, pink, purple, red or red with white tips and are branched so that with polyps expanded the animals form a complete network for capturing animals in the water passing through the colony.

Stylaster roseus is found widely in the Caribbean, the Bahamas and southern Florida at depths of 6 m to at least 30 m.

References: Roos 1971, Boschma 1955.

CLASS SCYPHOZOA

The members of the Scyphozoa have the medusoid portion of the life cycle predominating over the polypoid stage, which may even be absent. The medusa is almost invariably larger in size than the polyp of any species. There are five orders of Scyphozoans, two of which are considered further.

ORDER SEMAEOSTOMAE

The margin of the flat to dome-shaped bell of the Semaeostomae is scalloped into numerous lappets. The length of the tentacles around the margin of the bell varies, but the corners of the mouth, on the undersurface of the bell, have four frilly oral arms. The largest coelenterate, *Cyanea*—a medusa with a diameter reaching well over 2 m, is a member of this order.

Aurelia aurita (Linnaeus)
Moon jelly

This large medusa is one of the most commonly observed animals above Caribbean coral reefs. The bell has a

The fan-like colonies of *Iciligorgia schrammi* can be quite large, being over 2 m across, and are found protruding from reef faces where there is some current. It is usually found only in the clear water areas of deep reefs. Jamaica, Discovery Bay, 36 m depth.

Plexaura homomalla is a bushy gorgonian with a flattened growth form which is fairly common on Caribbean reefs. Jamaica, Spring Gardens, 12 m depth.

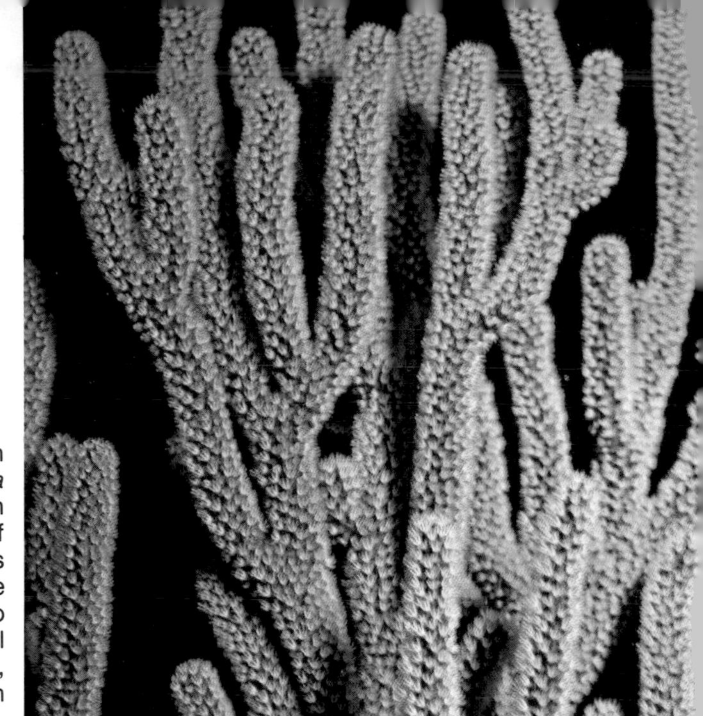

The gorgonian *Plexaura homomalla* has a high concentration of prostaglandins in its tissues which have made it of interest to the pharmaceutical industry. Jamaica, Discovery Bay, 15 m depth.

The projecting spindles of *Muricea pinnata* are visible just below the polyps of the gorgonian in this closeup view. Puerto Rico, La Parguera, Laurel Reef, 12 m depth.

fringe of a great many very short tentacles and four oral arms of varying length around the mouth. The bell can reach over 40 cm in diameter and be quite thick. *Aurelia aurita* is 95 to 96% water, and the pulsations of the bell along with oxygen consumption vary directly with temperature. The moon jelly can perceive light and moves away from brightness.

Aurelia aurita is world-wide in the tropics and temperate areas and can occur in sizable numbers over reefs on occasion.

References: Hyman 1940, Mayer 1910.

ORDER CUBOMEDUSAE: SEA WASPS

The Cubomedusae have a four-sided bell with a tentacle or group of tentacles at each posterior corner. Several species of sea wasps can occur in Caribbean waters, and some of the species can give a dangerous and painful sting.

Most Caribbean species are members of the genus *Carybdea,* with the bell only a few to several centimeters in length. Surprisingly, the largest species are not necessarily the most dangerous to humans. *Carybdea alata* Reynaud seems to be the most dangerous and could under certain circumstances produce a fatal sting, although no definite records of such exist. The trailing tentacles can be contracted so they are only a few centimeters in length or extended to a length of well over 2 m. Species of *Carybdea* can swim surprisingly fast by pulsations of the bell; with tentacles extended they are reminiscent of a fishing boat trolling lines astern.

Sea wasps are more abundant in certain locations and at certain times. Along the northern coast of Jamaica where there is often a layer of slightly turbid lower salinity water (due to freshwater outflow from the island), the sea wasps will be found mostly within a few meters of the surface at night. Exactly where they spend the day is not conclusively known, and the occurrence of sea wasps near the surface seems to vary considerably with wave conditions and moon phase. They are somewhat attracted to lights.

When diving at night in areas where *Carybdea* is sighted or may occur, caution will easily avoid an unfortunate encounter. Wear clothing or a wet suit covering as much of the body as possible. Particularly vulnerable are heavily vascularized areas (such as the front and side of the neck) and the face. If sea wasps are sighted, try to stay below the upper few meters of water as they are generally concentrated in that area. With a light sweeping the area ahead, a diver can proceed cautiously. If switching from regulator to snorkle (or vice versa), check that the free mouthpiece has not fouled any tentacles (this has nearly brought one diver to grief).

If stung badly by sea wasps, exit the water as soon as possible. Do not rub or touch the affected area as this serves only to discharge nematocysts not initially released. Monosodium glutamate (meat tenderizer) should be sprinkled on the area of contact. This material is a proteolase and serves to break down much of the complex protein molecules of the toxin on the skin and in undischarged nematocysts. Reaction to Cubomedusae stings varies with individuals. If a general reaction occurs (pain all over, difficulty in breathing) medical attention should be sought rapidly; in most cases anti-histamines are administered.

The life cycles of some Cubomedusae are known. The species are widely distributed in the Caribbean, but the dangerous species seem to be rare in the southern Florida area.

References: Cutress and Studebaker 1972, Mayer 1910.

CLASS ANTHOZOA

The third class of Coelenterata has its life cycle restricted to the polypoid phase exclusively, no medusoid stage occurring. They usually attach to a substrate and have the oral end expanded into a flattened oral disk. The mesoglea is cellular in many species and the tentacles hollow. The gastrovascular cavity is divided into compartments by complete or incomplete septa. Generally one or more siphonoglyphs, a ciliated groove on the margin of the mouth, occur.

A calcareous skeleton may be produced. The gonads develop on the septa; anthozoans may be mon- or dioecious. A planula larva may be produced which is capable of being transported some distance by ocean currents.

Left: Gorgonians of the genus *Muricea* are "scratchy" if run over the skin due to the projecting spindles formed of spicules. Bahamas, Cat Island, 20 m depth.

Right: The genus *Eunicea* is one of the most common reef gorgonian genera, but identification of a specimen to species usually requires examination of the spicules. Bahamas, Cat Island, 10 m depth.

A typical view of a species of *Eunicea* on the reef. Bahamas, Cat Island, 10 m depth.

A typical candelabrum-like colony of a species of *Eunicea*. Bahamas, Cat Island, 12 m depth.

SUBCLASS OCTOCORALLIA (ALCYONARIA)

The polyps of the Octocorallia are fairly uniform in their construction, the diversity of the subclass being expressed in the structure and organization of the colony and variation in the skeletal elements (spicules). This colonial subclass includes the gorgonians and sea fans of shallow reefs plus a number of other organisms not found on reefs or only below diving depths. As their name implies, the polyps almost without exception bear eight tentacles which in turn usually possess small projections termed pinnules. Spicules of calcium carbonate (calcite) are found in the tissue of the polyps and are incorporated with a "horny" epidermal secretion into the skeleton.

The Octocorallia are generally divided into six orders, one of which (Gorgonacea) is extremely important and abundant on Caribbean reefs. A second order (Telestacea) is of minor importance. A third order, the Alcyonacea, is abundant on reefs in the Indian and Pacific Oceans, but the few species in the western Atlantic occur below diving depths.

The classification of the octocorals essentially is subjective due to the lack of knowledge concerning the source (environmental or genetic) of variation in the spicules and morphology of the "species." There is no real definition as to what constitutes an octocoral species and practically no information on the effects of environmental conditions on taxonomic characters.

References: Bayer 1961, Bayer and Owre 1968.

ORDER TELESTACEA

Family Telestidae

Telesto riisei (Duchassaing and Michelotti)

The colonies of *Telesto riisei* are organized around a long axial polyp at the terminus of a main stem. Side branches may arise from this main stem but are often short and bear terminal axial polyps. The polyps are white and often expanded during the day. The skeleton is often coated by a variety of other organisms, particularly algae. Colonies may reach over 30 cm in extent and may be the only octocoral of any significance as a fouling organism.

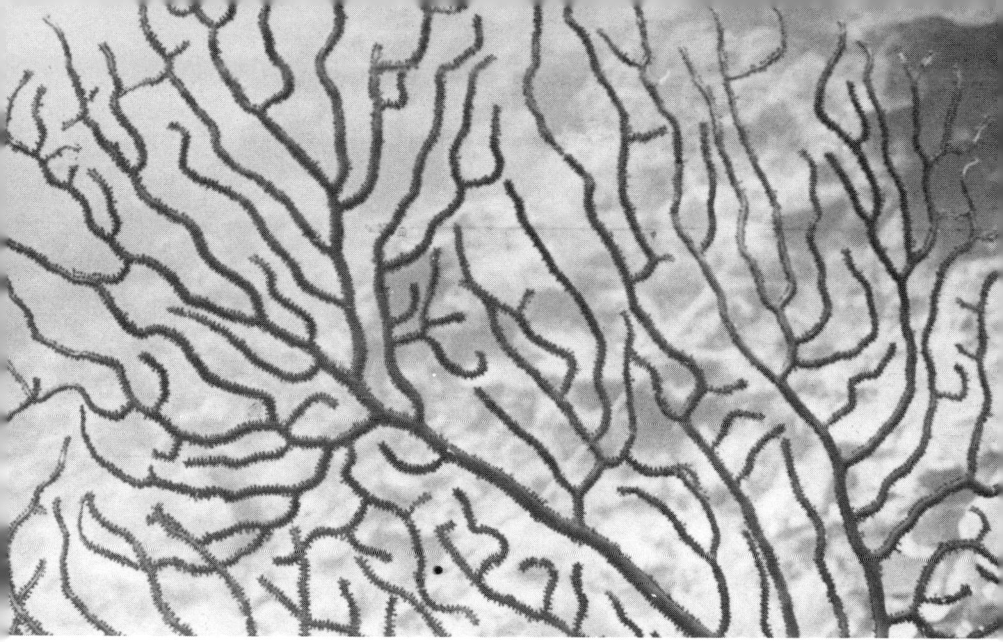

A portion of a colony of the gorgonian *Iciligorgia schrammi* photographed at 40 m depth on the fore reef, Discovery Bay, Jamaica. The polyps of this typical outer reef face gorgonian are found only on the flattened edges of the branches.

An unidentified deep-reef gorgonian photographed at 70 m depth, Discovery Bay, Jamaica. Sponges of the genus *Agelus* are also visible.

Most species of the genus *Pseudopterogorgia* have tall, plume-like branches. Puerto Rico, Desecheo Island, 12 m depth.

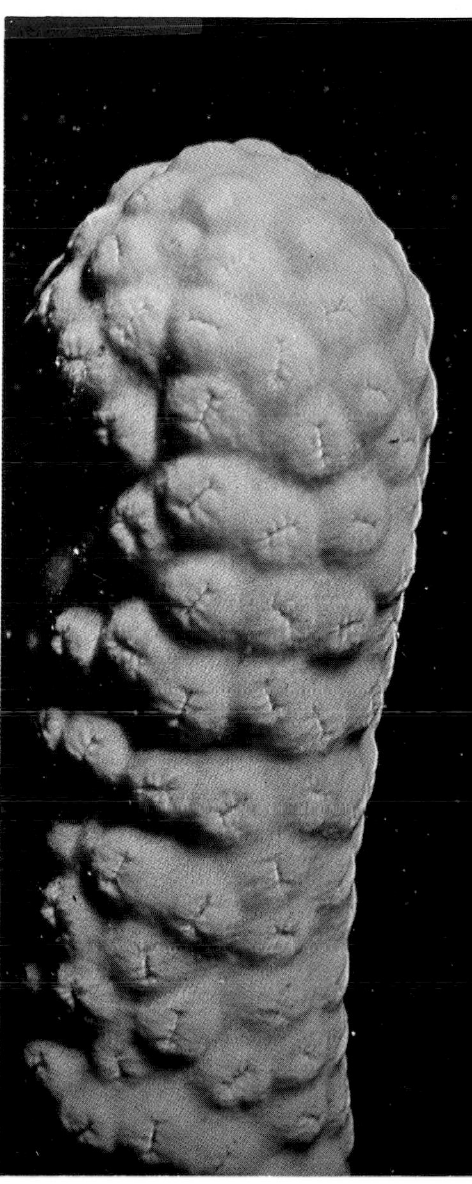

Left: Gorgonians of the *Plexaurella* appear quite "bushy" when the polyps are expanded. Bahamas, Bimini, Turtle Rocks, 11 m depth.

Right: When the polyps of species of *Plexaurella* are contracted the branches have a drastically different appearance then when they are expanded. Puerto Rico, La Parguera, Laurel Reef, 12 m depth.

Close-up view of a gorgonian, probably of the genus *Eunicea*. Photographed at 10 m depth, Discovery Bay, Jamaica.

Small portion of a sea fan, *Gorgonia ventalina*. The darkened area in the center of the photo is an area where all the polyps have been killed due to the feeding activities of a flamingo tongue, *Cyphoma gibbosum*.

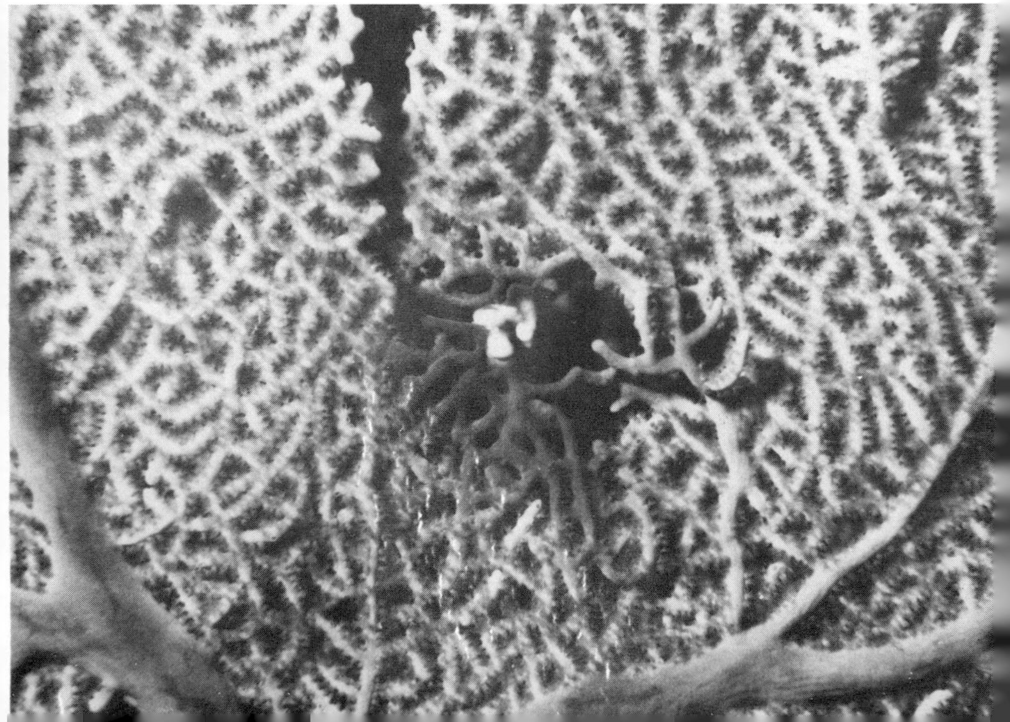

The species is known from 0 to 55 m from southern Florida to Brazil. *Telesto riisei* is found in areas of moderate turbidity where reefs can occur but is seldom observed on reefs with extremely clear oceanic water.

References: Bayer 1961, Rees 1969, 1972.

ORDER GORGONACEA: GORGONIANS AND SEA FANS

This order is the major group of Octocorallia occurring on Caribbean reefs and generally occurs in an abundance far beyond that found on Indo-Pacific coral reefs. The reasons for this "flowering" of the Gorgonacea on Caribbean reefs is not understood, but its existence cannot be denied.

All colonies possess an axial skeletal structure of either a horny or calcareous central cylinder (suborder Holaxonia) or a zone of tightly bound spicules (suborder Scleraxonia). Most species have an erect skeletal structure attached to a solid substrate by a holdfast, but a small number of species may occur as an encrusting mat. Many species possess zooxanthellae similar to those occurring in the stony corals.

Identification of Gorgonacea is generally based on the morphology of the colony and the size, shape, location and orientation of the calcareous spicules (requiring the use of a microscope). These identifying characters can be determined from dried colonies, the most common method of preserving Gorgonacea. While this is convenient in many instances, knowledge of the soft parts of gorgonians has suffered because of this fact. The spicules are colored, giving the colony its color, but color itself is generally a poor character on which to base an identification since it is quite variable and subject to environmental control in some cases.

The Gorgonacea serve as a substrate for many other organisms. Fire coral, *Millepora* spp., may encrust entire colonies, particularly the sea fans of the genus *Gorgonia.* Bivalve molluscs, sponges and algae may grow upon dead sections of gorgonian skeletons; whether these organisms simply take advantage of already dead substrate or themselves kill a portion of the gorgonian is not known. The gastropod mollusc

This gorgonian of the genus *Pseudopterogorgia* has a flamingo tongue, *Cyphoma gibbosum*, climbing on it at night. The flamingo tongue feeds largely on gorgonians. Bahamas, Acklins Island, 15 m depth.

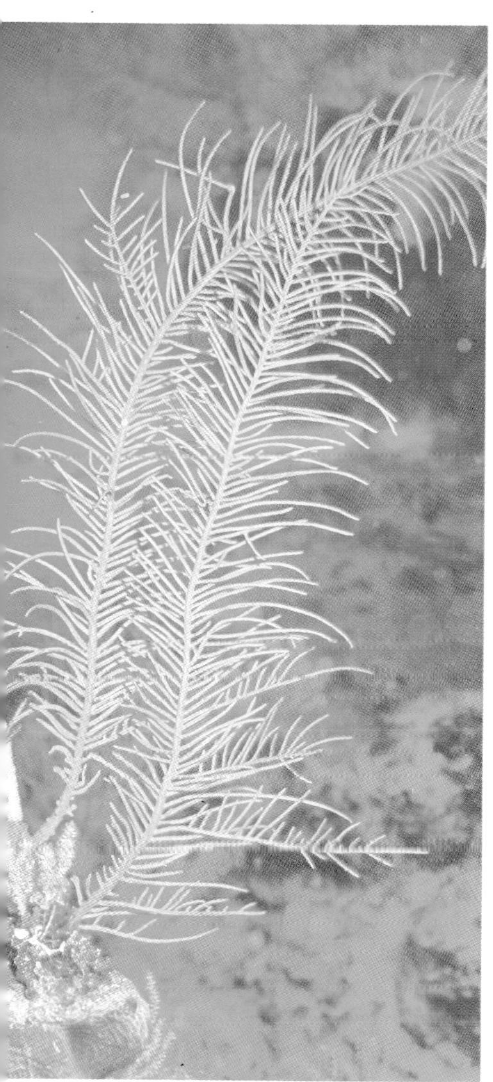

Left: The plumes of these colonies of *Pseudopterogorgia* sp. are over 1 m in height and are growing on the sides of a finger of reef on the outer slope. Bahamas, Exuma Island, 10 m depth.

Right: Gorgonians of the genus *Pseudopterogorgia* have relatively small polyps which are normally contracted and inconspicuous. Puerto Rico, La Parguera, Laurel Reef, 12 m depth.

Cyphoma gibbosum feeds on gorgonian polyps by crawling slowly over the skeleton, grazing at will. Other organisms, such as basket starfishes and brittlestars, climb tall gorgonians to reach a position more advantageous for filter-feeding in reef areas.

References: Bayer 1961, Hyman 1940, Opresko 1973, Preston and Preston 1975, Goldberg 1973a, 1973b, Rees 1973.

SUBORDER SCLERAXONIA

Family Briareidae

Briareum abestinum (Pallas)

This common gorgonian is generally encrusting, with its large polyps expanded. It may occasionally be lobate or finger-like and has even been found growing on the axis of a dead gorgonian. The polyps are purplish gray, packed with zooxanthellae and sway with wave surge.

Briareum abestinum can be one of the most abundant gorgonians from back reef lagoon areas to fore reefs at moderate depths. It is particularly abundant in areas of *Acropora cervicornis* mounds where the somewhat unstable substrate is not well suited to the gorgonians with a rigid central axis (suborder Holaxonia).

The species occurs in the Caribbean, the Bahamas and southern Florida at depths from 0 to 32 m but generally is found above 20 m.

References: Bayer 1961, Kinzie 1973, Opresko 1973, Goldberg 1973.

Family Anthothelidae

Iciligorgia schrammi Duchassaing

This distinctive large gorgonian is branched in nearly one plane and can occur in groups of colonies. The skeleton is brownish to tan and the polyps, which are located on margins of the somewhat flattened branches, are somewhat lighter in color. This species may reach 2 m across. The fine ends of the branches (termed twigs) are expanded in size, somewhat like a fist on the end of an arm. The polyps of *I. schrammi* lack zooxanthellae.

Iciligorgia schrammi is characteristic of deep reefs with clear water and good current flow and can be the most common gorgonian in such situations. It is often found protruding from vertical faces and seems to prefer slightly overhanging surfaces in the shallow portion of its depth range. It is known from southern Florida, the Bahamas, the Caribbean and South America to the Amazon River. Its known depth range is between 1.5 and 50 m, but it is common only below about 10 m.

References: Bayer 1961, Kinzie 1973, Goldberg 1973.

Family Plexauridae

Plexaura homomalla (Esper)

This bushy, flattened gorgonian has recently been the subject of much study since it was discovered to contain high amounts of a type of chemical (prostaglandins) valuable in the pharmaceutical industry. Investigations were made of the potential to harvest from natural populations or to culture in the ocean sizable quantities of this gorgonian. Advances in chemical synthesis of prostaglandins have now made such considerations less important. The species is tan in color and can reach nearly 1 m in height. The smaller branches tend to grow only from the upper surface of the larger branches.

There are two forms of *Plexaura homomalla*, one in shallow water (about 0-30 m) and a second, with more slender branches occurring more densely, in deeper water (about 15-50 m). The species is known from southern Florida and Bermuda to the Caribbean.

Two other species of *Plexaura* occur in the Caribbean, distinguished from *P. homomalla* by their colony form and spicules.

References: Bayer 1961, Kinzie 1973.

Muricea spp.

Six species of *Muricea* occur in the Caribbean area. All have tall, pointed calyces (area of the polyp) with a projecting spindle or spindles formed of large spicules. They are

Gorgonia ventalina is one of the common "sea fans" of the Caribbean and can vary widely in color. Bahamas, Exuma Islands, 9 m depth.

Gorgonia flabellum is most often found in shallow water with heavy wave action. It is quite similar to the other species of *Gorgonia* and identification of the species is often a problem. Bahamas, Exuma Cays, 3 m depth.

Gorgonia mariae is the smallest of the sea fans and is most often found in deep water. Whether it really deserves ranking as a species is a subject of some disagreement among specialists. Puerto Rico, Desecheo Island, 15 m depth.

Gorgonians of the genus *Ellisella,* usually red in color, are found only in deep-reef environments on Caribbean reefs. They usually have relatively few, if any, branchings. Bahamas, Crooked Island, 30 m depth.

never purple in color and have fairly stout branches. Members of *Muricea* are prickly or scratchy when run over the skin due to the projecting spindles.

Muricea pinnata Bayer has a large terminal spike on the lower portion of each polyp. This spike is somewhat more apparent than in most of the other species and is easily seen or felt in live colonies.

The species of *Muricea* are found widely in the tropical western Atlantic or along the southern United States coast. Not all species commonly occur on reefs.

References: Bayer 1961, Kinzie 1973.

Eunicea spp.

The genus *Eunicea* is an important group of reef-dwelling alcyonarians. Well over 10 species occur in the Caribbean. Identification to species usually requires examination of the spicules. They may be tall, bushy or candelabrum-like in form. The calyces are usually prominent and often protrude a considerable amount, but not invariably so. Most species occur from a few meters depth to a maximum of about 30 m.

References: Bayer 1961, Kinzie 1973.

Plexaurella spp.

The genus *Plexaurella* has several Caribbean species, with identification to species requiring examination of the spicules. They are often tall and sparsely branched with thick branches. The polyps are large and usually expanded, giving the colony a "fuzzy" appearance.

Most Caribbean species commonly occur from about 10 to 50 m.

References: Bayer 1961, Kinzie 1973.

Family Gorgonidae

Pseudopterogorgia spp.

There are several species of *Pseudopterogorgia* on Caribbean reefs. Most are tall, plume-like colonies. On the leeward side of some islands in the Caribbean, a zone of dense growth of these species can occur at 7-10 m, with colonies reaching heights over 1.5 m. They are pinnately branched, with no interconnections between branches, and some are slimy to the touch with abundant mucus.

Pseudopterogorgia bipinnata (Verrill) produces planulae in Jamaica in late January and early February. Unlike stony coral planulae, those of *P. bipinnata* do not contain zooxanthellae. In the laboratory they settle 11 days after release and must acquire their initial zooxanthellae from the environment, as these plant cells are abundant in the adult colonies.

References: Bayer 1961, Kinzie 1973, 1974, Goldberg 1973.

Gorgonia ventalina Linnaeus
Sea fan

This species shares the common name "sea fan" with the following two species. The complexes of anastomosed branches, generally compressed in one plane, are found on nearly every reef and are a characteristic part of this environment in the Atlantic. The similar appearance of the three Atlantic sea fans makes them somewhat tricky to distinguish. The shape of the spicules is the most positive method of doing so, but distinction can be made in the field with accuracy.

Generally in *Gorgonia ventalina* the individual branches are compressed somewhat in the plane of the fan, particularly those branches connecting ascending branches. The ascending branches are, however, occasionally somewhat compressed at a right angle to the fan plane. *Gorgonia flabellum* has the branches, particularly the ascending branches, strongly compressed at a right angle to the plane of the fan. The third species, *G. mariae*, does not have the branches compressed at a right angle to the fan plane, is a much smaller species and has other distinctions.

Color is a poor character on which to base any identification. It is quite variable within a species due to pigments produced in the spicules. *Gorgonia ventalina* can be yellow, purple or occasionally whitish, while *G. flabellum* shares some of these colors.

This species has the widest distribution, both on the reef

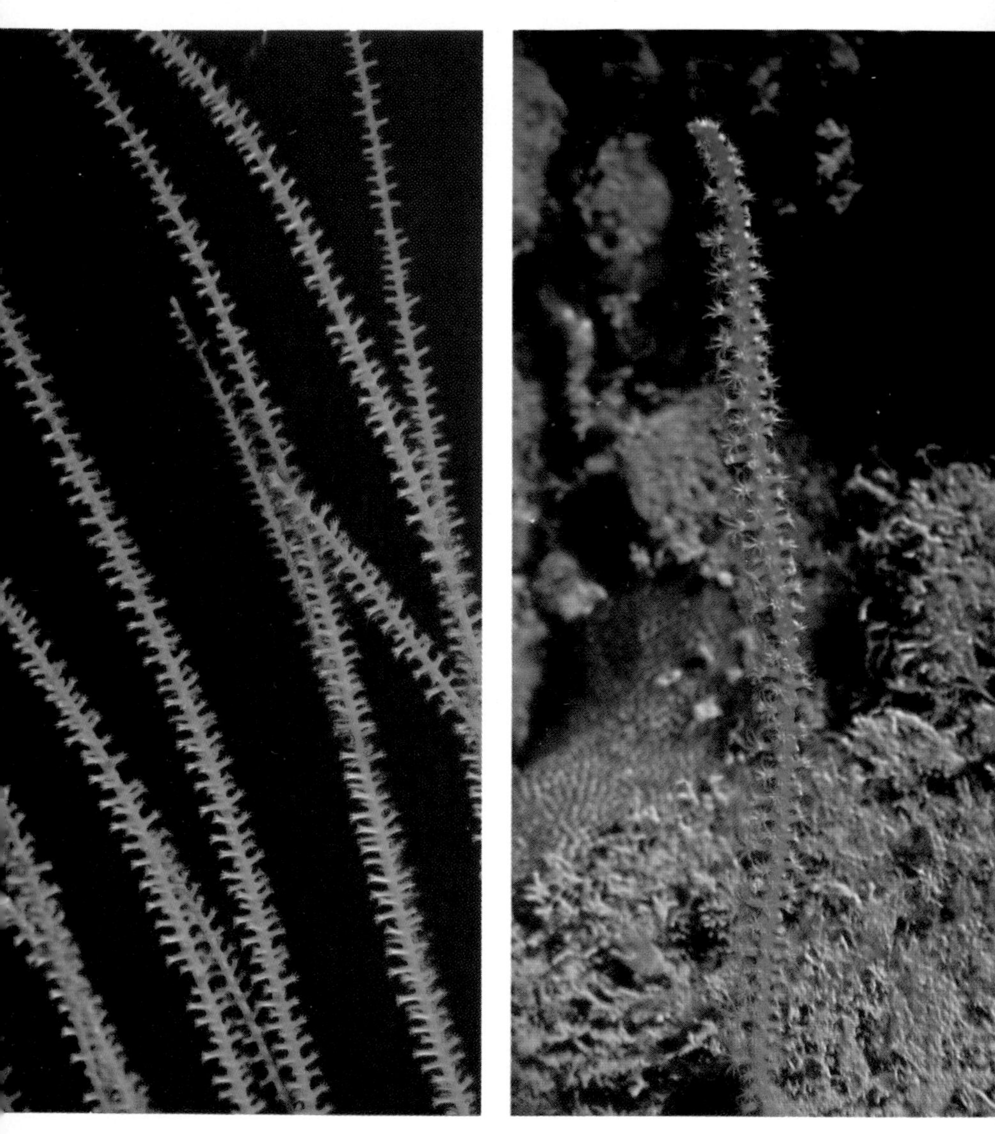

Left: The polyps of this colony of *Ellisella* sp. are clearly visible and are expanded during the day. Caicos Islands, Providenciales, 30 m depth.

Right: Some specimens in the genus *Ellisella* can consist of a single, whip-like red or orange filament. Puerto Rico, Rincon, 40 m depth.

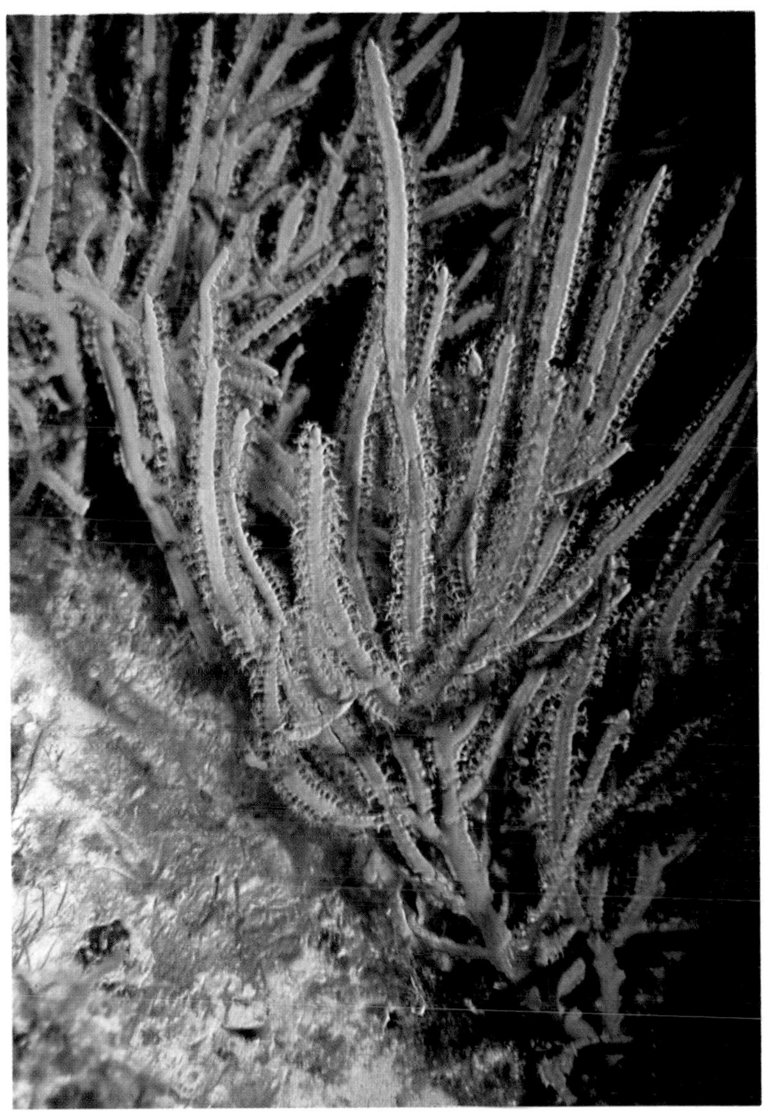

The polyps of *Pterogorgia citrina* are found only on the sides of its greatly flattened branches. Puerto Rico, Mona Island, Playa Sardinera, 15 m depth.

and geographically, of any of the species of *Gorgonia.* It can occur near shore in areas of extreme wave action and on deeper outer reefs at 15 m or more in depth. It can reach a height of nearly 2 m and shows a somewhat "clumped" (non-random) distribution of individuals on a reef. The plane of the fan is normally oriented perpendicular to the direction taken by incoming waves. If colonies with established fans are turned 90° so that the plane is parallel to the wave direction, they often die. Those that live may develop new accessory fans at a right angle to the original.

The species is known from Bermuda to Curacao, including the Florida Keys and western Caribbean. It is not known from the Gulf of Mexico coast of Florida.

References: Wainwright and Dillon 1969, Bayer 1961, Kinzie 1973.

Gorgonia flabellum Linnaeus
Sea fan

This species of sea fan has the branches connecting the ascending branches of the fan strongly compressed at a right angle to the plane of the fan. Small lateral branchlets may occur on either side of the fan in this and the other species, probably in response to local environment conditions. *Gorgonia flabellum* can be grayish white, pale lavender or yellow in color.

The skeleton is composed of calcite (a form of calcium carbonate), fibrous "gorgonin" which contains a collagen-like component and some other materials. Like other sea fans, the plane of the fan occurs perpendicularly to the direction of incoming waves (parallel to the line of the crest or trough). Growth in this plane possibly reduces the forces which tend to twist the base of the fan out of this alignment.

Gorgonia flabellum is often restricted to shallow water with very strong wave action. It occurs in areas generally somewhat shallower and rougher than *G. ventalina* where the two occur in the same geographic area. It is seldom found below 10 m depth and can reach sizes near those of *G. ventalina*.

Its known geographic distribution is somewhat odd. It is

abundant and easily distinguished from *G. ventalina* in the Bahamas, but becomes scarce and less distinctive in Florida and the Lesser Antilles. The distinctions based on spicule shape remain, but external differences between this species and *G. ventalina* are less pronounced. It is not known to occur in Bermuda.

References: Wainwright and Dillon 1969, Bayer 1961.

Gorgonia mariae Bayer
Sea fan

This is the smallest of the sea fans, the fan-like form reaching only about 30 cm in height. The anastomosing mesh of the branches is large, the branches are not strongly compressed and branches off the ascending members are pinnate. Small colonies show few connections of the branches until strong lateral branches have been developed. Colonies are generally pale or yellow, often with some violet near the base. There are two other growth forms of this species. One has short free branchlets from one or both faces, while the bright yellow plumose form, which may reach 40 cm in height, has the inner and lower branches anastomosed but the terminal branches free.

This is generally a deeper water species than the preceding two species and has been taken as deep as 47 m although it has been recorded as shallow as 5 m. Pale specimens translocated from 30 m depth to 1 m often died; those surviving developed within two weeks an increase in colored spicules. Known from Cuba, Jamaica, Puerto Rico, the Virgin Islands and the northern Lesser Antilles.

References: Bayer 1961, Kinzie 1973.

Pterogorgia citrina (Esper)

The greatly flattened branches of this gorgonian, with the polyps occurring only on the edges, are quite distinctive. The colonies are often greatly branched and not in a single plane. The color varies from pale yellow to yellow and the species is common on rocky bottoms on shallow reefs. Each polyp has its own slit-like aperture into which it can contract along the edge of the branches.

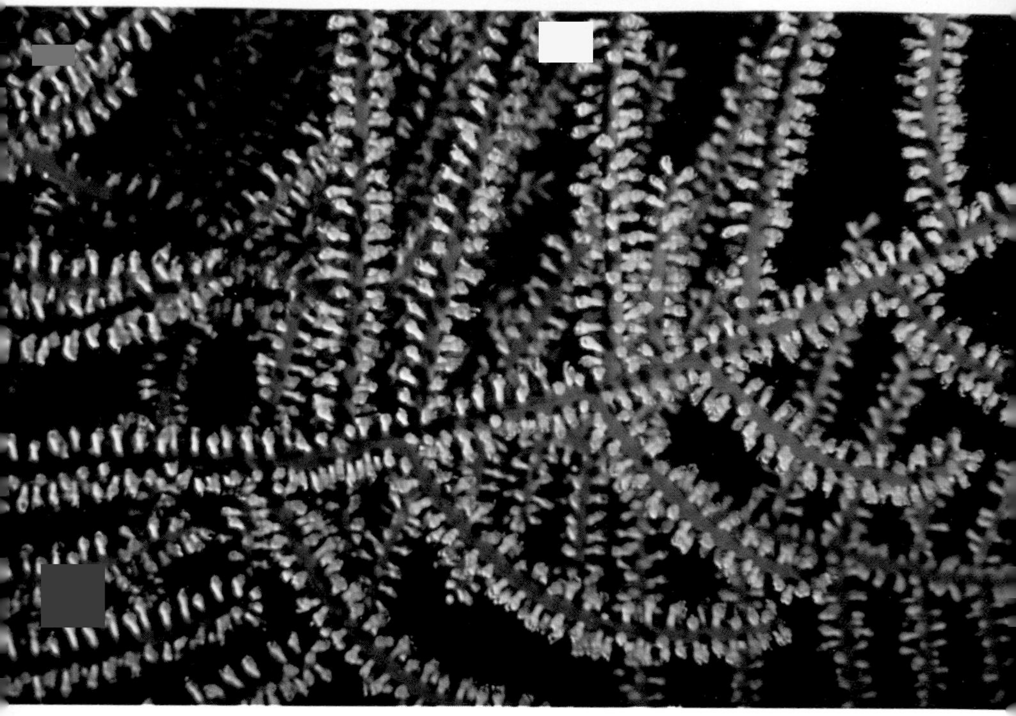

Unidentified gorgonian. Bahamas, Eleuthera Island, 20 m depth.

Unidentified gorgonian. Bahamas, Eleuthera Island, 15 m depth.

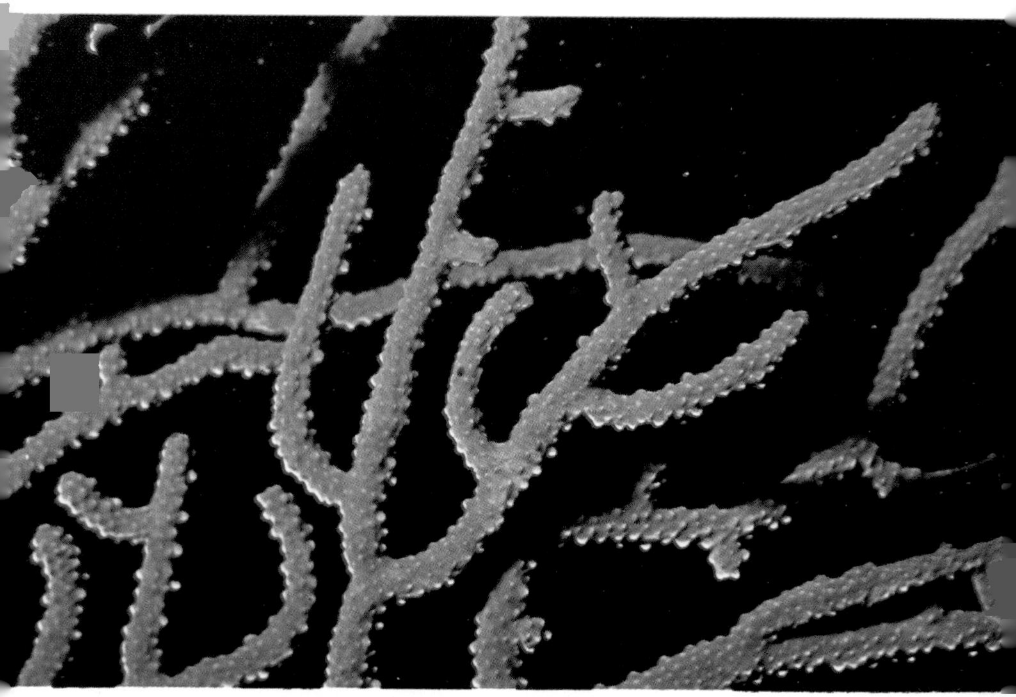

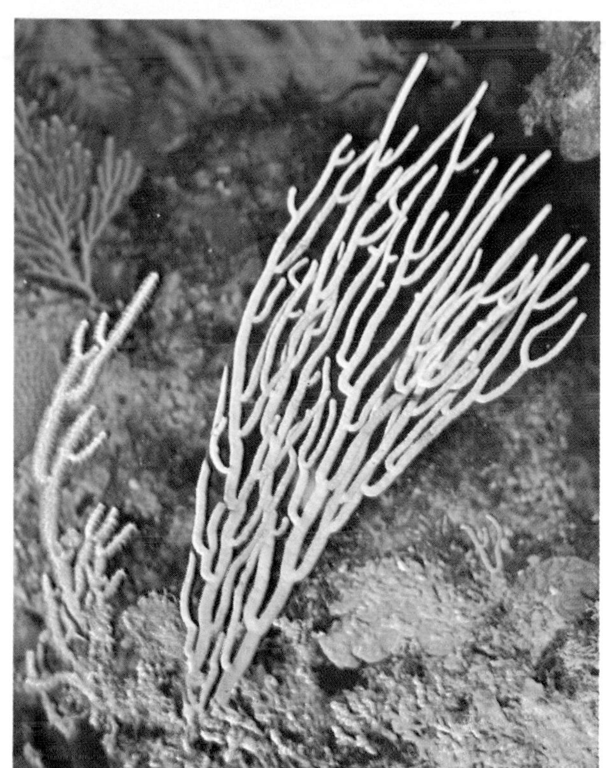

Unidentified gorgonian. Bahamas, Cat Island, 15 m depth.

Unidentified gorgonian. Puerto Rico, La Parguera, 22 m depth.

Pterogorgia citrina occurs from Bermuda and southern Florida to the southern Caribbean. Two other species of *Pterogorgia, P. anceps* (Pallas) and *P. guadalupensis* (Duchassaing and Michelotti) also occur in the Caribbean and have the polyps retracted into a common groove, rather than individual slits, on the edges of the branches. *Pterogorgia anceps* may have the branches y- or x-shaped in cross section with three or four edges; *P. guadalupensis* has wide branches, often few in number, usually in a single plane. All three species may occur together in one area.

References: Bayer 1961, Kinzie 1973.

Family Ellisellidae

Ellisella spp.

The members of the genus *Ellisella* are deep reef gorgonians generally red in color and unbranched or with a few slender whip-like branches. They lack zooxanthellae and have the polyps clearly protruding from the sides of the branches.

There are at a minimum six species in this genus occurring in the western Atlantic. Two species, *Ellisella barbadensis* (Duchassaing and Michelotti) and *E. elongata* (Pallas), reach sizes of nearly 2 m and can occur in dense stands on rocky, often vertical substrates at about 20 to at least 250 m. Three other smaller species may also occur within diving depths on deep reefs. Most species have wide geographic ranges, generally from southern Florida to the Caribbean. At least one species occurs in Bermuda.

References: Bayer 1961, Kinzie 1973.

SUBCLASS HEXACORALLIA (ZOANTHARIA)

While the subclass Octocorallia tends to be octamerously symmetrical, the Hexacorallia are somewhat hexamerously symmetrical, particularly in their internal structure. Their skeleton, if the species produces one, may be calcareous or horny and is never constructed of spicules, which members of the Hexacorallia lack. The group is very heterogeneous in structure; the polyps differ in morphology, a great diversity of nematocysts is encountered and colonial structure takes many different forms.

There are five orders of Hexacorallia, all occurring on Atlantic reefs and some of prime significance in reef structure. The stony corals (Scleractinia) need no comment regarding their importance on reefs. The sea anemones (Actiniaria), zoanthids (Zoanthidea), black corals (Antipatharia) and tube anemones (Cerianthidia) contribute little or nothing to reef structure but are important faunal elements of reef communities. The hexacorals may be solitary (sea anemones, cerianthids) or colonial (some stony corals, black corals, others).

ORDER ACTINIARIA: SEA ANEMONES

The sea anemones are solitary. They may occur in small groups but are capable of surviving without any other members of their species adjacent. They also completely lack a skeleton, contributing nothing to the carbonate structure of the reef.

These are ecologically important animals. Within the protection of their nematocyst-armed tentacles is found a sizable assortment of organisms, particularly crustaceans, which occur nowhere else. The relationship existing between the sea anemones and their associates can vary from mutualism, where both species benefit, to parasitism, where the anemone is robbed of captured food or actually preyed upon itself.

The sexes in actinians may be in separate polyps or within a single polyp. Planula larvae are produced sexually and are capable of dispersing some distance.

Family Aliciidae

Lebrunia danae (Duchassaing and Michelotti)

The tentacles of this unusual anemone are seldom visible because the greatly ramified pseudotentacles arise below the rim of the oral disk and form a dense radiating branchwork effectively hiding the rest of the anemone. The pseudoten-

Unidentified gorgonian, possibly *Swiftia exserta.* Puerto Rico, Rincon, 22 m depth.

Unidentified encrusting gorgonian. Puerto Rico, La Parguera, Laurel Reef, 12 m depth.

Closeup view of polyps of unidentified gorgonian. Bahamas, Cat Island, 10 m depth.

The branched pseudotentacles of *Lebrunia danae* comprise most of what can be seen in this photograph of an entire anemone. Puerto Rico, Desecheo Island, 30 m depth.

tacles which are very light to dark brown and have whitish oval or circular bumps (called acrorhagi) which are filled with nematocysts capable of stinging humans. The entire anemone may be 30 cm across, with the column generally hidden in a crevice.

Lebrunia danae is known from Bermuda, the Bahamas, the Caribbean and Brazil at depths of 2 to 60 m and occurs generally in reef environments. The Pederson cleaning shrimp, *Periclimenes pedersoni*, occasionally occurs with it, particularly in the deeper portion of its depth range.

References: Correa 1964, 1973, Fisher 1973.

Lebrunia coralligens (Wilson)

This anemone occurs in crevices of coral heads, where only the ends of the pseudotentacles are visible. The tan pseudotentacles are much less ramified than they are in *Lebrunia danae* and have enlarged, often double-lobed tips which are often ringed in white. This species may occur singly or in groups of a few individuals.

Lebrunia coralligens is known from the Bahamas, the Caribbean and Brazil from 1 to 10 m.

References: Correa 1964, 1973, Fisher 1973.

Family Aiptasiidae

Aiptasia tagetes Duchassaing and Michelotti

This small anemone is common in a wide variety of habitats, and its hardiness is attested to by its ability to colonize and flourish via its planula in flowing sea water aquaria. The tentacles, which can number nearly 100, are translucent brown. The oral disk can be bluish white or nearly colorless and reaches only a few centimeters in diameter. Individuals may occur singly or in small groups.

Aiptasia tagetes occurs widely in the Caribbean at depths to at least 15 m, and its congener, *A. pallida*, is found from North Carolina to Texas.

References: Duerden 1901, Fisher 1973, Carlgren and Hedgepeth 1952.

A small sea anemone, *Lebrunia danae,* growing on a mat of an unidentified zoanthid. Photographed at Discovery Bay, Jamaica, 12 m depth.

These small specimens of *Lebrunia danae* show the extreme branching of the pseudotentacles. Coral of the genus *Agaricia* nearly surrounds the anemones. Jamaica, Discovery Bay, 54 m depth.

Normally, the pseudotentacles of *Lebrunia coralligens* are the only portions of this anemone visible. It is typically found within crevices in coral heads. Puerto Rico, La Parguera, 3 m depth.

Aiptasia tagetes is a small anemone found often in great numbers. It is usually quite translucent except for the oral disk which may be bluish-white in color. Puerto Rico, Aguadilla, 12 m depth.

The anemone *Heteractis lucida* has whitish knobs along its tentacles which contain the batteries of nematocytes. Jamaica, Discovery Bay, 15 m depth.

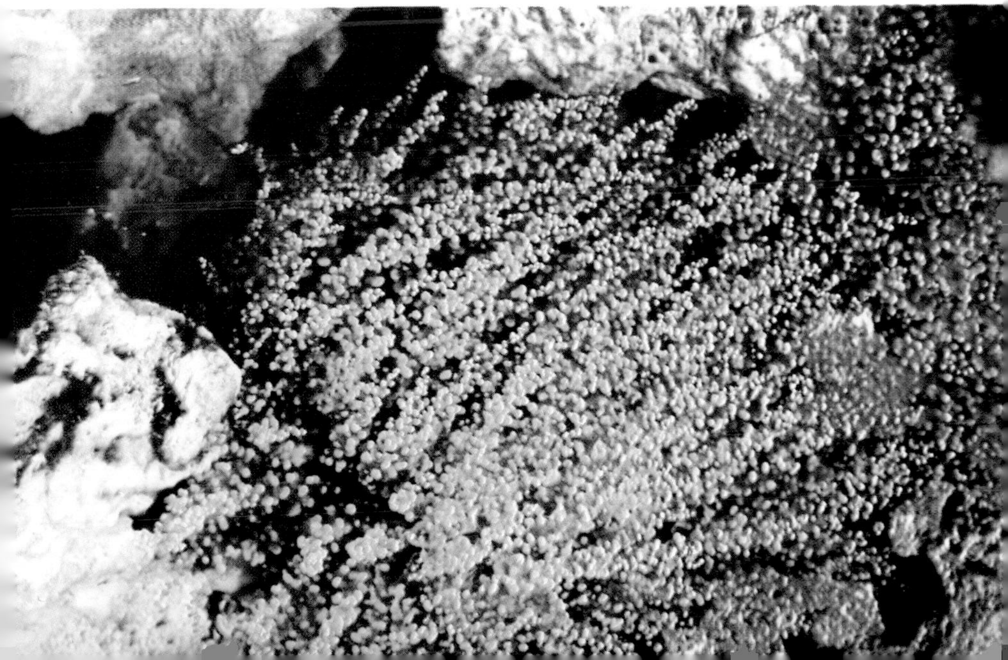

The sea anemone *Stoichactis helianthus* resting on some knobs of the coral *Montastrea annularis*. The parts of the coral colony adjacent to the sea anemone are dead with the white skeleton exposed. Probably the sea anemone is able to attack the coral nearby by a feeding response and thus is able to kill it. Photographed at 12 m depth at Discovery Bay, Jamaica. This is fairly deep for *S. helianthus* to occur.

Heteractis lucida Duchassaing and Michelotti

This anemone is easily distinguished by the whitish knobs containing nematocysts located all along its pale brown, translucent tentacles. The tentacles may number in the hundreds and reach over 10 cm in length. *Heteractis lucida* occurs with the column in crevices and only the tentacles exposed; it is most common in clear water fore reef environments. The species is capable of mildly stinging humans and only rarely has any commensal crustaceans associated with it.

Heteractis lucida is known from the Bahamas and the Caribbean at depths to at least 30 m.

Reference: Fisher 1973.

Family Actiniidae

Condylactis gigantea (Weinland)

The largest and most spectacular of the Caribbean sea anemones is *Condylactis gigantea*, which has many large, often colorful, tentacles. The tentacles may reach over 10 cm in length, taper gradually and end in a rounded tip. The tips may be yellow or purple in color, with the remainder of the tentacle whitish to greenish.

The base and column are usually hidden in a crevice. The anemone is found in both fore reef and lagoonal areas. It can be common in certain areas but does not occur in concentrated groups in reef environments. Various commensal shrimp such as *Periclimenes pedersoni* and *Thor amboinensis* may occur with it.

Condylactis gigantea is known from Bermuda and southern Florida to Brazil and occurs throughout the Caribbean at depths of 0 to at least 30 m.

References: Correa 1964, Fisher 1973.

Family Sagartidae

Bartholomea annulata (Lesueur)

This is perhaps the most common anemone in the Caribbean. It may have nearly 200 long tentacles with whorls and incomplete spirals of nematocysts which are readily apparent on the tentacles. They range from nearly transparent pale

The nematocyst batteries of *Heteractis lucida* readily distinguish it from nearly every other Caribbean anemone. Puerto Rico, La Parguera, Laurel Reef, 18 m depth.

Condylactis gigantea is one of the most delicately colored Caribbean anemones and its large size makes it probably the most commonly observed on the reef. Puerto Rico, Mona Island, Playa Sardinera, 15 m depth.

Bartholomea annulata is an extremely common anemone and this individual, its tentacles protruding from a crevice in the bottom, is typical in coloration. The antennae of a snapping shrimp, *Alpheus armatus,* can be seen protruding from the tentacles on the left. Cayman Brac, 5 m depth.

This specimen of the anemone *Phymanthus crucifer* has the column buried in the sediment with only the oral disk and tentacles exposed. Jamaica, Discovery Bay, 3 m depth.

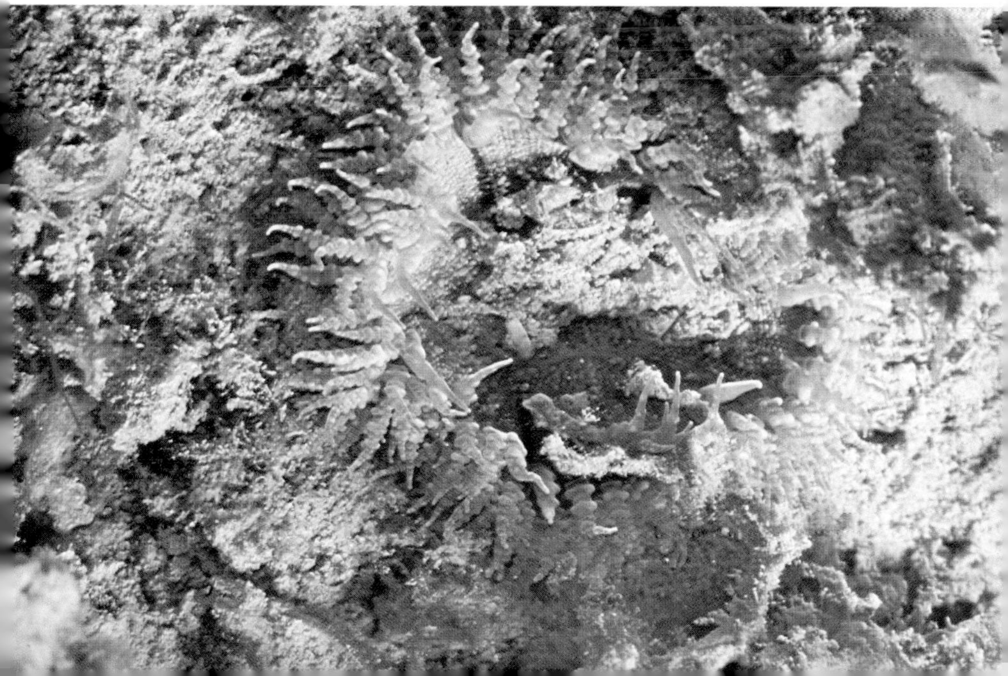

brown with a few opaque patches of nematocysts to dark brown with the abundant nematocyst patches cream colored. Much of the ground color of the anemone is produced by zooxanthellae in the tissues. *Bartholomea annulata* is large, with the expanded mass of tentacles reaching over 30 cm in diameter.

The base and much of the column are hidden in crevices or holes of reefs, rocky areas or isolated solid substrates in grass beds. They can occur in the opening of dead *Strombus gigas* shells. When expanded the tentacles are completely exposed, but when contracted the entire anemone is drawn back into its hole, often out of sight.

A wide variety of crustaceans occur associated with *B. annulata*, a community which has not been well studied. Its depth range is from 1 to at least 40 m and it probably occurs somewhat deeper. It is found from Texas and Bermuda to the northern coast of South America.

References: Carlgren and Hedgepeth 1952, Clarke 1955, Duerden 1901, Mahnken 1972.

Family Phymanthidae

Phymanthus crucifer (Lesueur)

This anemone has a large number of small tentacles around the rim of the oral surface. The oral disk may be 15 cm in diameter and has a number of folds on its peripheral edge. Its surface may have white flecking. The 200 or more banded, striped or granulated tentacles occur on the margin of the oral disk.

Generally only the oral disk and the tentacles are visible. The column is usually buried in sand or is located in a crack in rock and the oral surface is level with the surface of the bottom. *Phymanthus crucifer* occurs in back reef areas in water less than 7 m deep. It is known from the West Indies and the Bahamas.

Reference: Duerdon 1901.

Family Stoichactidae

Stoichactis helianthus (Ellis)

This large flattened anemone is one of the most easily

identified of Caribbean actinians. It has hundreds of stubby tentacles with rounded tips covering its entire oral surface exclusive of a small area around the large oval mouth. Its column is short, expanding greatly away from the base, and the anemone can reach at least 25 cm in diameter. The species varies in color and is occasionally whitish to gray with a pinkish tinge.

Stoichactis helianthus can occur in large numbers, particularly in shallow back reef areas. Individual anemones can be close together in limited areas. This is one of the few Caribbean anemones which can sting humans. The surface feels sticky to the touch due to the penetration of nematocysts into the skin, and a sharp sting is felt if uncalloused skin (like a bare stomach) is applied to the anemone. This pain persists for some time and the skin may blister in the area of contact.

Stoichactis helianthus has various crustaceans (a small unidentified species of *Periclimenes, Thor amboinensis* and *Mithrax cinctimanus*) occurring with it. The amphinonid polychaete *Hermodice carunculata*, a predator of corals and gorgonians, has been observed attacking it. This anemone is generally found in water 0.5 to 2 m deep. No records exist of it from Bermuda, but it is common in the Bahamas and the West Indies.

References: Duerden 1901, Lizama and Blanquet 1975.

Family Bunodactidae

Bundosoma granulifera (Lesueur)
Warty anemone

The warty anemone is common in shallow water and has light and dark columnar bands. Generally there are 96 fairly short conical tentacles arranged in multiple rows around the large oral disk. The tentacles often have white patches and may be pink or purplish on their oral side. The oral disk has radiating lines, and the column may be red with lighter striping. The anemone reaches 8 cm in diameter.

Bundosoma granulifera occurs on rocky substrate, often in holes. It is known from the Bahamas and the Caribbean at depths above 10 m.

Reference: Duerden 1901.

This closeup photograph shows the tentacles on a portion of the oral disk of the anemone *Phymanthus crucifer.* There can be several hundred tentacles which are variously marked. Puerto Rico, La Parguera, 2 m depth.

This individual of *Phymanthus crucifer* was photographed at night in a back-reef area and was found growing close to the corals *Agaricia agaricites* (left) and *Acropora cervicornis* (right). Jamaica, Discovery Bay, 5 m depth.

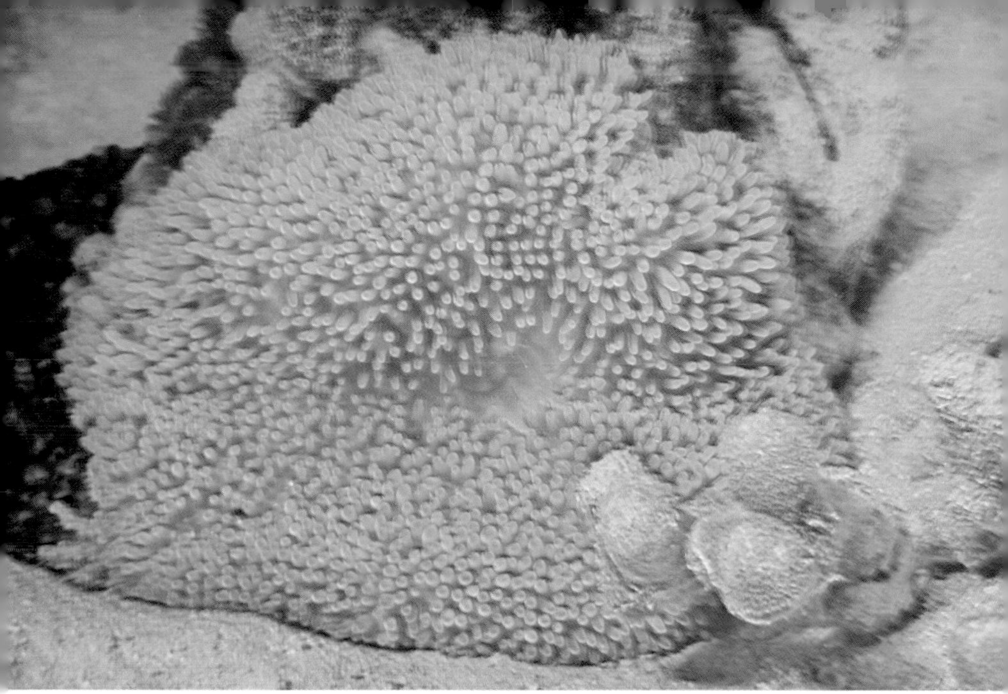

The large anemone *Stoichactis helianthus* has short, rounded tentacles and can be very abundant in back-reef areas in shallow water. Its sting is painful to humans and contact should be avoided if possible. Puerto Rico, La Parguera, Mario Reef, 1 m depth.

The tentacles of *Stoichactis helianthus* can shelter a variety of commensal crustaceans and other species can be found beneath the rim of the oral disk. Puerto Rico, La Parguera, Laurel Reef, 1 m depth.

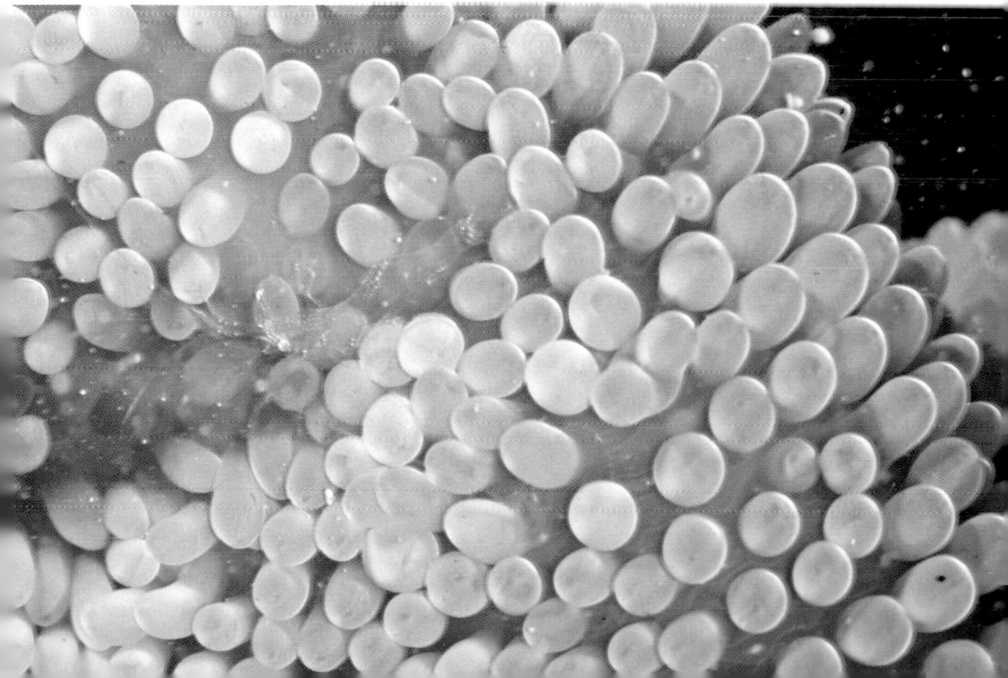

ORDER CORALLIMORPHARIA: CORAL-LIKE ANEMONES

The corallimorpharians are a small order of Hexacorallia. They lack a skeleton and a siphonoglyph. They may form sheet-like colonies or can occur singly. Although a few species are present on Caribbean reefs, they are of only minor importance.

Family Actinodiscidae

Rhodactis sanctithomae (Duchassaing and Michelotti)

These large corallimorpharians are often found in groups on rocky substrates. They may reach 10 cm in diameter and are translucent greenish or brown with numerous small unusual tentacles on the central portion of the oral disk. The rim is greatly expanded laterally at night and has a series of very short pointed tentacles radiating out.

Rhodactis sanctithomae often is found on vertical rocky areas shaded by a flattened coral head, on dead branches of the coral *Acropora cervicornis* and on the dead upper surface of a colony of *Montastrea annularis*. It has also been observed on a sponge symbiotic with the zoanthid *Parazoanthus parasiticus*. It is known from 2 to 30 m from the Bahamas, the Caribbean and Bermuda.

References: Correa 1964, Fisher 1973.

Paradiscosoma neglecta (Duchassaing and Michelotti)

This corallimorpharian anemone resembles an inverted umbrella. It has irregular short tentacles around the rim and reaches over 8 cm in diameter. *Paradiscosoma neglecta* is brown to dark green in color, occasionally with wide rings of varying color on the oral disk. Green specimens may also have a number of white streaks radiating on the oral disk from the protruding mouth.

This species can occur on rocky substrates in shaded

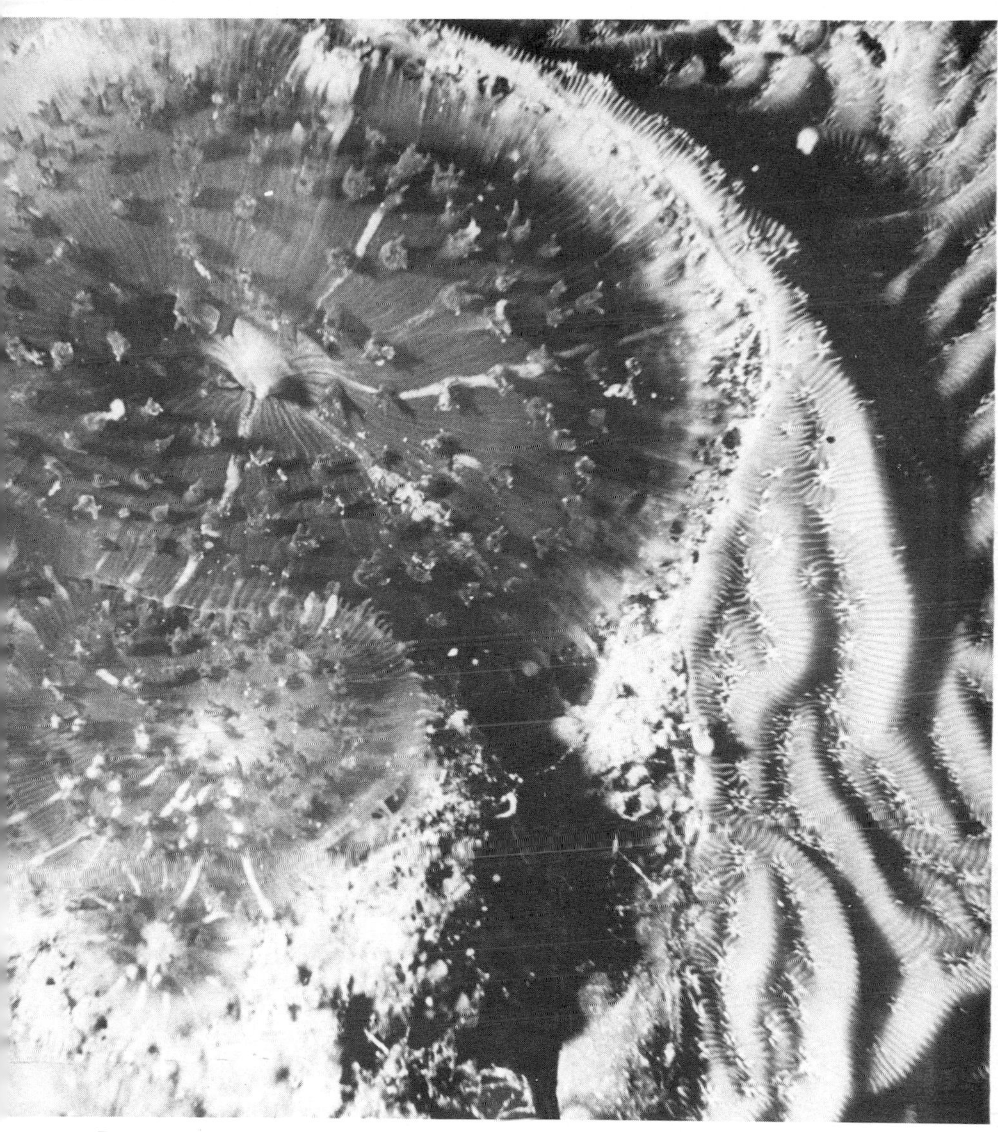

The corallimorpharian *Rhodactis sanctihomae* near a colony of *Agaricia.* It is difficult to determine from the photo but it appears that the corallimorpharian may have killed portions of the coral colony. At least it seems to be able to prevent the coral from overgrowing it.

The warty anemone, *Bundosoma granulifera,* is found on rocky substrates, often in small holes, in shallow water. Bahamas, Exuma Cays, 1.5 m depth.

The rim of these corallimorpharians, *Rhodactis sanctithomae,* is expanded at night. They are growing on a cylindrical sponge, possibly *Gelloides ramosa.* Bahamas, Eleuthera Island, 15 m depth.

The corallimorpharian *Ricordea florida* can occur in huge numbers covering large areas of substrate. The colors are often spectacular with the tentacles colored differently than the remainder of the animal. These individuals are growing on a dead branch of the coral *Acropora cervicornis*. Puerto Rico, La Parguera, San Cristobal, 12 m depth.

This cluster of *Ricordea florida* is growing on a rocky substrate and is only a small part of a much larger mass. Puerto Rico, La Parguera, Mario Reef, 1 m depth.

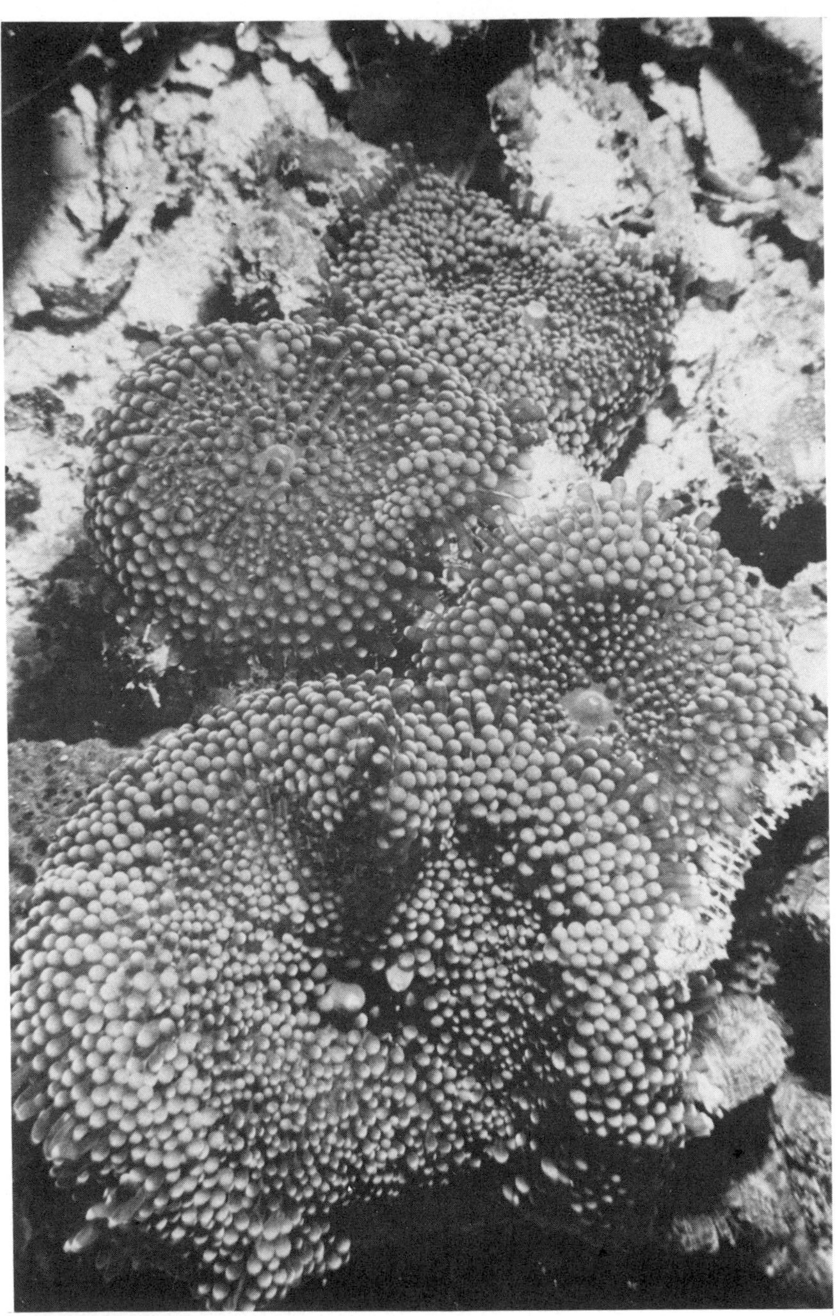

Several individuals of the corallimorpharian *Ricordea florida*, Discovery Bay, Jamaica, 15 m depth.

locations or in reef caves. It is known from the Bahamas and the West Indies at 12 to at least 30 m.

Reference: Correa 1964.

Family Corallimorphidae

Ricordea florida (Duchassaing and Michelotti)

Groups of hundreds of polyps of this corallimorpharian occur covering large areas of rocky substrates. Although no more than 5 cm across, *Ricordea florida* has extremely short, rounded tentacles over its entire oral surface. Color is quite variable in this species, the tentacles being light green, green, brown, green-brown or blue-gray, often with their bases a different color. The mouth and the outer rows of tentacles may be colored differently than the rest of the tentacles.

Ricordea florida is found in both back reef and fore reef areas where rock substrate is available. It is known from the Bahamas and the Caribbean from 0 to 20 m.

Reference: Fisher 1973.

Genus and species unknown

This distinctive corallimorpharian probably represents an undescribed species. The numerous tentacles are most unusual, being nearly clear with enlarged bright orange balls on the tips. The column is pale orange, reaching about 5 cm in diameter, and the tentacles may reach 5 cm in length.

This species is known from Puerto Rico at depths of about 10-30 m and is expanded at night.

ORDER ZOANTHIDEA: ZOANTHIDS

The zoanthids are generally small, anemone-like in form and may be colonial or solitary, symbiotic or free-living. They have two cycles of tentacles, a single siphonoglyph and both complete and incomplete septal pairs.

Both symbiotic (with sponges and other invertebrates) and free-living species occur on Atlantic reefs, and six species are considered further.

References: Walsh 1967, West 1971.

The corallimorpharian *Paradiscosoma neglecta* resembles an inverted umbrella with irregular short tentacles around its rim. Jamaica, Discovery Bay, 30 m depth.

This unusual corallimorpharian is only seen at night when its clear tentacles with orange balls at the tips are expanded. It can not be identified to genus or species. Puerto Rico, La Parguera, Caracoles Reef, 12 m depth, night.

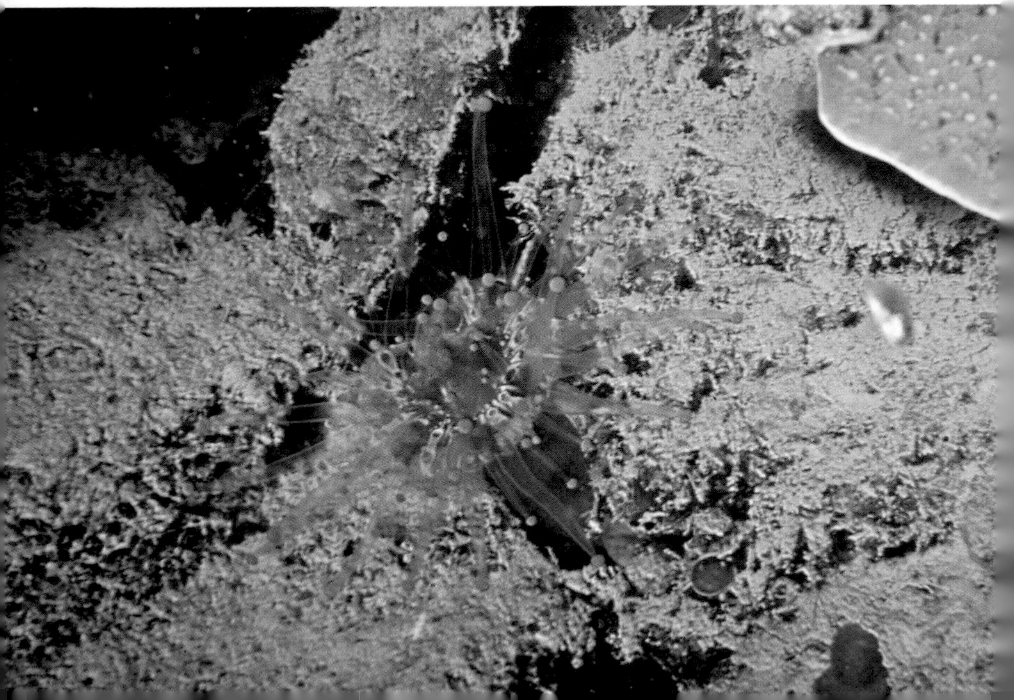

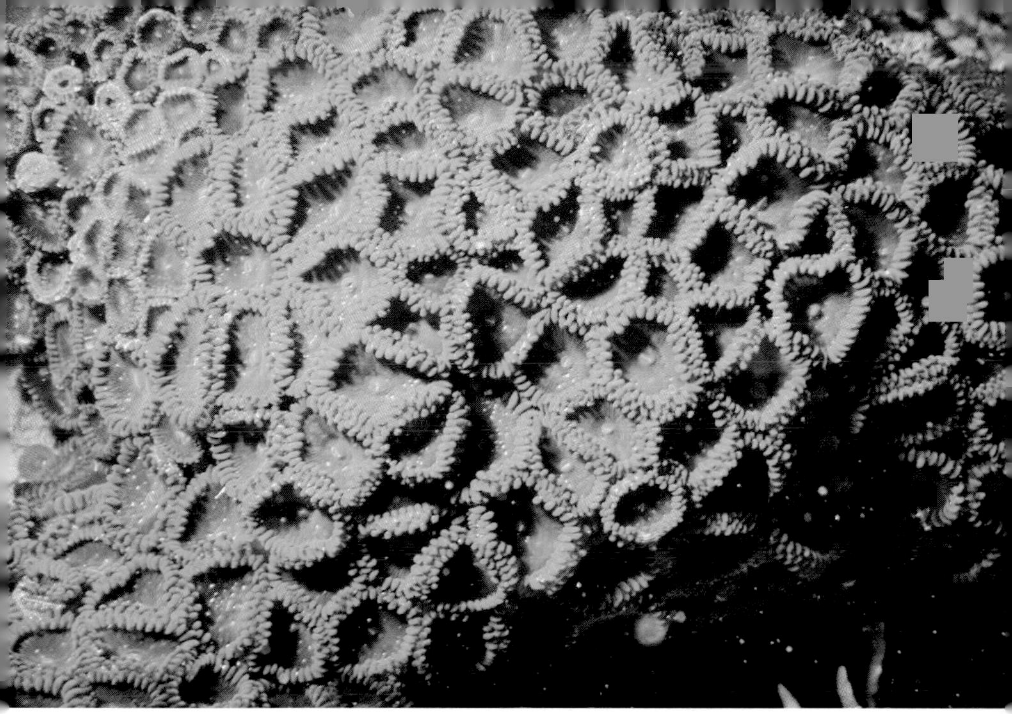

The green zoanthid *Zoanthus sociatus* can form dense mats completely covering the bottom in shallow water. Puerto Rico, La Parguera, Mario Reef, 1 m depth.

Palythoa caribbea is another zoanthid which forms dense mats and occurs in areas where there is some wave action. Puerto Rico, Aguadilla, 1.5 m depth.

Zoanthus sociatus Ellis

This green zoanthid forms resilient mats which can cover extensive areas in shallow water, excluding other organisms. They are particularly abundant on the back side of shallow reef flats. They have two circles of about 30 stubby tentacles each and a large oral disk. The polyps are interconnected with branching stolons and reach 2 to 5 cm in height.

Several other species of *Zoanthus* are reported to occur in the West Indies. *Z. sociatus* is restricted to water less than 5 m in depth.

References: Duerden 1901, Kinzie 1973.

Palythoa caribbea Duchassaing

This zoanthid forms mats, like *Zoanthus sociatus*, but is yellow to golden brown in color. *Palythoa caribbea* has short rounded tentacles surrounding a large oral disk. It occurs in shallow water where some wave action is felt, such as on shallow fore reef areas.

It is West Indian in distribution; several other species of *Palythoa* reportedly occur in this region.

Reference: Duerden 1901.

Palythoa grandis Verrill

This is a sizable zoanthid, reaching at least 3 cm (over one inch) in diameter, occurring singly or in clumps of a few to several polyps. The individual polyps resemble flattened mushrooms with a large greenish brown oral disk. The edges of the polyp are often curled inward, with the short curved tentacles pointed outward or toward the mouth, and are lighter in color than the oral disk. Often portions of the edge are flattened, particularly adjacent to neighboring polyps, giving the polyps the appearance of being polygonal in shape.

The species is known from the Caribbean, seeming to prefer rocky substrate on deep reefs at depths of 20 to 40 m.

Reference: Pax 1910.

Parazoanthus swiftii (Duchassaing and Michelotti)

This colonial zoanthid is symbiotic with at least two species of sponges. It occurs abundantly with the branching sponge *Iotrochota birotulata*, where it may cover as much as one half the surface area of the sponge with up to 200 polyps in band-like rows winding around the branches of the sponge. The yellow to yellow-orange polyps with pale yellow tentacles (up to 26 in number) contrast markedly with the dark green color of the sponge. The zoanthids, extracts of which are toxic to fish, may discourage sponge-eating fishes such as the rock beauty from sampling their hosts, and the distinctive color of the zoanthid on the sponges may serve to advertise its presence.

The reproduction of any one colony is asexual by budding, usually at the ends of the series of polyps. *Parazoanthus swiftii* generally occurs at 10-20 m depth, but it has been found in less than 6 m of water. It is known from several localities in the Caribbean including Jamaica, Puerto Rico and the Virgin Islands.

Reference: West 1971.

Parazoanthus parasiticus (Duchassaing and Michelotti)

This symbiotic zoanthid is known to occur with a wide variety of sponges, including *Cliona delitrix, Cliona* spp., *Spheciospongia* sp., *Gelloides ramosa* and *Callyspongia vaginalis.* Unlike *P. swiftii*, it occurs as single polyps or groups of at most two or three polyps. Multiple individuals are usually recently budded, and eventually individuals seem to spread themselves out evenly on the surface of the sponge.

The column of the polyp is white due to calcareous sand encrustations, and the tentacles are brown due to the presence of zooxanthellae. While not as brightly colored as *P. swiftii*, this species is also conspicuous on its sponge hosts; the coloration probably prevents predation by sponge-feeding fishes. The sexes are separate, but it appears that reproduction on a single sponge is usually by asexual budding.

Parazoanthus parasiticus is known from the West Indies, the Bahamas and Bermuda at depths of 10 to 33 m.

Reference: West 1971.

Palythoa grandis is a large zoanthid which does not form large mats. It occurs in somewhat deeper water than the mat-forming species and has tiny tentacles around its convoluted oral disk. Jamaica, Discovery Bay, 25 m depth.

The symbiotic zoanthid *Parazoanthus swiftii* occurs with sponges. The polyps are found in band-like rows as shown here and are in the expanded state. Puerto Rico, La Parguera, Mario Reef, 8 m depth.

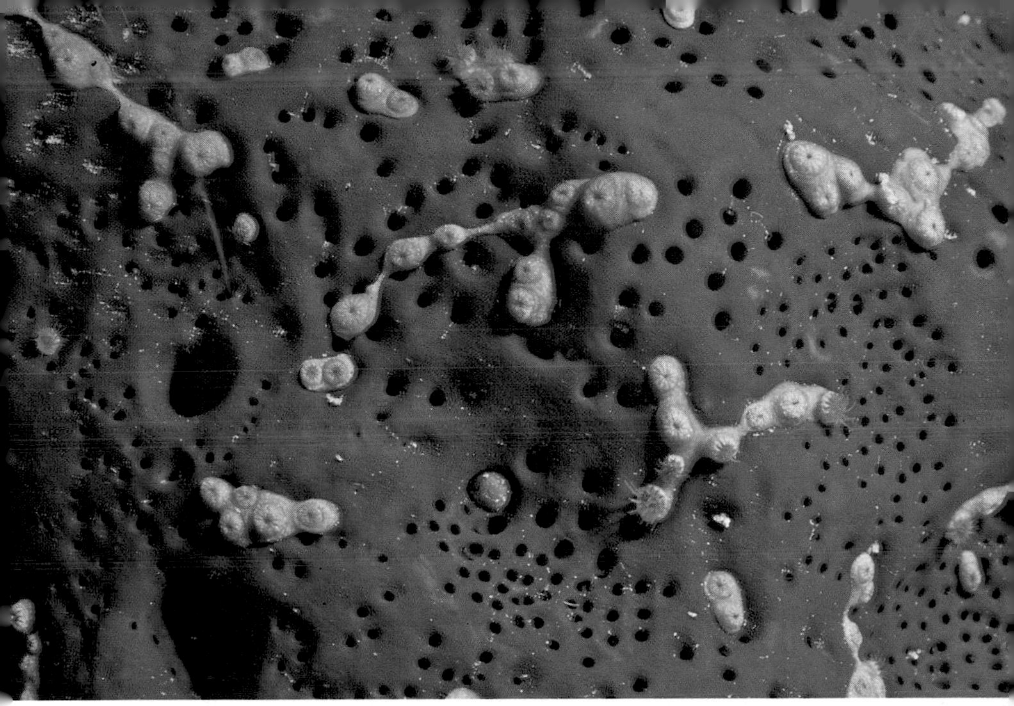

The polyps of *Parazoanthus swiftii* on an unidentified sponge in this photograph are in a contracted state. The small ostia on the surface of the sponge are also visible. Jamaica, Discovery Bay, 20 m depth.

The zoanthid *Parazoanthus parasiticus* is another zoanthid which occurs symbiotically with sponges. In this case the polyps are expanded and the tentacles visible. Puerto Rico, La Parguera, Mario Reef, 12 m depth.

Parazoanthus tunicans Duerden

This poorly known zoanthid was observed encrusting on the skeleton of a hydroid beneath an overhang on a deep patch reef in the Bahamas. The individual polyps are small, with numerous slender tentacles, and are orange in color. The column appeared encrusted with white material.

The species is known from the Caribbean and the Bahamas at moderate depths (30 m).

Reference: Duerden 1901.

SUBCLASS HEXACORALLIA
ORDER SCLERACTINIA: STONY CORALS

The scleractinians are the stony corals of reefs, the organisms with which coral reefs are identified and one of the major (if not the major) contributors of calcium carbonate to the reef structure. The polyps of stony corals are somewhat similar to those of sea anemones but produce a calcium carbonate cup (the corallite) and are usually colonial, producing a massive calcareous skeleton (the corallum) from the many corallites. Often scleractinians are considered in two informal groups, the hermatypic or reef-building corals (those making a significant contribution to reef structure) and ahermatypic or non-reef-building corals (often small, soliary species without large skeletons).

The coral polyps possess nematocysts similar to those of sea anemones on tentacles which generally number six or multiples of six. Each polyp has a single mouth (monostomodeal) which lacks the siphonoglyph (a ciliated groove along one edge).

Many stony corals, particularly those that are hermatypic, contain small unicellular plants called zooxanthellae (Dinoflagellata) in their gastrodermis. These zooxanthellae are pigmented, giving corals most of their color, and play a role in the production of calcium carbonate by the coral polyp. The exact nature of their contribution is not known and seems to vary within species of corals. Generally, however, ahermatypic corals lack zooxanthellae while hermatypic species possess large numbers. The zooxanthellae can be expelled by a coral (usually termed "bleaching") when under stress. Most reef corals maintained in complete darkness for a number of months will "bleach," and expulsion of zooxanthellae has been noted in corals on reefs after the passage of hurricanes over the reef.

It is believed that the requirement of light for the zooxanthellae is the reason why coral reefs are limited to fairly shallow waters. With increasing depth below about 30 m corals are generally less heavily calcified than in shallower water and the ability to form reef structures is much less than in shallow water. Reef corals may occur to depths approaching 90 to 100 m in extremely clear water, but below 45 to 50 m their constructional abilities are severely limited and may be surpassed by those of other groups of organisms such as the sclerosponges.

The living polyps occupy only the outer surface of the often massive skeleton. Other organisms can also occupy the skeleton, often deleteriously. Endolithic algae are usually found in a band or bands in the skeleton just below (to 15 mm depth) the living coral surface. Boring sponges and clams occur in the skeleton and weaken it by their mechanisms of removing calcareous material.

Scleractinian coral polyps are usually contracted into the corallite during the day, with a few exceptions noted under individual species. They expand, however, at night to resemble the typical polyp and are capable of capturing zooplankton with their tentacles and nematocysts. The role of zooplankton (animal plankton) in coral nutrition is poorly understood, particularly in combination with the contribution of the zooxanthellae, although most species have the potential for capturing such food.

Within a colony all reproduction is asexual. New polyps are budded from other polyps as the colony increases in diameter or length. The rate of growth is variable between species, with branched species generally growing faster than massive species, and is strongly influenced within each species by environmental conditions. Sexually produced larvae, termed planulae, result in the establishment of new colonies.

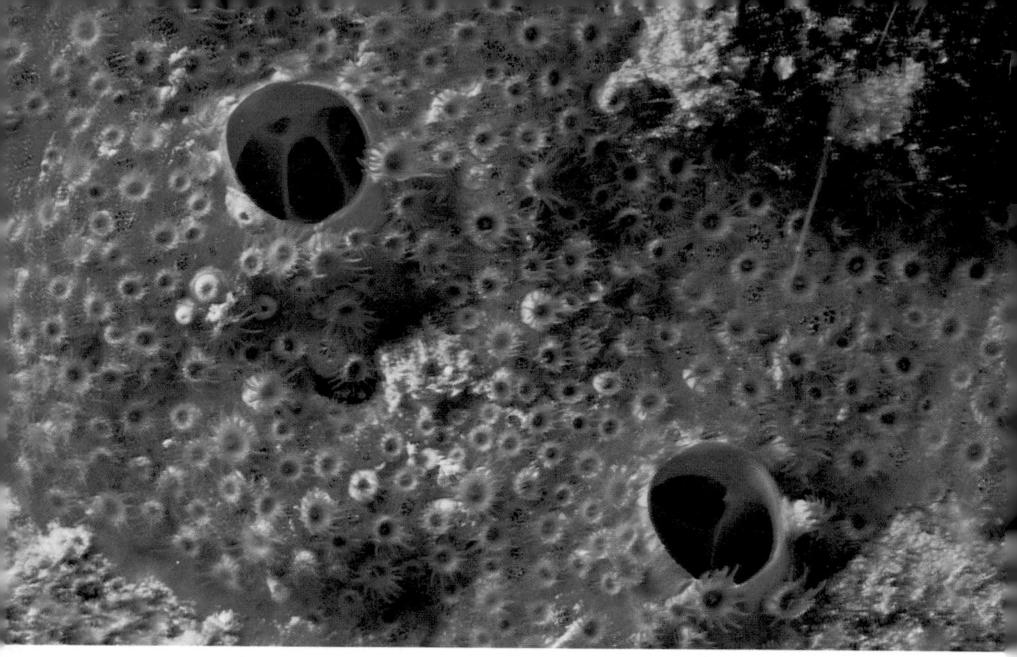

Parazoanthus parasiticus also occurs with boring sponges such as this *Cliona delitrix* with the small incurrent ostia of the sponge visible between the zoanthids. The large excurrent ocula of the sponge are also present. Jamaica, Discovery Bay, 15 m depth.

The zoanthid *Parazoanthus tunicans* has been observed with an unidentified hydroid as is shown here. This relationship is poorly known. Bahamas, Cat Island, 27 m depth.

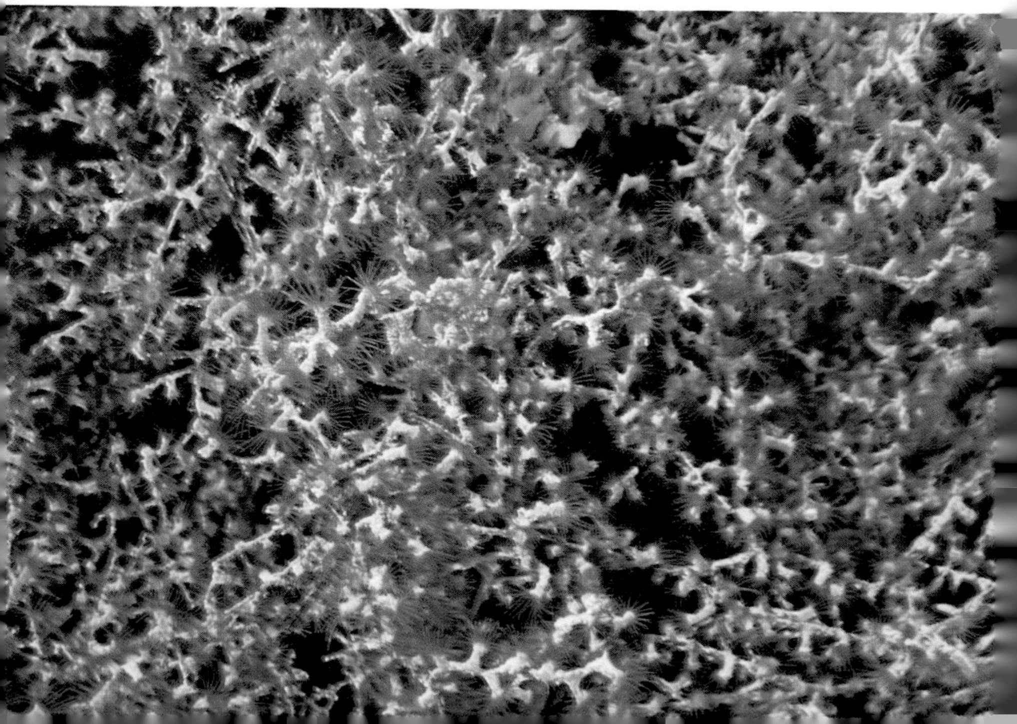

This unidentified zoanthid forms small mats like some shallow-water species but occurs in the deep-reef habitat. Jamaica, Discovery Bay, 36 m depth.

The coral *Stephanocoenia michelinii* often has an unidentified symbiotic organism associated with it. The red dots on the coral colony are this "symbiont," perhaps a zoanthid or polychaete worm, which contract at the approach of a diver. Jamaica, Discovery Bay, 24 m depth.

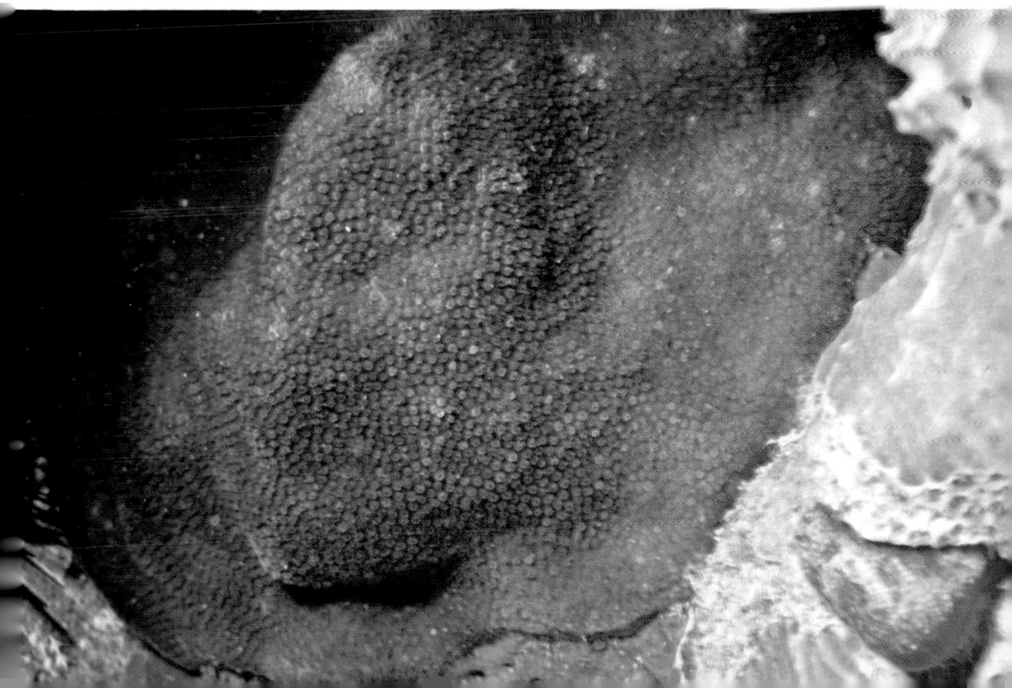

Larvae may either swim (entering the plankton and covering large distances) or crawl (staying close to the parent) until they attach to the bottom to initiate a new colony.

A number of organisms prey directly on corals. Certain fishes pick polyps from the surface of the colony (butterflyfishes) while others ingest or scrape portions of skeleton with their attached polyps (puffers, parrotfishes). Often predation by fishes leaves characteristic marks on corals, such as the scrapes produced by parrotfish beaks. Some gastropod molluscs feed on coral polyps by inserting their proboscis into the polyp, and a few polychaete worms feed on branched corals by engulfing the tip of a branch in their mouth.

The number of species of scleractinians present in various locations in the western Atlantic varies somewhat. Where conditions are near optimal and reef development best, the highest number of species is encountered. In most locations in the Caribbean, where there is no extensive fresh water run-off, forty to sixty species of reef-building corals can be expected to occur. Jamaica has more species recorded than any other locality, but this is probably due to the greater collecting effort expended there.

In areas of less optimal conditions, the number of corals occurring is lower. In the northern Gulf of Mexico eighteen species occur. Twenty-six are known from Palm Beach County, Florida, and Bermuda has a fauna of about twenty species. In the Gulf of Caiacou in Venezuela the coral fauna is reduced for a Caribbean locality, probably due to fresh water run-off. The coral fauna of Brazil is not as well known as some Caribbean localities but appears to be considerably less than that of the Caribbean. There are at least ten species in common between the two areas.

There are no extensive reefs in the eastern Atlantic off the African coast, but some corals do occur there and at least seven species are in common with the western Atlantic.

References on scleractinian geography: Bermuda—Laborel 1966, 1970, Bullock 1972; *Florida*—Goldberg 1973, Smith 1971; *Bahamas*—Squires 1958, Storr 1964; *Gulf of Mexico*—Tresslar 1973, Rannfeld 1972, Kornicker *et al.* 1959; *Cuba*—Duarte-Bello 1961, Kuhlman 1970, 1971,

1974; *Jamaica*—Goreau and Wells 1967, Wells and Lang 1973; *Puerto Rico*—Almy and Carrion-Torres 1963; *Curacao and Netherlands Antilles*—Roos 1971; *Bonaire*—Scatterday 1974; *Barbados*—Lewis 1960, 1965; *Venezuela*, Olivares 1973, Olivares and Leonard 1971; *Colombia*—Pfaff 1969, Antonius 1972; *Panama*—Porter 1972; *Brazil*—Laborel 1969.

References on Scleractinia: Bayer and Owre 1969, Hyman 1940, Yonge 1973.

Family Astrocoeniidae

Stephanocoenia michelinii Milne-Edwards and Haime

This species forms small to moderately large heads and is often simply encrusting. It is not particularly distinctive or conspicuous and, although common on many reefs, it could never be considered abundant. Brown or light brown in color, it may have zoanthid-like symbionts protruding from its surface which retract on the approach of a diver, leaving only small reddish spots. In specimens from deeper water the corallites of the colony are larger and spaced wider apart than those from shallow water, a phenomenon which occurs in some other corals. Perhaps this allows a greater surface area for the zooxanthellae of one polyp to receive the necessary light for photosynthesis.

The species occurs throughout the Caribbean, the Bahamas, the offshore banks in the northern Gulf of Mexico, southern Florida and Bermuda. Its bathymetric range is wide, from 1 to 95 m, but it is most common at 3 to 50 m.

References: Roos 1971, Squires 1958, Goreau and Wells 1967, Tresslar 1973, Scatterday 1974.

Family Pocilloporidae

Madracis decactis (Lyman)

This species is often encrusting, with raised knobs in a tight bunch, but can also be massive. It is greenish to brown in color and is the most widespread member of the genus on reefs. It can occur from back reef areas to the deep reefs and

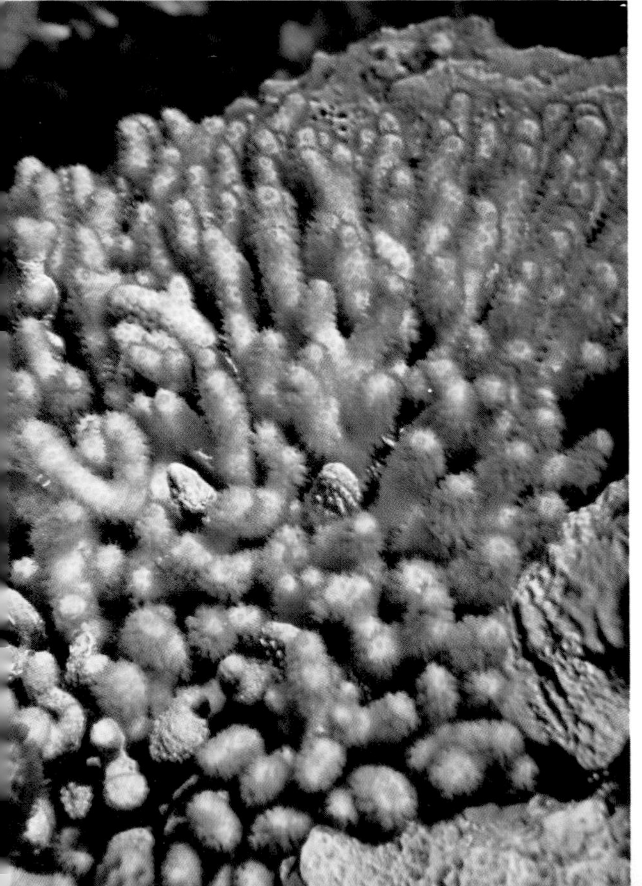

Madracis decactis is usually the most widespread member of the genus on reefs. It grows close to the substrate and does not tend to form large heads as some of the other species of *Madracis* do. Puerto Rico, Desecheo Island, 12 m depth.

The delicately branched coral *Madracis mirabilis* can form huge clumps several meters across and is often found at the edge of steep drop-offs which occur at fairly shallow depth (15-30 m). Bahamas, Cat Island, 20 m depth.

Madracis pharensis grows as small nodular branches arising from a dead basal mass in areas of reef caves which are quite dark. Jamaica, Pear Tree Bottom, 23 m depth, cave.

Madracis formosa is a deep-reef species of the genus which forms large clumps in deep water. Jamaica, Discovery Bay, 35 m depth.

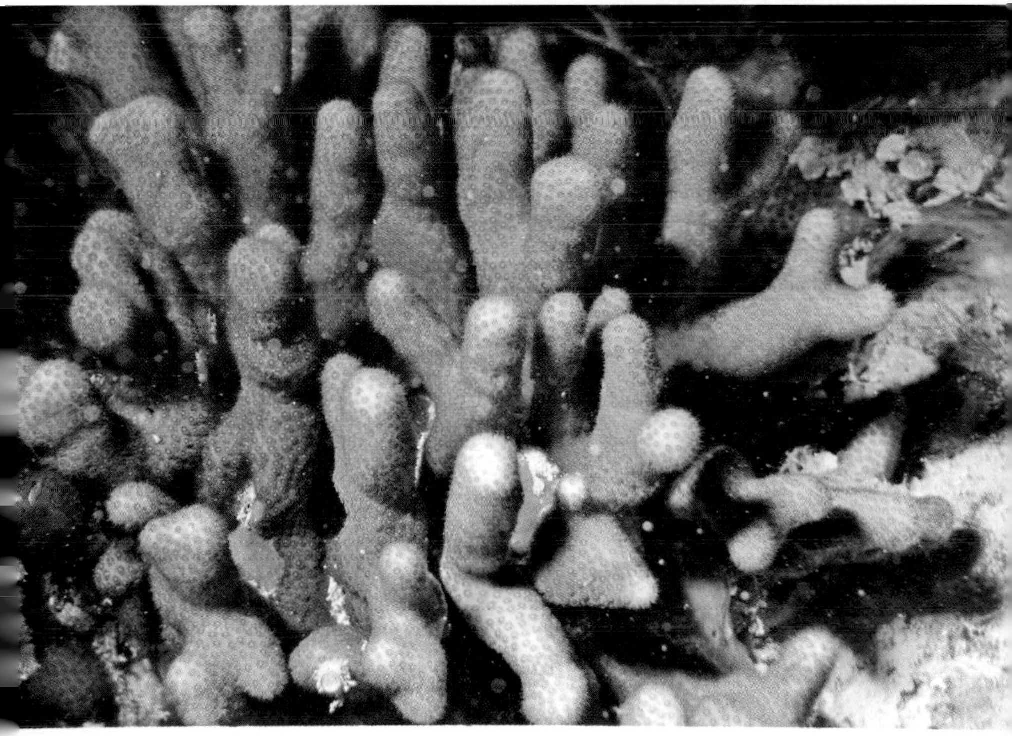

is a minor contributor to reef construction. In shallow water (less than 10 m depth) it often occurs in cavities out of direct light.

Madracis decactis occurs throughout the Caribbean, the Bahamas, southern Florida, Bermuda and in the eastern Atlantic. It is the geographically most widely distributed member of the genus *Madracis* in the Atlantic.

References: Wells 1973a, Roos 1971, Scatterday 1974.

Madracis mirabilis Duchassaing and Michelotti

This is a branching form which can occur in huge masses several meters across. The individual branches project outward and only the outer few centimeters possess living polyps. These masses are quite fragile and are easily broken by a careless diver. The species may range in color from pale cream to bright yellow.

There is some confusion between this species and *M. asperula* Milne-Edwards and Haime. *Madracis mirabilis* is definitely known from several localities in the Caribbean and Bermuda. It is found on outer reefs in water at least 10 m deep, but can also occur somewhat shallower in slightly turbid water.

References: Wells 1973a, Goreau and Goreau 1973, Scatterday 1974.

Madracis pharensis (Heller)

This coral has two different growth forms, the first an encrusting form which occurs strictly on the deep reef, but in open areas. The second form has nodular proliferations arising from a basal, often dead, encrusting mass and occurs in cavernous, dark situations. The cavern-dwelling form can occur considerably shallower than the other, but a light is needed to observe it in these caverns. The species often has a rose or pinkish tinge to the polyps. The species is known from various localities in the Caribbean.

References: Wells 1973a, Goreau and Goreau 1973.

Large clumps of *Madracis mirabilis* on the tops of buttresses at Spring Gardens, Jamaica, depth 15 m. The branches of this coral are quite fragile and while the clumps are large, they are easily damaged.

Elkhorn coral, *Acropora palmata* is characteristic of most shallow water Caribbean reefs. It is often so abundant at depths above 6 m that the zone of the reef is termed the *"Acropora palmata"* zone. Jamaica, Rio Bueno, 4 m depth.

The cylindrical branches of staghorn coral, *Acropora cervicornis,* are characteristic of moderate depths on Caribbean reefs. They can form tight clumps or cover such large areas horizontally that they resemble thickets. Bahamas, Exuma Islands, 9 m depth.

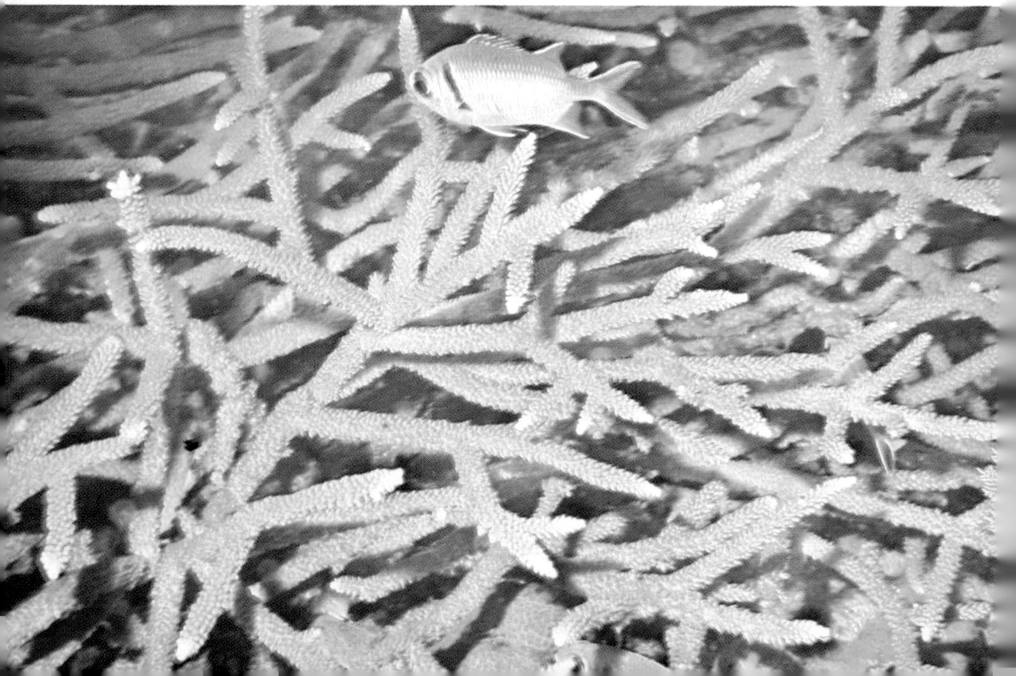

The apical polyps of *Acropora cervicornis* lead the outward growth of each branch. As they grow the branches increase relatively little in diameter but increase greatly in length. Normally the basal portions of the branches are dead, lacking living polyps, but growth continues at the tip of the branch. Jamaica, Discovery Bay, 10 m depth.

Acropora prolifera has fusion between cylindrical branches forming a flattened plate-like branch. Bahamas, Cat Island, 12 m depth.

Extremely dense stand of *Acropora palmata* which reaches nearly to the surface at Buck Island, St. Croix, Virgin Islands.

Ends of the branches of elkhorn coral, *Acropora palmata,* with the individual polyps visible. The tips of the branches are lighter in color and are the most actively growing region.

Madracis formosa Wells

This is, like *M. mirabilis,* a branched species forming sizable clumps. It is distinguished from all other western Atlantic species of *Madracis* in having the septa of the corallite arranged in eights rather than in tens. The branch tips are blunt and the species is generally pale beige in color. The polyps, like other species of *Madracis,* are often somewhat expanded during the day.

Madracis formosa is restricted to outer reef locations and is most abundant below 35 m, where it often forms clumps over 1 m in diameter and height. It is known presently only from Jamaica.

References: Wells 1973a, 1973b, Goreau and Goreau 1973.

Family Acroporidae

Acropora palmata (Lamarck)
Elkhorn coral

If any coral can be considered as representative of shallow water Caribbean reefs, it is surely *Acropora palmata.* It is the most abundant stony coral in shallow water areas, often growing up to low water levels, and its impressive form, irrespective of its abundance, would certainly earn it a special regard.

The "*Acropora palmata*" zone is a characteristic component of most West Indian reefs, and it thrives where wave conditions are rough. Normally the yellow-brown flattened branches grow oriented toward the incoming surf in areas of consistent wave direction. When there is extremely consistent heavy surf the coral may be reduced to an encrusting form without its characteristic branches. In more protected areas the branches may grow into "antler" type formations.

Severe storms such as hurricanes can have disastrous effects on reefs of *A. palmata.* Entire reefs may be reduced to rubble, much of this transported over the reef crest or piled above low water levels. Large colonies may be overturned and often renew their growth in the inverted position.

Acropora palmata is strictly a shallow-water coral. Seldom are colonies found below 15 m, and its greatest abun-

Undersurface of a plate of the coral *Agaricia* with sponges, bryozoans, serpulid worm tubes and algae exposed. Normally these undersurface areas are heavily shaded with the plate in growth position and constitute a poorly known reef habitat which is quite abundant. Jamaica, Discovery Bay, plate from 50 m depth.

Agaricia agaricites can assume a wide variety of forms such as this knobby form shown here. It has the widest range of forms and the widest bathymetric distribution of any of the species of *Agaricia*. Bahamas, Bimini, Turtle Rocks, 8 m depth.

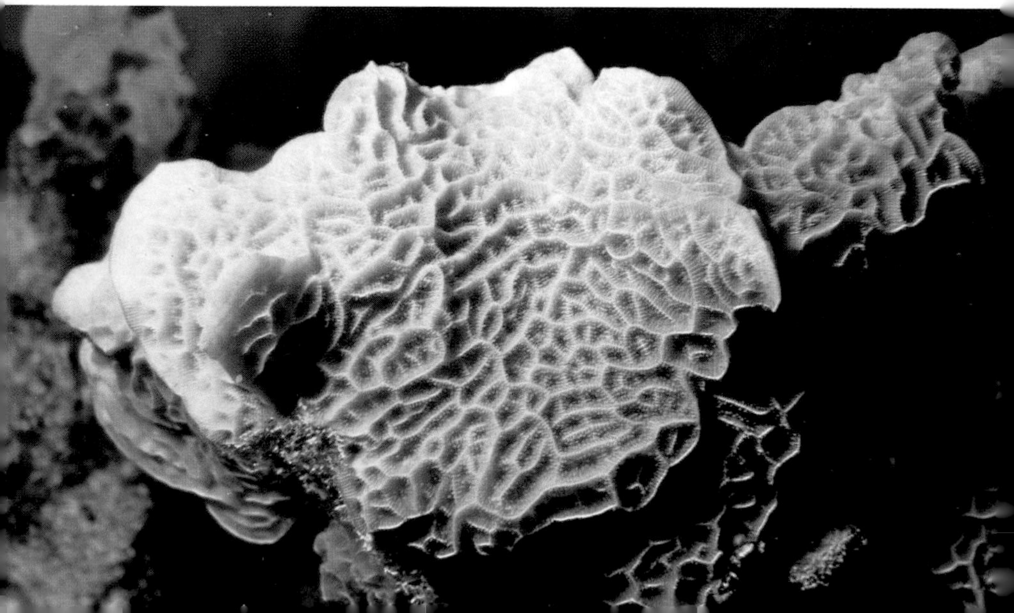

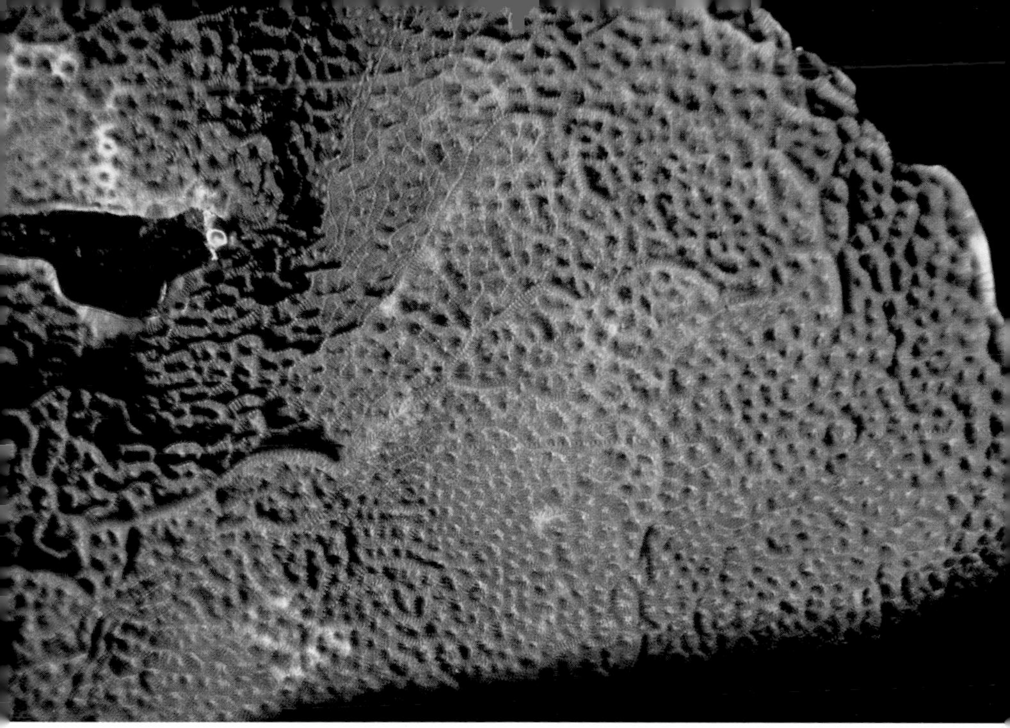

Agaricia agaricites can be more plate-like and can add significantly to reef structure in some areas. Jamaica, Discovery Bay, 15 m depth.

The ridges of *Agaricia agaricites* are often relatively high as compared to other species of *Agaricia.* Jamaica, Discovery Bay, 10 m depth.

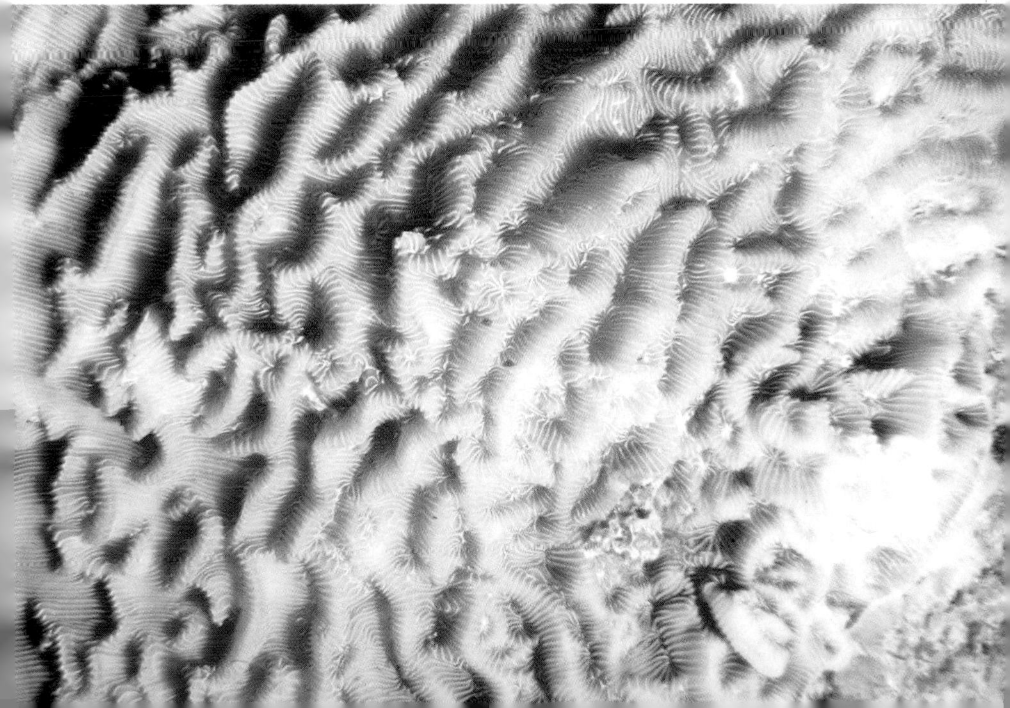

dance is in the top 6 m of water. It can occur in surprisingly turbid water but may be limited in some areas by low winter temperatures. It, like all other species of *Acropora*, does not occur in Bermuda and is uncommon along the Florida Keys and the western edge of the Great Bahama Bank. Entire barrier reefs, with no adjacent reef flat, may be built of this coral. The famous barrier reef at Buck Island, St. Croix is an excellent example of such a situation, but similar reefs are found in many areas of the Caribbean.

Occasionally the branches of *A. palmata* will have lumpy growths of polyps, termed "neoplasms," on the normally flattened branches. If any portion of the coral surface dies this provides a site of attachment for a wide variety of organisms, and branches of *A. palmata* with algae, hydroids, and actinians in sections have been observed. Certain crabs, such as *Domecia acanthophora*, form cavities in the junctions of branches by preventing the coral from growing in these areas.

References: Goreau 1959, Goreau and Goreau 1973, Roos 1964, 1971, Scatterday 1974.

Acropora cervicornis (Lamarck)
Staghorn coral

Just as the large flattened branches of *Acropora palmata* are characteristic of the shallow reef facing the open sea, the delicate but often lengthy cylindrical branches of *A. cervicornis* are characteristic of reefs below 6 to 9 m on these same seaward reef faces. The branches with the polyps protruding on all sides can form dense clumps or thickets covering areas of many square meters. In these thickets usually only the upper layers of branches support living polyps. Those lower down are dead and are encrusted by algae or other organisms. Boring sponges, other erosive organisms, the effect of waves and the ever increasing weight of the growing coral mass above cause the bottom layer of such thickets to be constantly crumbling into rubble. The growth of the coral above, which helps cause the demise of the lower branches, must keep pace with the destruction below to remain at a stable level. *Acropora cervicornis* is one of the most rapidly growing corals. Length increases of nearly 30 cm per year have been recorded for single branches under optimal conditions. The tip of each branch has an apical polyp which leads in the outward growth, and the coral is light yellow or tan in color.

The species can also occur in shallow, quiet back reef areas where the water is fairly clear, forming clumps of branches. In more exposed conditions the coral occurs deeper and cannot stand up to the vigorous wave action of the "*Acropora palmata*" zone. It occurs from low water to 50 m but is most common at 12 to 22 m. Like *A. palmata, A. cervicornis* is moderately aggressive or capable of attacking by an extracoelenteric feeding response a number of immediately adjacent corals. Other more aggressive species are not affected by the attack of *A. cervicornis* and can attack it successfully, if close enough, by the same method.

The species is found throughout the Caribbean and the Bahamas but is relatively uncommon in southern Florida. The temperature conditions of the reefs of southern Florida may be marginal for this species. It does not occur in Bermuda.

References: Goreau 1959, Goreau and Goreau 1973, Scatterday 1974, Roos 1971, Lang 1973.

Acropora prolifera (Lamarck)

This species resembles *A. cervicornis* in that its individual branches are cylindrical with a single apical polyp, but it is more branched. Often fusion between branches occurs, producing flattened branches in which the original cylindrical elements can still be discerned. In this sense it is somewhat intermediate in appearance between *A. palmata* and *A. cervicornis.*

The masses of *A. prolifera* are much smaller than those of *A. cervicornis* and often the species assumes a somewhat vase-shaped aspect of fused branches. It is reportedly more brown than *A. cervicornis* in color.

Generally this species is not as abundant as the preceding two species of *Acropora* but is most common on the seaward slope of outer reefs, often with or near *A. cervicor-*

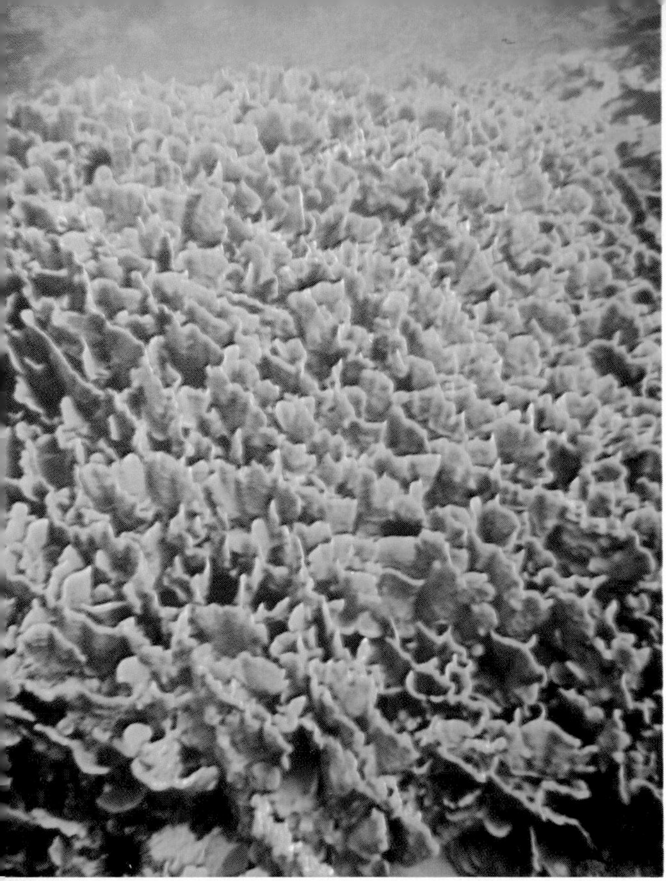

Agaricia tenuifolia can form massive box-like structures on reef buttresses, producing a tremendous maze inhabited by large numbers of other reef organisms. Jamaica, Discovery Bay, 12 m depth.

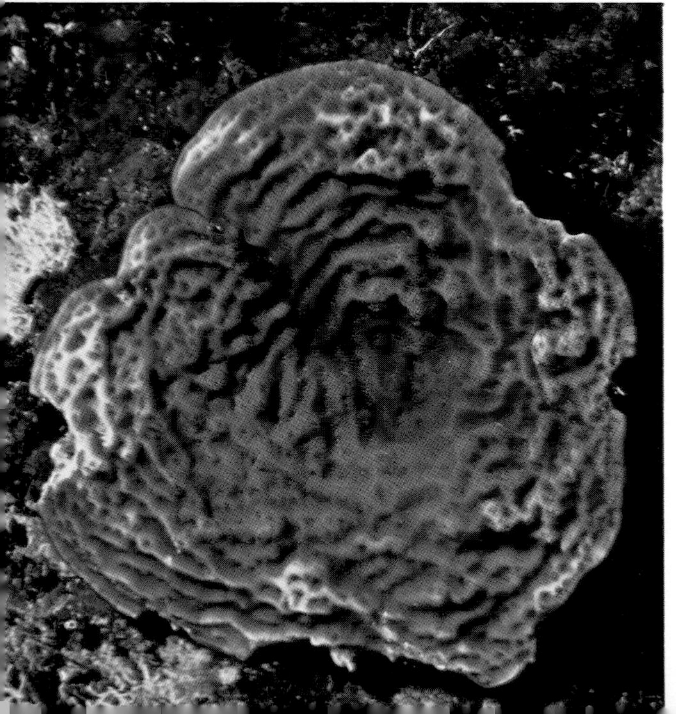

Agaricia fragilis is a delicate plate-like form which is the only species of *Agaricia* that occurs in Bermuda. Bermuda, south shore reefs (patch reefs), 15 m depth.

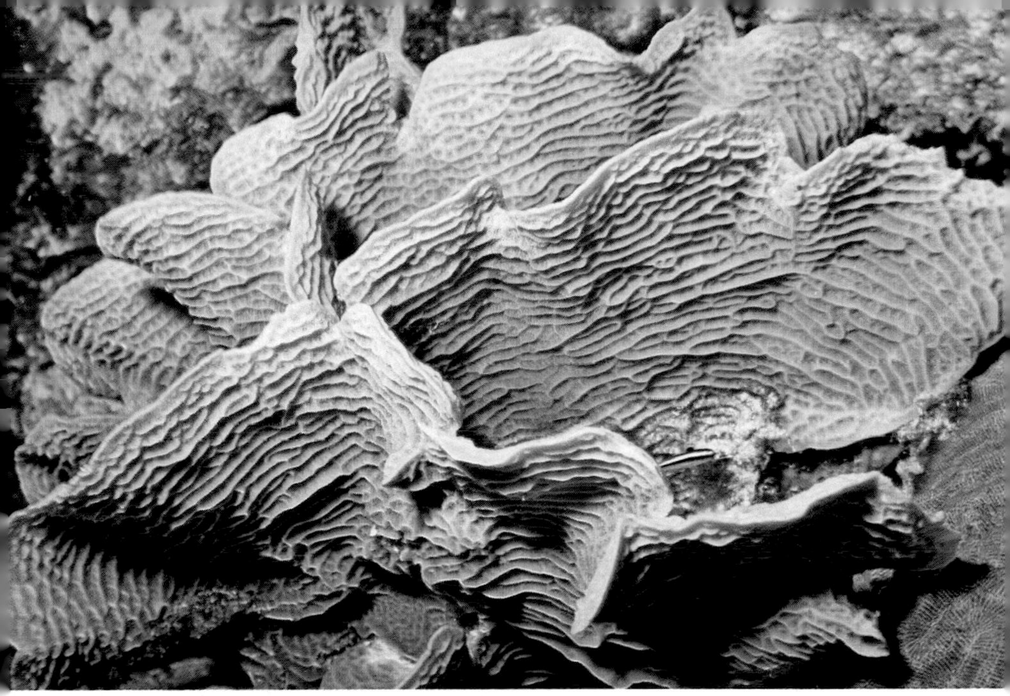

The plates of *Agaricia tenuifolia* which make up the box-like structure are bifacial, having living polyps on both surfaces of the plate. Puerto Rico. Mona Island, Playa Sardinera, 15 m depth.

Agaricia undata has fragile, convoluted blades which are unifacial, having living polyps only on one side. The side without living polyps is usually away from the light and a diverse community of invertebrates can grow in this habitat. Jamaica, Discovery Bay, 36 m depth.

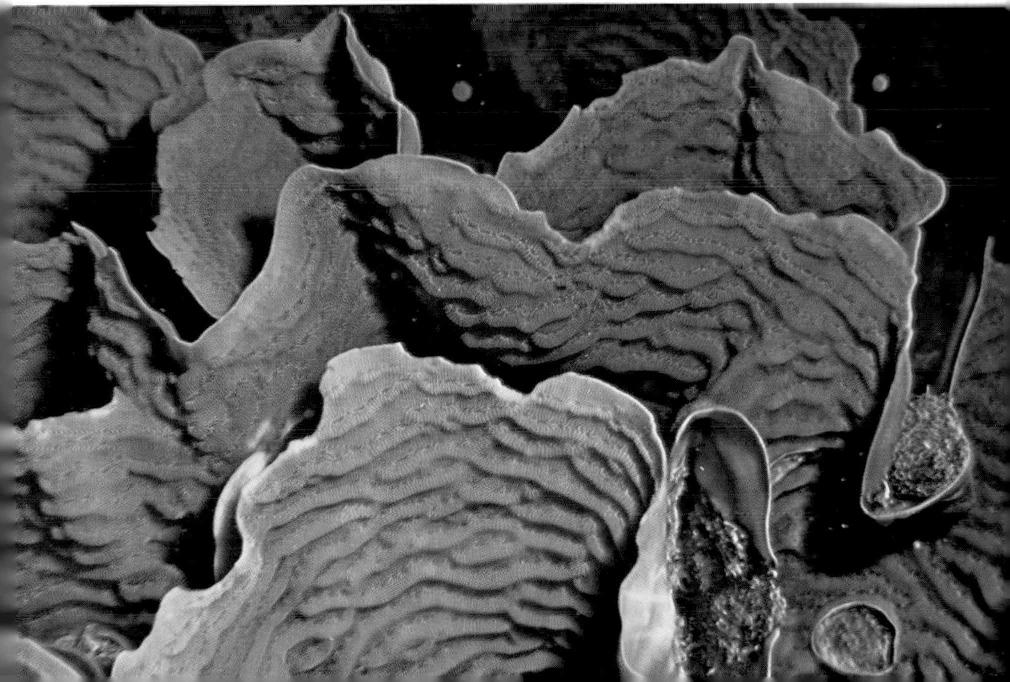

nis. It reaches 30 m in depth but is most common at about 7 m. It is known from several locations in the Caribbean, the Bahamas and southern Florida.

References: Goreau and Wells 1967, Scatterday 1974, Goreau and Goreau 1973.

Family Agariciidae

The lettuce corals (Agariciidae) are among the most lovely and fragile corals occurring on Atlantic reefs. Their delicate beauty is not really indicative of the important role they play in the construction of reefs, particularly in deeper sections. Various species are also important elements of the shallow reef environment, their foliaceous form well adapted to deep reef situations where light is limited.

They are low on the aggressive hierarchy, attacking only a few other genera (*Siderastrea, Porites, Madracis, Stephanocoenia*) by extracoelenteric digestion, but compensate with their ability to compete by overgrowth and cutting off their competitors' light.

The thin plates of most agariciids cover a large area in relation to their weight and can be "primary hermatypes" under optimal conditions. They also produce an important environment on their dark, shaded undersurfaces where extremely high densities of brachiopods, molluscs, and coralline sponges are found.

It is often necessary to have a specimen in hand to make a positive identification. Only about one-half of the eight western Atlantic species of *Agaricia* are dealt with specifically, the rest being considered as a unit.

References: Lang 1973, Jackson *et al.* 1971, Wells 1973a.

Agaricia agaricites (Linnaeus)

Agaricia agaricites has at least five different growth forms, each of which is termed a subspecies by most authorities. These subspecies have overlapping but slightly different depth ranges; *A. agaricites* as a whole ranges from 1 to at least 75 m. All of the growth forms are fairly common under

Right: A large "head" of *Agaricia,* probably *A. tenuifolia* or *A. agaricites,* but a definite identification is not possible. This colony is supported only by a narrow stalk, the remainder of the base probably having been attacked by boring organisms. Photographed at 15 m depth, windward shore of Hogsty Reef, Bahamas.

Below: Large colony of *Agaricia* on a vertical wall at Acklins Island, Bahamas, depth 30 m. It is difficult to determine which species is represented by this colony.

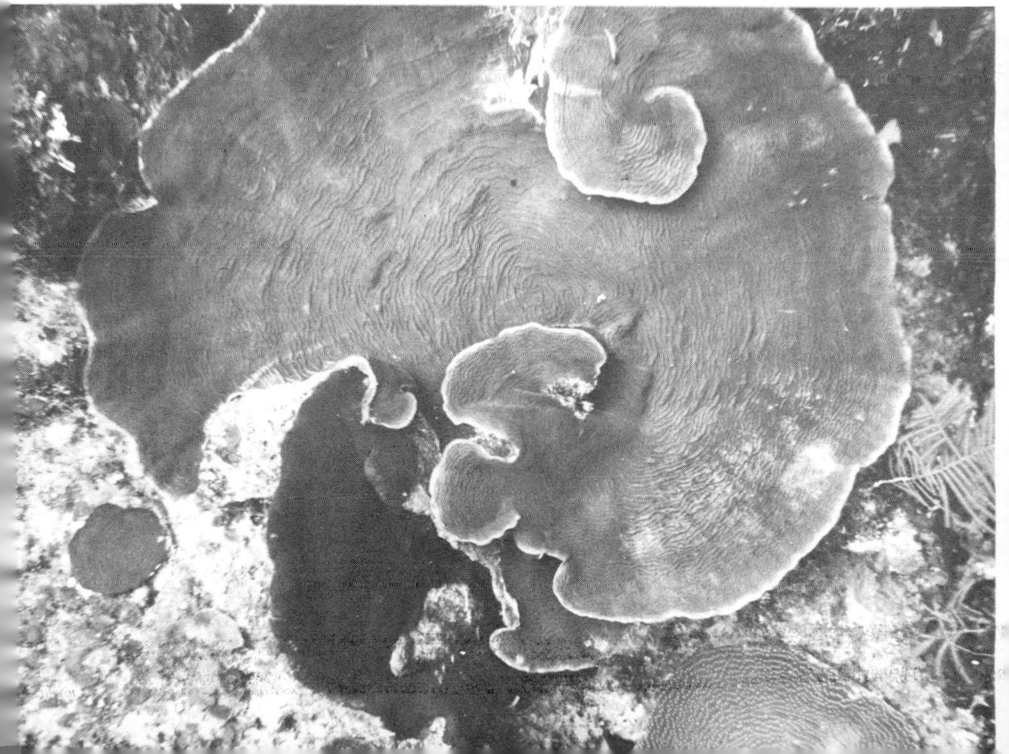

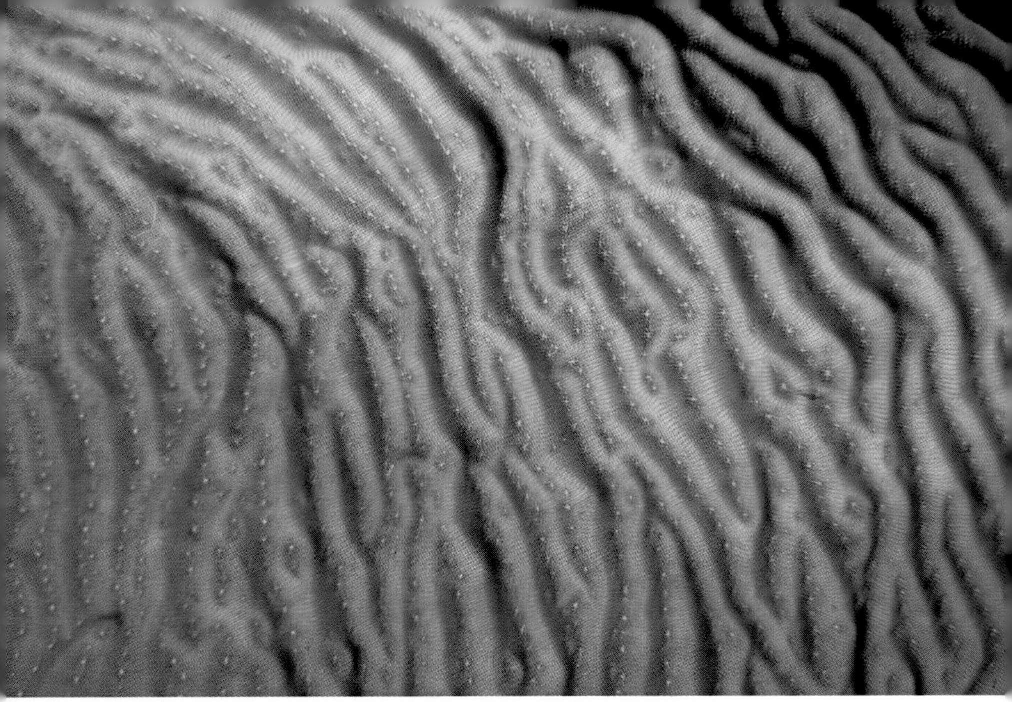

Agaricia sp., possibly *A. lamarcki*. Closeup view showing the corallites and ridges. Jamaica, Discovery Bay, 36 m depth.

Agaricia sp., possibly *A. lamarcki.* This species and *A. grahamae* form broad, thin plates which can cover large areas of deep-reef substrate. Their attachment is often poor and they are easily damaged by divers. Jamaica, Discovery Bay, 30 m depth.

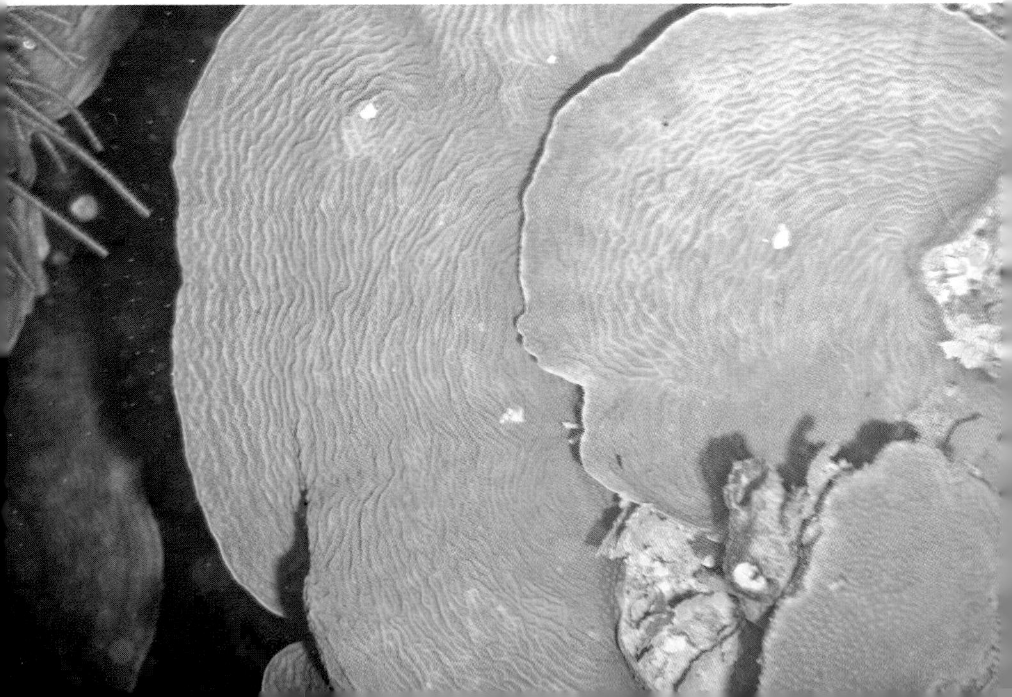

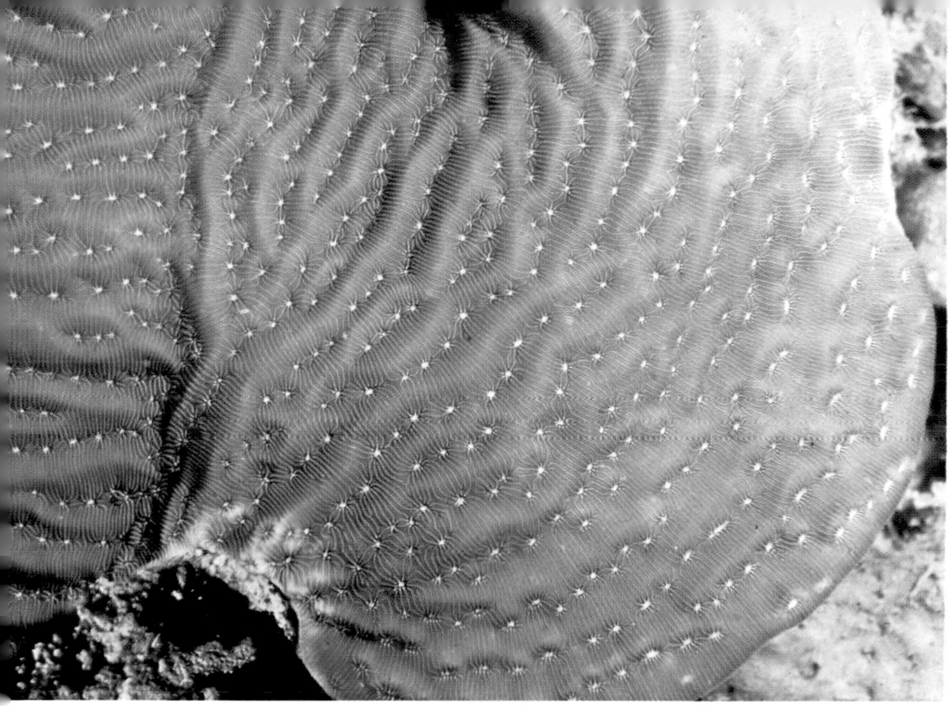

Agaricia sp., possibly *A. grahamae.* The white mouths of the polyps are easily visible in this colony. The deeply shaded undersurface of corals, such as this *Agaricia,* provides an unusual reef habitat. Jamaica, Discovery Bay, 56 m depth.

Helioseris cuculata forms lovely, delicate cup-like masses which do not reach great size as compared to the colonies of *Agaricia.* Jamaica, Spring Gardens, 18 m depth.

Plate-like corals on a deep-reef face at 45 m depth, Discovery Bay, Jamaica. The corals visible in the photograph are mostly *Montastrea annularis* and various species of *Agaricia.* The dark, calm environment created by these overhanging plates is quite evident in this photo. This under-plate environment has a fauna which does not occur elsewhere.

the proper conditions and may be correlated with light and wave conditions.

The species has been reported as occurring on mangrove roots on occasion. Generally it is brown to tan with some forms having a blue iridescence. On the coral buttresses of the northern coast of Jamaica *A. agaricites* is very abundant, sometimes being the co-dominant species of coral. The extreme hardness and toughness of this coral has been reported as responsible to some degree for the resistance to storm damage of the buttresses.

The distribution of colonies in some areas is non-random and may be produced by associative settlement of planulae or erosion and separation of large colonies into several small ones.

References: Squires 1958, Roos 1971, Goreau and Wells 1967, Lewis 1974b.

Agaricia tenuifolia Dana

The vertical fronds of *Agaricia tenuifolia* can form clumps several meters across which are surprisingly thick. This species does not extend nearly as deep as other species of *Agaricia*, having been recorded only as deep as 18 m, with its greatest development in water only 5 to 10 m deep. This species and *A. agaricites* are the only members of this genus which make a significant contribution to the construction of shallow water reefs.

References: Wells 1973a, Goreau and Goreau 1973.

Agaricia undata Ellis and Solander

The often vertical, folded blades of *Agaricia undata*, a deep-water species, make this one of the most attractive of the lettuce corals. Its depth range is between 16 and 80 m, with large concentrations at the 30 to 50 m level. It forms large colonies; the undersides of the plates on the outer edge of the colony are thrust up and plainly visible. The light tan colonies are often closely associated with other species of *Agaricia*.

Reference: Wells 1973a.

The surface of colonies of *Helioseris cuculata* has a radiating series of lines and concentric groups of polyps. Jamaica, Discovery Bay, 36 m depth.

Siderastrea siderea forms large masses and occurs in a wide variety of reef habitats. Jamaica, Discovery Bay, 3 m depth.

Siderastrea radians also occurs in a number of reef habitats except for the deeper reefs. Puerto Rico, Aguadilla, 6 m depth.

The finger-like branches of *Porites furcata* form dense stands in many back-reef areas, such as is shown in this photograph, where wave action is very light. Only the outer few centimeters of the branches have living polyps, the basal portions are dead and encrusted by a variety of organsisms. Puerto Rico, La Parguera, Mario Reef, 1 m depth.

Agaricia spp.

The remaining species of *Agaricia* are very similar in appearance and almost require the presence of a specimen in hand for a positive identification. The importance of these species in the ecology of deep reef areas cannot be emphasized enough.

Agaricia fragilis Dana has two growth forms and reaches to at least 80 m depth. It is the only species of *Agaricia* known to occur in Bermuda.

Agaricia grahamae Wells and *A. lamarcki* Milne-Edwards and Haime are very similar in appearance and can cover entire reef faces with their foliate, slightly overlapping plates. The valleys form concentric rows around the initial portion of the colony as the colony grows outward, producing a very flattened plate. The outer edge of the plate is usually a lighter color than the rest of the colony. *Agaricia grahamae* occurs from 9 to 76 m while *A. lamarcki* is known from 4 to 46 m. The coral *Mussa angulosa* has been observed to "attack" an individual of *A. grahamae* which was beginning to overgrow the more aggressive coral. The *M. angulosa* expanded over a period of hours and its mesenterial filaments digested a small portion of the *A. grahamae.* The portion of the coral killed by such an attack is seldom regenerated and quickly becomes overgrown with algae and sponges.

References: Wells 1973a, Lang 1973.

Helioseris cuculata Ellis and Solander

This fragile, cup-like agariciid coral is common in much of the West Indies at depths between 8 and 50 m. It can occur as shallow as 3 m and is known to a depth of at least 90 m. Although it may grow in other forms, the thin cup-shaped colonies are most attractive, with each polyp having a greenish mouth surrounding the brown remainder of the polyp.

Colonies of *H. cuculata* reach only 10-20 cm in greatest dimension and do not nearly approach the size or importance of the members of *Agaricia* in the deep reef community. This coral can be found on vertical faces on rough shores, such as along the northern coast of Mona Island, but in such locations the colonies are small in size.

Reference: Wells 1973a.

Family Siderastreidae

Siderastrea siderea (Ellis and Solander)

The massive form of this species can reach a size of 2 m across and may be encrusting when young. It deserves careful comparison with the following species, *S. radians. Siderastrea siderea* reaches a much larger size than *S. radians.* Its polyps are somewhat larger and the depth of the pit in which the polyp rests is greater. Aside from differences in skeletal structure, the best method of differentiating the species is that *S. siderea* seems to have the sides of the corallite forming a deep pit in its center, which often shows as darker than the surrounding area, while *S. radians* lacks such a distinct depression in the center of the corallite. The species is usually a light reddish brown in color.

Siderastrea siderea occurs in all types of reef environments but not in muddy habitats or tide pools as does *S. radians.* It reaches a depth of at least 40 m but is most common near a depth of 10 m. It occurs widely in the Caribbean, the Bahamas, the northern Gulf of Mexico, southern Florida and Bermuda.

References: Almy and Carrion-Torres 1963, Roos 1971, Squires 1958.

Siderastrea radians (Pallas)

While the rounded masses of *Siderastrea radians* reach a much smaller size (maximum about 30 cm diameter) than *S. siderea*, this species is more widely distributed in marine environments. It can be greenish to brown in color but is generally less reddish than *S. siderea.* It occurs in all reef environments except the deeper reefs and can tolerate the extreme conditions in muddy bays and tide pools. Small egg-shaped colonies bearing polyps on their entire surface have been found rolling freely around the bottom. Evidently the polyps on the bottom are changed often enough by the colony rolling slightly that they are able to survive. This species can also occur as a thin encrusting growth.

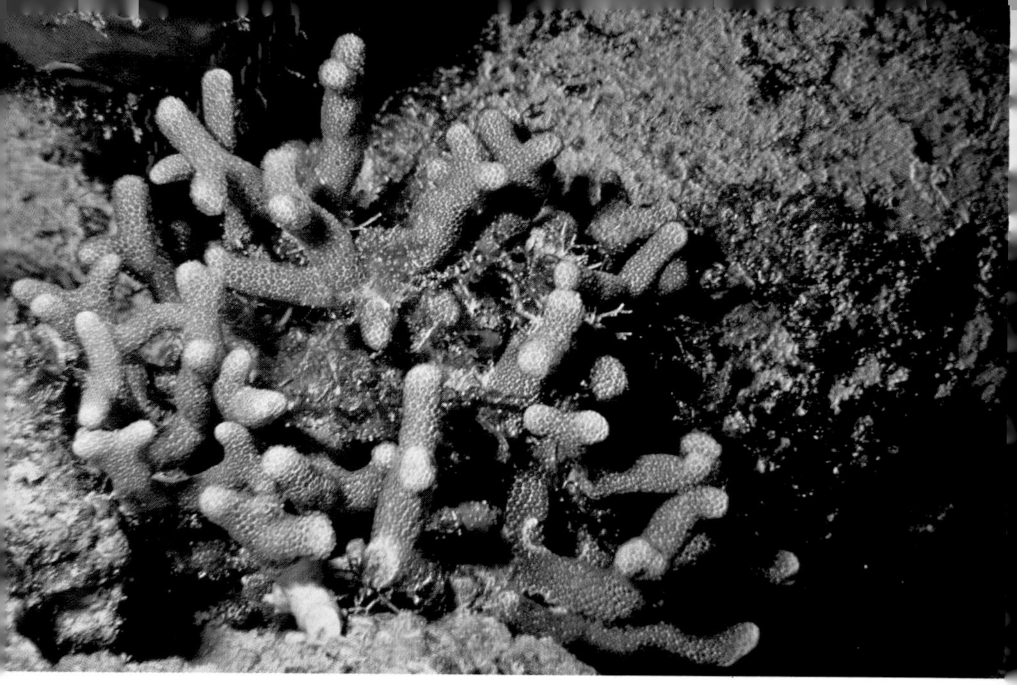

Porites divaricata has thinner branches than the other species of *Porites* and does not form clumps as large in size as do some other species. Puerto Rico, Mona Island, Playa Sardinera, 15 m depth.

The branches of *Porites porites* are often so densely packed that when the polyps are expanded it appears that they are a single, knobby head. Jamaica, Spring Gardens, 18 m depth.

The branches of *Porites porites* are larger in diameter than any other of the branched species of the genus in the Caribbean. Bahamas, Exuma Islands, 10 m depth.

Porites asteroides is the only Caribbean species of the genus which does not form branched colonies. Rather it occurs as heads or plates with raised mounds on the surface of the colony as is illustrated here. Jamaica, Discovery Bay, 20 m depth.

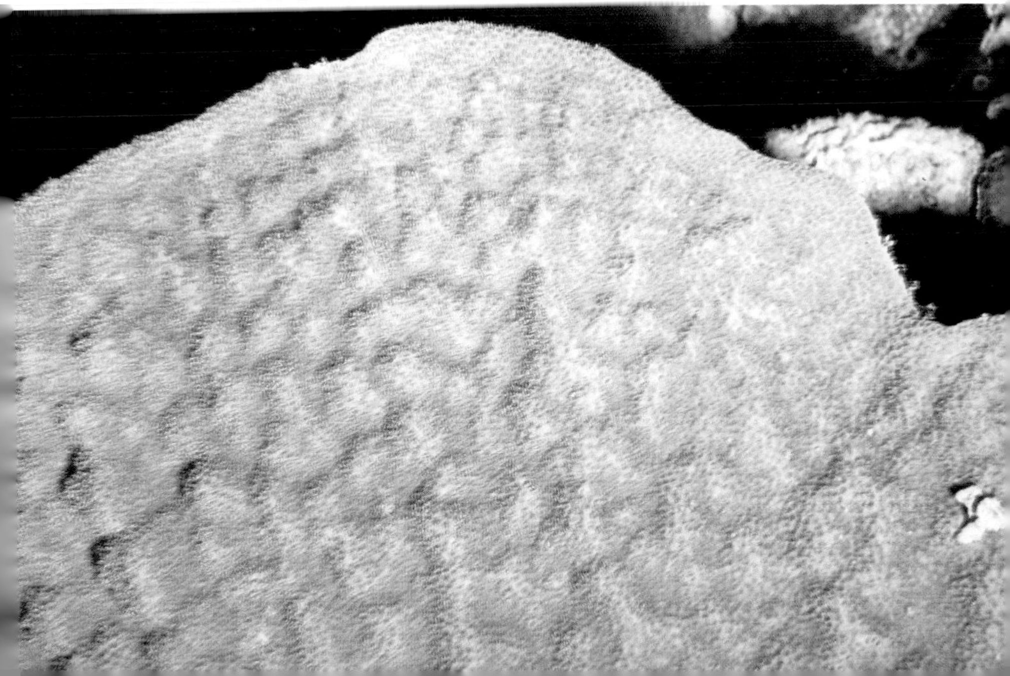

The tentacle of S. *radians* terminates in a knob of nematocysts, an unusual arrangement for a stony coral. The initial formation of the calcareous skeleton by a newly settled planula is best known from studies of this species.

The species is uncommon below 10 m depth. It occurs throughout the tropical western Atlantic, including Brazil, and in the eastern Atlantic.

References: Duerdan 1904, Laborel 1974, Almy and Carrion-Torres 1963, Roos 1971, Squires 1958.

Family Poritidae

Porites furcata Lamarck

This thin-branched *Porites* can form extremely dense stands under the proper conditions. Off southwestern Puerto Rico it is the dominant coral in the areas leeward of the reef crest, where water temperatures and light intensity are high in the shallow depths. At 1.0 to 1.5 m depth these stands may be 50 cm high and many meters wide. Much coral rubble produced from dead *P. furcata* is also found in these areas. These dead areas may be produced during periodic low water conditions or by catastrophic storms. Blocks of larger, massive corals moved across the reef flat by hurricane intensity storms can produce "trenches" through the stands of *P. furcata*.

In other locations, *P. furcata* may be most abundant in deeper fore reef areas such as on the northern coast of Jamaica.

The branches of *P. furcata* are thinner than *P. porites* and they can be more densely concentrated. The distance between branches can be quite small in both species. Typically only the upper portions of a branch of *P. furcata* are alive; the lower portions are probably killed by being shut off from free water circulation by the growth of the upper portions. The branches provide shelter for a wide variety of organisms. Ophiuroids (brittlestars), chitons, sea urchins, algae and others are found here.

The color of *P. furcata* varies from yellow to buff-brown. There are occasionally purple tones on the lower portion of colonies.

The species is known throughout the Caribbean, the Bahamas and southern Florida, but has not been definitely recorded from Bermuda. It reaches a depth of at least 20 m and can be found as shallow as mean low water. The earliest known fossils of *P. furcata* are from Florida and are of Pleistocene age.

References: Glynn 1973 a, Squires 1958.

Porites divaricata Lesueur

This is the most delicate Atlantic species of *Porites.* The branches are only about 6 mm in diameter and form at most a small clump with widely spaced branches. The species is yellowish or pale greenish. The colonies are typical of back reef areas in shallow water, but occur rarely as deep as 15 m. The small clumps of *P. divaricata* may form beds in some areas, such as southern Biscayne Bay.

Porites divaricata is widely distributed in the Caribbean to southern Florida. Many authors have considered it as a form or subspecies of *P. porites.*

Reference: Squires 1958.

Porites porites (Pallas)

The branches of *Porites porites* are thicker than in the preceding two species, often 25 mm in diameter. The branches resemble fingers less than in the other species and are stubbier. The species occurs as groups of short branches sometimes forming a tight clump but more often widely spaced. The polyps are often expanded somewhat during the day. Colors range from pale beige (most common) to purplish with pale tips on the branches.

Porites porites can occur in many reef situations including back and clear water fore reefs. It is rare below 20 m but can occur as shallow as low water levels. It is recorded from throughout the Caribbean, the Bahamas and southern Florida.

References: Squires 1958, Glynn 1973a.

This small colony of *Porites asteroides* has the sponge *Mycale laevis* growing on its undersurface. An orange fringe of the sponge is visible around the colony and the oscula are quite prominent. Bahamas, Cat Island, 15 m depth.

Favia fragum is a small coral which most often occurs as a spherical colony on rocky substrates as is shown in this photograph. Jamaica, Discovery Bay, 5 m depth.

Colonies of *Favia fragum* seldom become more extensive than that illustrated here. This particular individual may well be produced by the fusion of two previously separate colonies by the narrow strip of living coral. Bermuda, south shore reefs, 10 m depth.

The polyps of *Diploria strigosa* are expanded only at night and have the tentacles along the sides of the valleys. Puerto Rico, Mona Island, Playa Sardinera, 15 m depth.

Porites asteroides Lesueur

This coral can occur in a variety of growth forms. In shallow waters or those exposed to heavy surf it is encrusting, while at 8-10 m colonies are rounded, hemispherical or spherical. Deeper than 10 m they are flattened with the surface directed toward the source of light. In caves or crevices their flattened surface is directed toward the opening through which light enters. Large rounded colonies are often broken into several smaller patches of living coral by death and erosion of portions of surface. This results in what appears to be clumped colonies of *P. asteroides*, but examination of the rock in between reveals it to be dead *P. asteroides*.

Colonies are yellow or greenish yellow with small polyps, and often the surface of the colony appears to be "lumpy" with low mammillate projections. Fan worms often occur with *P. asteroides* and the sponge *Mycale laevis*, which grows on the undersurfaces of certain corals, can also be associated with it.

Asexual reproduction is accomplished in two manners. In extratentacular budding new mouths appear outside the ring of tentacles; this occurs principally during early growth of the colony. Intratentacular budding, where the new mouth arises within the tentacular ring of the polyp, occurs almost exclusively in older, massive colonies.

Porites asteroides occurs abundantly in nearly all reef zones to depths of over 50 m. It is known throughout the tropical western Atlantic including Bermuda and the banks off the Texas coast.

References: Roos 1967, 1971, Lewis 1974b.

Family Faviidae

Favia fragum (Esper)

While attaining only a small size, the nearly spherical colonies of *Favia fragum*, consisting of certainly less than 200 polyps in the largest examples, are common in certain areas of the shallow reef. Probably more is known about the early life history of this coral than any other Atlantic species.

In Barbados *F. fragum* was found to have a non-ran-

Plates of *Porites asteroides* on an "island" of massive corals in a dense bed of *Acropora cervicornis* at Discovery Bay, Jamaica at 10 m depth. The knobby surface of the plates is very characteristic of *P. asteroides*. *Agaricia agaricites* is also visible on this coral clump.

Clump of *Porites porites* among the knobby form of *Montastrea annularis*. Eleuthera Island, Bahamas at 20 m depth.

Diploria clivosa has straighter valleys than its relative, *D. strigosa*, and does not occur as deep on the reef. Puerto Rico, Desecheo Island, 12 m depth.

Diploria clivosa has the septa relatively close together and they are hidden to a great extent by the tissue of the polyps. Puerto Rico, La Parguera, 20 m depth.

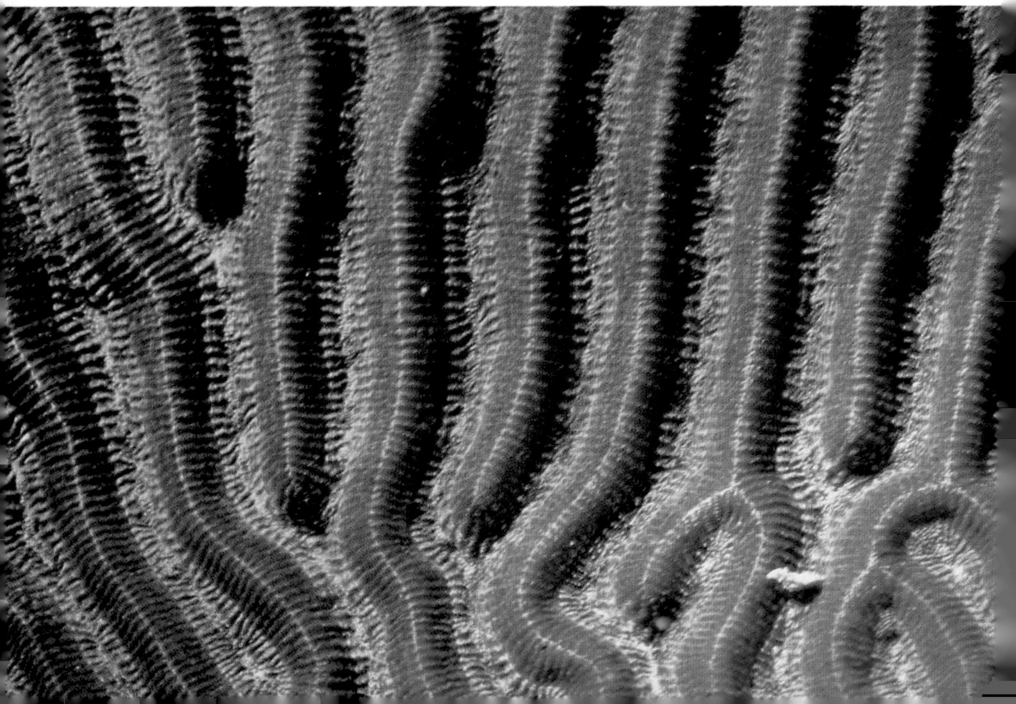

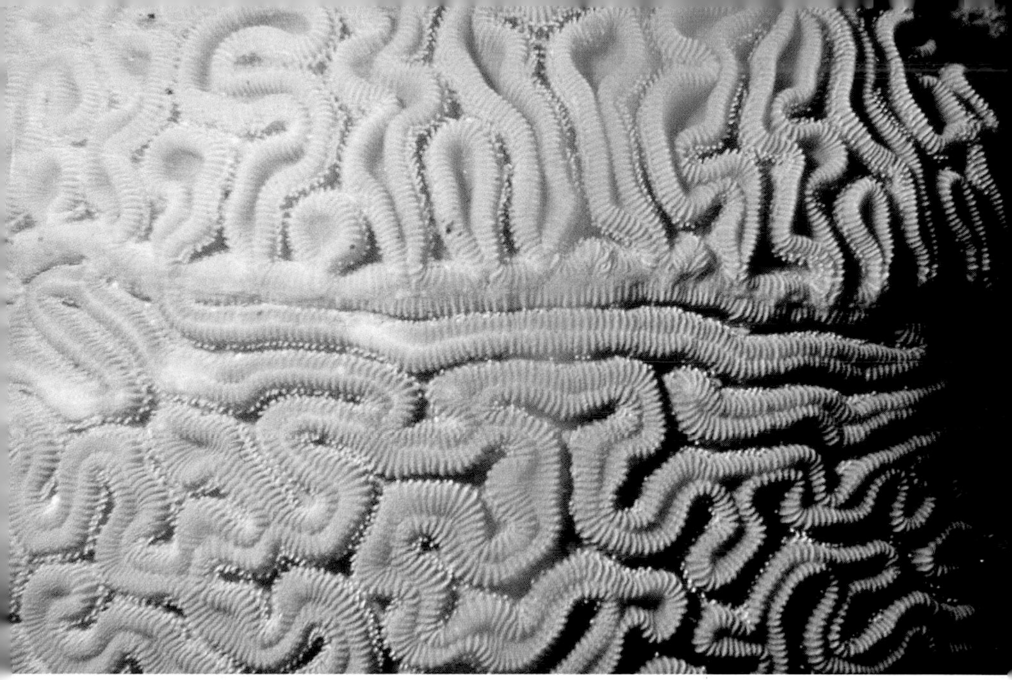

Diploria labyrinthiformes has a deep groove between the ridges which is variable in width. Here two colonies have grown together to form a single head, but the polyps and the resulting pattern of ridges indicate that they are acting as separate individuals. Bermuda, south shore reefs, 10 m depth.

An extreme example of groove formation in *Diploria labyrinthiformes*. The groove is wider than the valleys containing the polyps and its width is probably determined largely by environmental factors. Bahamas, Cat Island, 10 m depth.

A nearly perfect specimen of brain coral, *Diploria strigosa*, photographed at Buck Island, St. Croix, Virgin Islands at 10 m depth. This coral could be confused with *Colpophyllia natans,* but lacks an apparent groove on top of the ridges.

Diploria strigosa with several neon gobies on its outer surface. Photographed at Tierra del Bomba, Cartagena, Colombia at 8 m depth.

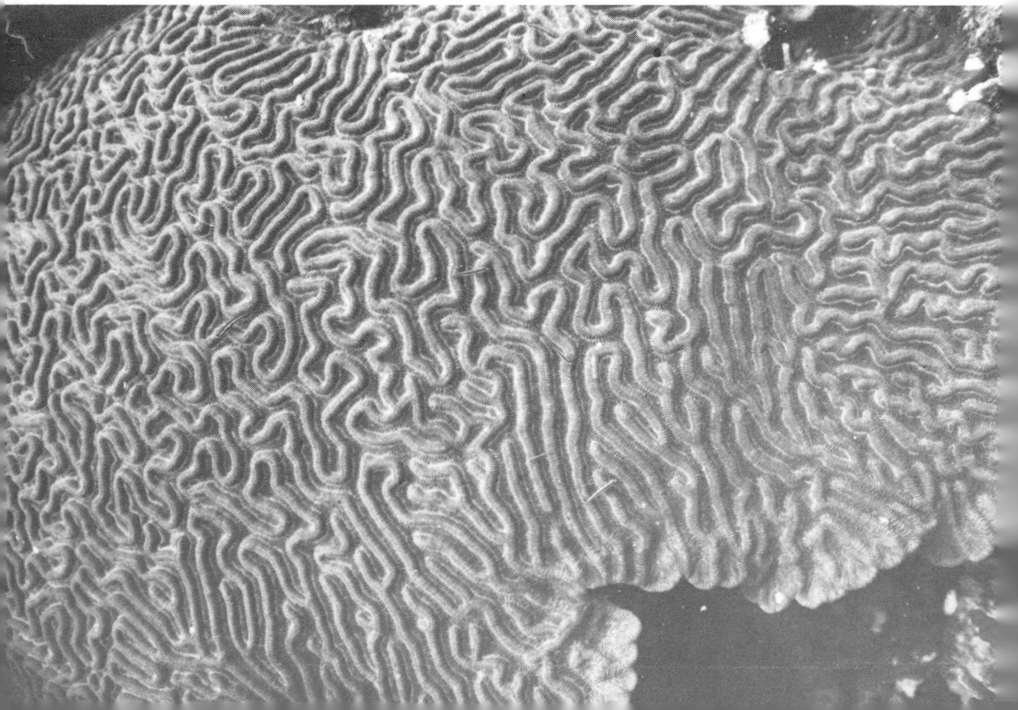

dom distribution in certain areas. Planulae are produced during the summer months, particularly a few days before and after the new moon. They are variable in shape, 1-2 mm long and may either swim or crawl away from the parent colony. Those that crawl move only a short distance and attach to the bottom where with luck they will grow, resulting in the clumping observed in this coral. The swimming larvae in the laboratory settle and metamorphose within 24-48 hours after release by the parent.

Evidence exists that the swimming planulae are able to recognize adult colonies and previously settled juveniles and tend to settle near them. Once settled, the young coral needs both food and light to grow (the planula carries its own supply of zooxanthellae), but if lacking either they can live for some time but not grow in size.

Favia fragum occurs widely in the tropical western Atlantic, including Bermuda.

References: Lewis 1974a, 1974b, 1974c, 1970.

Diploria strigosa (Dana)

The species of the genus *Diploria*, along with others in *Colpophyllia* and *Meandrina*, are commonly referred to as brain corals due to the resemblance of the convoluted surfaces of the hemispherical heads to the human brain. The polyps are found in single-file in the valley of the convolutions, a long valley containing many separate polyps. They are normally contracted during daylight but expand at night, at which time the tentacles lining either side of the valley can be seen. Ridges run between the valleys; in this species there is no groove on top of this ridge. The coral is yellow, brown or greenish brown, and often the grooves appear lighter than the ridges due to the presence of the contracted polyps.

This species can form immense heads well over 2 m across and is capable of making a significant contribution to reef structure. This species, like most brain corals, is slow growing. The annual increase of size of a head of *D. strigosa* has been estimated at up to 1 cm per year. This means a specimen 2 m in diameter would be at least 100 years old and

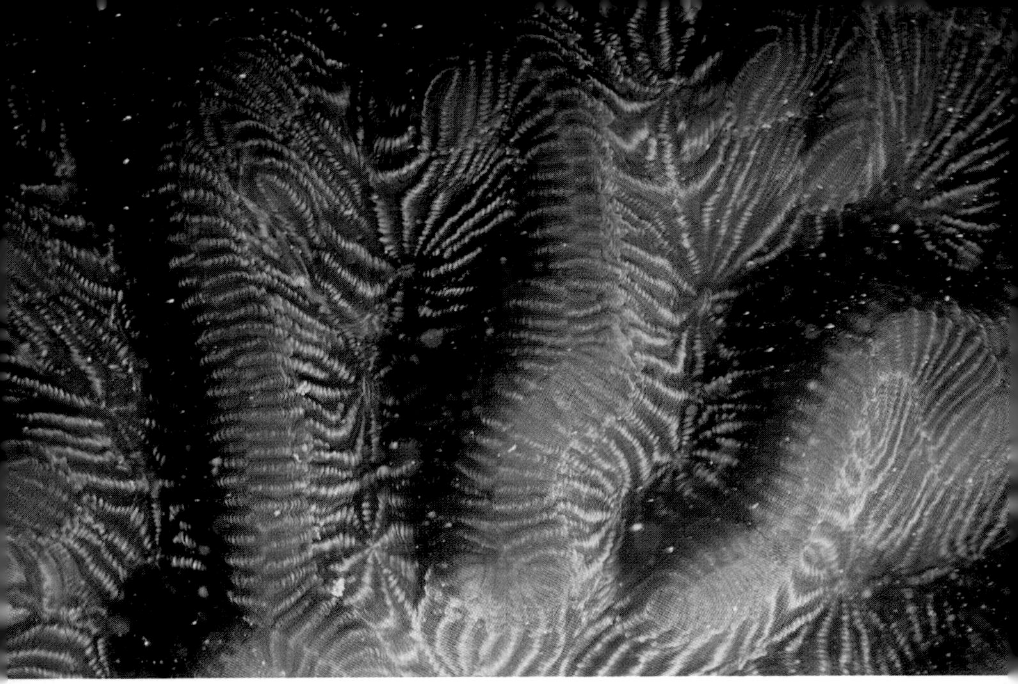

A closeup view of the head form of *Manicina areolata* in which the radiating septa of the individual polyps are visible. Puerto Rico, La Parguera, 18 m depth.

A portion of the rose coral form of *Manicina areolata* which is unattached to the substrate. These lovely little colonies are found in areas of sediment bottoms, often at some distance from true reefs. Jamaica, Discovery Bay, 5 m depth.

Colpophyllia natans is another of the so called "brain corals." It has a distinct groove on top of the ridges which distinguishes it from similar appearing species of *Diploria.* Bahamas, Grand Bahamas Island, 15 m depth.

At night the polyps of *Colpophyllia natans* are expanded but contract quickly if disturbed. Bahamas, Crooked Island, 15 m depth.

probably several hundred with all factors considered.

Diploria strigosa occurs from low water to at least 40 m but is most abundant above 10 m. It is perhaps the most widely distributed species of *Diploria* on the reef and has even been reported from muddy bays where few other corals grow. It occurs throughout the Caribbean, the Bahamas, southern Florida, the offshore banks of Texas and Bermuda. No species of *Diploria* is known to occur in the eastern Atlantic.

References: Roos 1971, Squires 1958.

Diploria clivosa (Ellis and Solander)

This species can form either hemispherical heads in quiet waters or flattened encrusting to plate-like forms in areas of wave action. *Diploria clivosa* is quite similar in appearance to *D. strigosa,* and difficulty may be encountered separating the two species. *Diploria clivosa* has more septa (plate-like structures on the side of and perpendicular to the ridges) per unit length (30 as opposed to 20 per centimeter), the grooves less interconnected, somewhat straighter generally and the entire corallum generally more flattened than *D. strigosa.* The color varies somewhat from yellow-brown, green and bluish gray to brown. Like other *Diploria* the polyps are retracted during the day and expanded at night. The ridges lack a groove on their upper surface.

In Bonaire it is one of the dominant corals on the leeward side of a fringing reef of *Acropora palmata* but is not as significant a constructor on reefs as are the other two species of *Diploria.* It does not occur as deep as *D. strigosa,* with its maximum depth being about 15 m and its distribution centered around 1 to 3 m. The species is recorded from southern Florida, the Bahamas and the Caribbean but does not occur in Bermuda, the northern Gulf of Mexico or the eastern Atlantic.

References: Squires 1958, Scatterday 1974.

Diploria labyrinthiformes (Linnaeus)

This is the easiest species of *Diploria* to identify and, although somewhat variable, it has certain characters which make it distinctive. It forms sizable heads over 1 m in diameter, usually nearly hemispherical, and has a groove of varying width and depth on the center of the ridges between the polyp-bearing valleys. This groove can be quite narrow or wider than the actual valley itself; the valleys are usually interconnected. The species is yellow to brown in color.

Diploria labyrinthiformes is a minor reef constructor on the seaward slope of reefs and is the most restricted species of *Diploria* in its distribution on reefs. The three Atlantic species of *Diploria* are only slightly "aggressive" in terms of their ability to attack other corals by an extracoelentric feeding response.

It occurs as deep as 43 m but is most common at 2 to 15 m. The species is found in the Caribbean, the Bahamas, southern Florida and Bermuda.

References: Roos 1971, Goreau and Wells 1967, Lang 1973.

Manicina areolata (Linnaeus)
Rose coral

Rose coral occurs in two very different growth forms, a small hemispherical head like form which may or may not be attached to the substrate and a cone-shaped form found growing unattached on muddy or sandy bottoms. It has a high tolerance for sediment and is an important lagoonal form although only a minor reef constructor. The yellow-brown, mustard or dark brown colonies generally have only one meandroid valley system, and the oral disk of the polyps may be green in color.

The unattached cone-like form can perform active movements; if turned over by wave action, the polyp can right itself by inflation of its tissues with water and rolling movements. This righting movement may take between one and two hours in small colonies to several days in large ones. Small individuals may be attached by a short stalk to a hard substrate if available, but become unattached later. The colony can actually "float" in the sediment by again inflating its tissues so that the skeleton and polyps have a specific gravity of about 1.3, which is less than that of the sediment. This coral is also able to readily remove sediment from its upper surfaces by sloughing mucus.

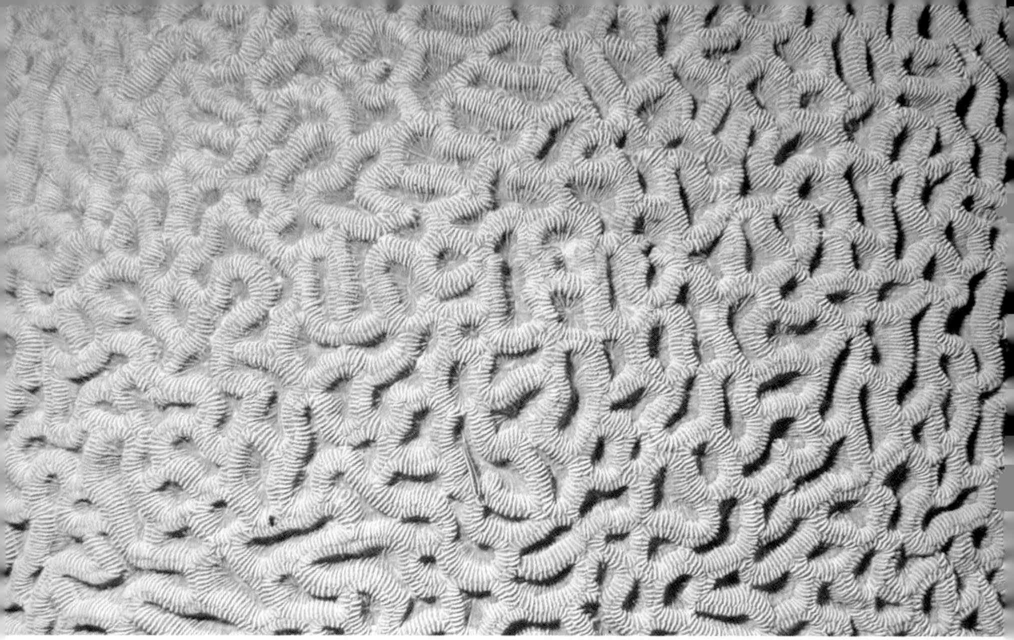

Colpophyllia breviserialis has short, often nearly circular, valleys with only one or a few polyps per valley. Jamaica, Spring Gardens, 18 m depth.

The polyps of *Cladocora arbuscula* are found only on the tips of the cylindrical branches and the species does not form masses large enough to be of significance in reef building. Puerto Rico, La Parguera, Laurel Reef, 9 m depth.

Montastrea annularis is one of the most important reef-building corals in the Caribbean. At depths of about 15 m and below it forms plate- or shingle-like structures. Bahamas, Whale Bay, 15 m depth.

Star coral, *Montastrea annularis,* can form sizeable heads which are often somewhat hollow inside. The caves formed by this coral open up a range of habitats for cryptic reef organisms.

Manicina areolata is evidently absent from Bermuda, but otherwise it is found widely in the tropical western Atlantic. It can occur in *Thalassia* beds or in lagoonal reef areas from near the surface and seems to be common in some deeper areas to as much as 65 m depth. The unattached form of *M. areolata* was encountered on a deep bank consisting of coral rubble coated with coralline algae and this form, often considered as typical of shallow waters, may be found just as deep as the attached forms.

References: Fabricius 1964, Roos 1971, Scatterday 1974.

Colpophyllia natans (Muller)

This is a common species of "brain" type coral found on the seaward sides of reefs. It forms sizable (over 1 meter diameter) heads and irregular, somewhat flattened masses. The valleys and intervening ridges are often differently colored, with the change occurring about mid-level on the side of the valley, where the septa show a distinct "step" and assume a less steep angle to the bottom of the valley. Some of the color combinations noted include green valleys with brown ridges, white with brown and yellow-green with green (yellow mouths to the polyps). The grooves are meandering and often continue for a considerable distance.

This species can be a major frame constructor on seaward reefs at depths of 6 to 20 m. It ranges from 0.5 to 55 m in the extreme. A variety of coral-dwelling fishes, usually small gobies or scaled blennies, can occur with this and other "head" corals. It occurs throughout the reefs of the Caribbean, the Bahamas, southern Florida and the banks off the Texas coast. It is not known from Bermuda, however.

References: Goreau and Goreau 1973, Roos 1971, Lang 1973, Almy and Carrion-Torres 1963.

Colpophyllia breviserialis Milne-Edwards and Haime

It was only recently that this species was recognized as distinct from *Colpophyllia natans*, but actually the two species are quite simple to differentiate. The short, often nearly circular or only slightly elongate, valleys of *C. brevi-*

serialis consist usually of only a single polyp or at most a very few polyps. The coral forms flattened heads of a size equalling those of *C. natans*, and the two species can occasionally be found growing adjacent to one another. The species is brown or with brownish ridges and green-brown valleys. The species could possibly be confused with *Isophyllastrea rigida* on the basis of a verbal description, but photographs show that these two species are quite distinct.

Colpophyllia breviserialis occurs on seaward reefs at 5 to 30 m. It is presently known from Jamaica, Colombia, Panama, the Cayman Islands and the Bahamas.

References: Wells 1973a, Pfaff 1969.

Cladocora arbuscula (Lesueur)

This delicate branching coral is easily distinguished from similar appearing species. The polyps of *Cladocora arbuscula* occur only on the ends of the branches, each terminal polyp occupying the entire end of the cylindrical, longitudinally ribbed branch. Although division of the branches occurs by splitting of the growing polyp, the diameter of the branch does not increase with increasing length. At first glance *C. arbuscula* is similar in appearance to *Oculina diffusa* Lamarck, but since *O. diffusa* has polyps at both the ends and along the sides of its branches, the two corals are easily distinguished. *Cladocora arbuscula* is typically brown while *O. diffusa* is yellowish.

This coral occurs in beds of turtle grass, *Thalassia testudinum*, and in high sediment reef environments. It may occur as clumps of branches reaching about 30 cm in diameter. It is uncommon on outer reefs with extremely clear water and is relatively unaggressive on the scale of extracoelenteric feeding responses of stony corals.

The species is known from the Caribbean, the Bahamas and southern Florida (but not from Bermuda) at depths of 0.5 to 21 m.

References: Goreau and Goreau 1973, Almy and Carrion-Torres 1963.

Montastrea cavernosa is an extremely common Caribbean reef coral forming heads or plates in a wide depth range of water. Puerto Rico, La Parguera, 22 m depth.

At night the polyps of *Montastrea cavernosa* expand and are capable of capturing zooplankton. They will contract at the slightest touch and the disturbance of one area of a head will cause all of the polyps to contract. Bahamas, Eleuthera Island, 20 m depth.

Astrangia solitaria is a tiny ahermatypic coral which occurs often on the undersurface of reef rubble. Also present in this photograph are the red tests of the foraminiferan *Homotrema rubrum.* Puerto Rico, La Parguera, Laurel Reef, 1 m depth.

The ahermatypic coral *Phyllangia americana* can occur beneath ledges on reefs where there is considerable shading. It often occurs in small groups, but they do not really constitute a colony. Puerto Rico, La Parguera, Laurel Reef, 12 m depth.

Montastrea annularis (Ellis and Solander)
Star coral

This and the following species of *Montastrea* are generally the most common species of coral on Atlantic reefs at moderate depths. *Montastrea annularis* forms massive boulders or heads reaching several meters across in shallow water (1-20 m) and flattened heads or plate-like colonies in deeper water (below 20 m). Columnar lobate heads are common and consist of large numbers of rounded knobs of *M. annularis* forming a larger rounded or spherical head. The corallites of *M. annularis* are raised above the surface of the coral, exhibiting a stellate or star-like pattern. The coral is brown to green in color.

There is great variation in this species, and much of it seems related to depth. The corallites vary in size from 1.5 to 5.0 mm in diameter, averaging about 3 mm. The distance between corallites and the density of the skeleton also vary and correlate roughly with depth. The calcification (production of the calcium carbonate skeleton) has been investigated somewhat in this species. The calcification rate is related to light intensity up to a certain level of light. Light beyond that level actually decreases the calcification rate. In summer, during the middle of the day, these limiting light intensities are reached at depths of 10-15 m in clear water and optimum calcification occurs at depths of 20-25 m. Considering seasonal light changes, the long-term optimum depth of calcification for *M. annularis* will be somewhat shallower, but certainly not less than 10 m in clear waters.

The species is slow growing compared to branching corals such as *Acropora cervicornis*, but rates of 1.0-2.5 cm per year increase in height have been recorded. Considering the actual amount of carbonate material accumulated per polyp, the growth of *M. annularis* must be more nearly comparable to *A. cervicornis*. *Montastrea annularis* in the warmer waters of the Caribbean seems to grow faster than in the cooler waters of Florida and Bermuda.

The species competes well against some species of branching and flattened coral by the extracoelenteric feeding response. Heads of *M. annularis*, termed "islands," are often found in the midst of huge thickets of *Acropora cervicornis*, a species which *M. annularis* can successfully attack.

Montastrea annularis is attacked by a wide variety of organisms other than corals. Boring sponges are quite abundant in this species, gastropod molluscs of the genus *Coralliophila* feed either on the polyps or on plankton ingested by the polyps and filamentous algae occur on areas where coral tissue was removed by mechanical action.

This coral occurs throughout the Caribbean, the Bahamas, the banks off the Texas coast, southern Florida and Bermuda. It reaches depths of at least 60 m.

References: Barnes and Taylor 1973, Lewis *et al.* 1968, Scatterday 1974, Squires 1958, Roos 1971, Milliman 1969.

Montastrea cavernosa (Linnaeus)

The corallites of this species of *Montastrea* are considerably larger (5-11 mm diameter) than those of the preceding species, *M. annularis. Montastrea cavernosa* forms sizable massive heads and plate-like structures becoming thinner near the lower depth limits of the species. In many localities at moderate depths it is the predominant species of coral present. The skeletons of individual polyps show considerable variation between specimens from different environments. The size of the corallites, distance from neighboring corallites and height of the corallite are evidently controlled by environmental conditions such as light and suspended sediment, but experimental investigation of environmental effects has not been undertaken. The polyps are quite variable in coloration, from green and brown to a definite light red. In some instances the polyps are able to fluoresce, which is why the coral can appear red at depths in the absence of red wave lengths from sunlight.

Either this species or *M. annularis* is generally the most common coral between 10 and 30 m in buttressed or sloping areas of Atlantic reefs lacking sizable thickets of *A. cervicornis. Montastrea cavernosa* has been reported as the dominant hermatype at moderate depths off Palm Beach County, Florida (20-30 m), the windward shore of Barbados and at various locations. In other locations, such as atolls in the wes-

Oculina diffusa forms bushy mounds of thin branches. It can be found in various shallow-water habitats, particularly those with high sedimentation. Puerto Rico, La Parguera, Mario Reef, 6 m depth.

At night the polyps of *Oculina diffusa* are greatly expanded, yet their pale coloration does not make them particularly apparent. Puerto Rico, La Parguera, Caracoles Reef, 10 m depth.

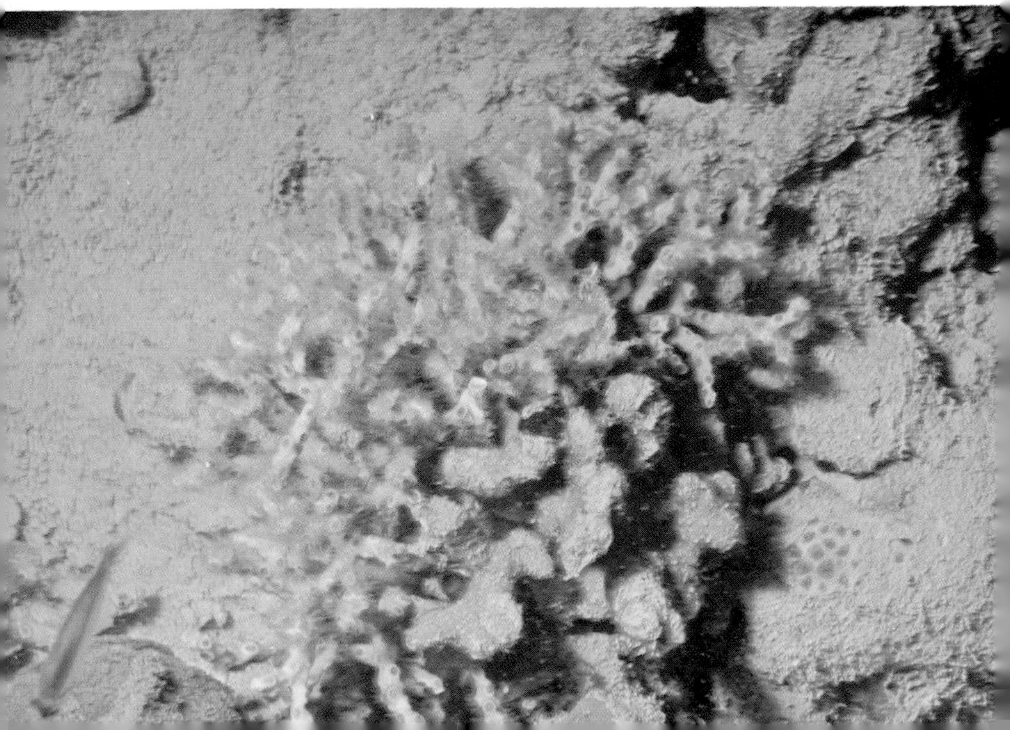

The septa of *Meandrina meandrites* are large and widely spaced. It can form small mounds, as shown here, or much larger heads. Jamaica, Discovery Bay, 15 m depth.

At night when the polyps of *Meandrina meandrites* are expanded it is almost impossible to even see the skeleton of the coral. The coral resembles some sort of anemone-like creature with densely packed tentacles. Puerto Rico, Mona Island, Playa Sardinera, 15 m depth.

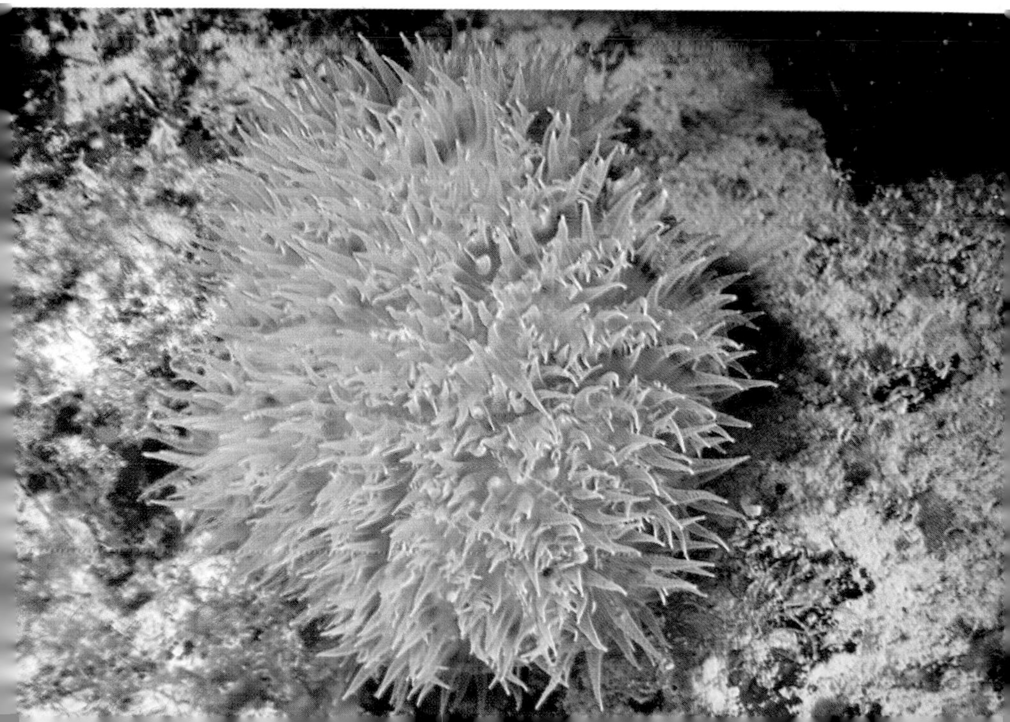

tern Caribbean, *M. annularis* predominates for undetermined reasons; the two species can also co-occur in nearly equal abundance.

Below 30 m *M. cavernosa* clearly predominates over *M. annularis*, but increasing importance of agariciid corals and sclerosponges in reef construction somewhat diminishes its contribution. *Montastrea cavernosa* is one of the most effective zooplankton feeders among the stony corals. A small percentage of the polyps, usually of the so-called "short" morphology, remain slightly expanded during the day, ready to feed on zooplankton. At night the entire colony of polyps expands to feed. Those polyps in locations where currents strike directly seem to capture more zooplankton than those in locations sheltered from direct contact with incoming water.

The species occurs in the Caribbean, the Bahamas, the banks off the Texas coast, southern Florida, Bermuda and Brazil. It also occurs off western Africa but is evidently absent from the Cape Verde Islands. It is one of the deepest occurring hermatypic corals, found at depths from only a few meters to at least 90 m.

References: Lewis 1960, Roos 1971, Goldberg 1973, Read *et. al.* 1968, Porter 1974a, Laborel 1974.

Family Rhizangiidae

Astrangia solitaria (Lesueur)

This small ahermatypic coral occurs as solitary polyps usually on the undersurface of rocks. It is common in occurrence but seldom seen due to its cryptic habit. The calices are cylindrical, reaching about 5 mm in diameter and 6 mm in height.

Astrangia solitaria ranges from Bermuda and southern Florida throughout the Caribbean. It is generally found in water 1-5 m deep, but has been collected as deep as 43 m.

References: Roos 1971, Goreau and Wells 1967, Wells 1972.

Phyllangia americana Milne-Edwards and Haime

Occurring under ledges and rocks, this small ahermatypic coral is found in small groups of polyps often somewhat connected. Each polyp measures about 10 mm in diameter by 10 mm in height.

Phyllangia americana is a shallow water species known from 0.3 to 17 m but is most common at about 3 m. It is scattered in its occurrence and is known from southern Florida to the southern Caribbean.

References: Goreau and Wells 1967, Roos 1971.

Family Oculinidae

Oculina diffusa Lamarck

This small coral is quite distinctive, occurring as clumps of tiny branches bearing widely scattered raised polyps. The skeleton, reaching over 30 cm across, has been described as "bushy," and the polyps with their tissue between the corallites (the coenosarc) are pale yellow. Considering the delicacy of the branches (about 10 mm maximum) this coral is surprisingly sturdy.

It occurs in a variety of shallow water habitats on reefs, including lagoonal, back reef and sloping bottoms. It generally occurs in areas of high sedimentation and contributes little to reef growth.

Oculina diffusa is known from southern Florida, Bermuda, the Bahamas, Cuba, Jamaica and Puerto Rico. Its occurrence in the southern Caribbean is uncertain. It is rare below 10 m in depth.

References: Squires 1958, Almy and Carrion-Torres 1963, Goldberg 1973.

Family Meandrinidae

Meandrina meandrites (Linnaeus)

The larger specimens of *Meandrina meandrites* occur as hemispherical or flattened heads, and small examples are attached to hard substrates or unattached on sediment bottoms. These cone-shaped unattached colonies resemble closely in form unattached specimens of *Manicina areolata*. *Meandrina meandrites* is distinctive because of its extremely large septa, spaced widely apart, and lack of a sizable groove on the ridges. In *M. meandrites* the septa on either side of a

Meandrina meandrites has one growth form, which can occur unattached on sediment bottoms, which is known as *M. meandrites* forma *braziliensis.* It resembles rose coral, *Manicina areolata,* but the two are easily distinguished by the size and prominence of their septa. Puerto Rico, La Parguera, Laurel Reef, 12 m depth.

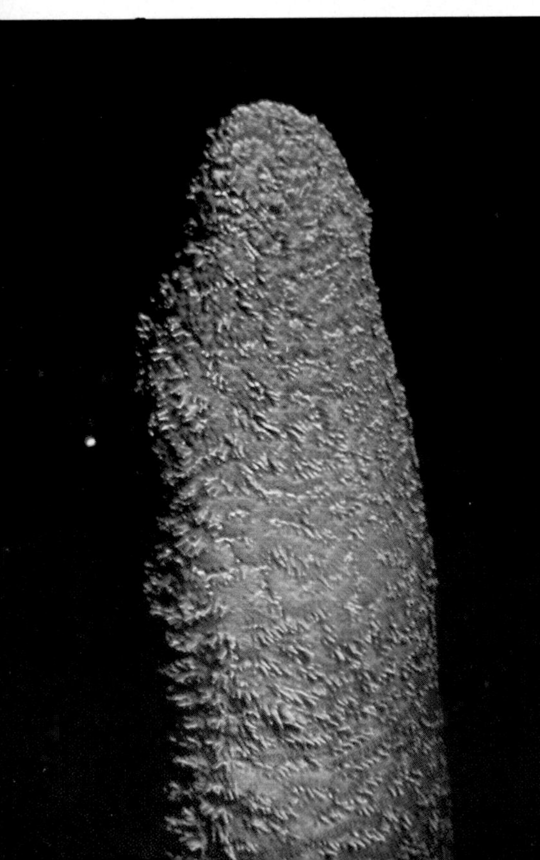

Goreaugyra memorialis has a pillar-like growth form and a broad groove between the ridges. It resembles both *Meandrina meandrites* and *Dendrogyra cylindricus* and is difficult to distinguish from the former. Bahamas, Grand Bahama Island, 15 m depth.

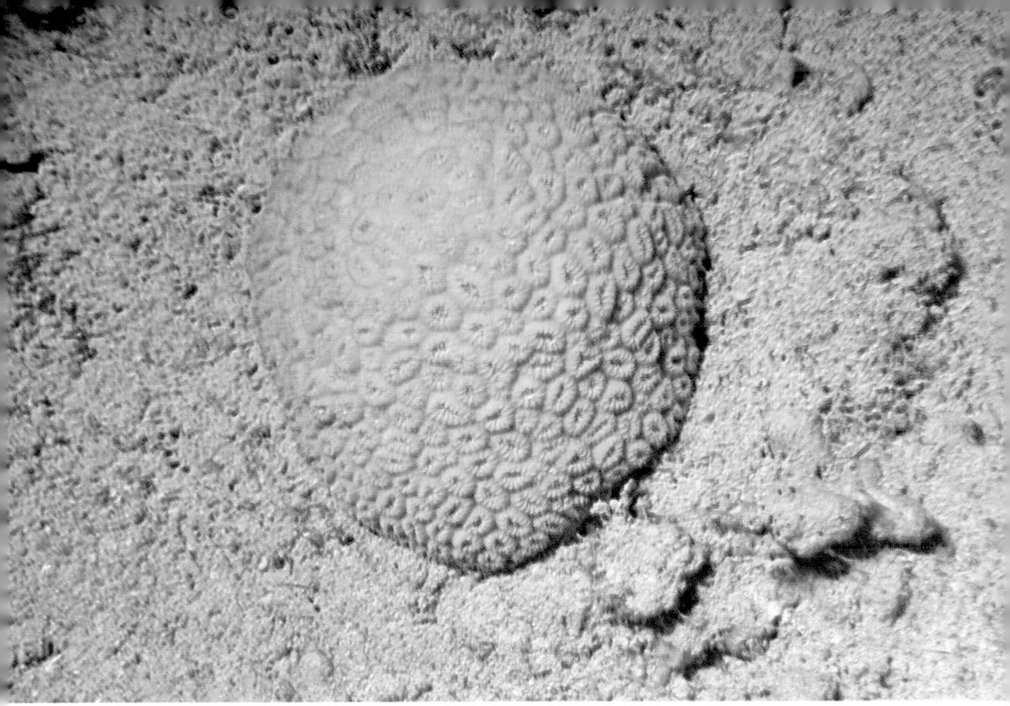

Dichocoenia stokesi is easily identified by its rounded heads and ellipsoidal or elongate polyps. Puerto Rico, Desecheo Island, 12 m depth.

The polyps of *Dichocoenia stokesi* are expanded at night and the tentacles cover nearly the entire surface of the colony. Puerto Rico, Mona Island, Playa Sardinera, 15 m depth.

ridge or valley are seldom the same in spacing and as such have an uneven appearance. The large colonies can reach nearly 1 m in diameter. The color varies from yellow or brown to white, and the polyps are contracted during the day. At night when the coral is expanded each tentacle is observed to have a distinct white tip.

The species is most common on reefs facing the open sea but does occur in back reef areas infrequently. Off Palm Beach County, Florida it was by far the most abundant coral below 30 m depth, where it was able to colonize a rubble bottom which was not suitable for other corals. Its known depth range is between 0.5 and 80 m, but it is most common between 8 and 30 m. The species occurs in the Caribbean, the Bahamas, southern Florida and Bermuda, but is reportedly absent from the Netherlands Antilles islands of Saba and St. Eustatius.

References: Roos 1971, Almy and Carrion-Torres 1963, Goreau 1960, Goreau and Goreau 1973, Goldberg 1973.

Goreaugyra memorialis Wells

This recently described pillar-shaped coral is presently known only from the northern Bahamas. The septa within the valleys are large and somewhat similar to those of *Meandrina meandrites*, but there is a broad groove, often considerably wider than the valley, of varying width between the valleys and their polyps. The growth form is similar to that of pillar coral, *Dendrogyra cylindricus*, but is easily distinguished in that *D. cylindricus* has only a small groove on the apex of the ridges. Some specimens which are intermediate between *G. memorialis* and both *M. meandrites* and *D. cylindricus* have been collected, and at present the status of this species must be regarded as uncertain. Its known depth distribution is between 18 and 30 m.

Reference: Wells 1973b.

Dichocoenia stokesi Milne-Edwards and Haime

This species forms small (to 50 cm diameter) hemispherical or rounded heads with corallites which are circular, elongate or y-shaped. The corallites are raised and the elongate ones may be 5 cm in the long axis while only 5 mm wide. This species could be confused with the one following, *D. stellaris*, but the two are distinguishable on growth form alone.

Dichocoenia stokesi can be whitish, cream, yellow or brown in color. If two individuals of different colors have grown together the tissue of the two original colonies will not merge although the two colonies together resemble a single rounded head. The polyps are contracted during the day and, while expanded greatly at night, are still clearly visible as separate polyps.

The species is common on both back and fore reef areas, but not on reef crests. There is one report of it occurring in a mangrove bay. It is found throughout the Caribbean, the Bahamas and southern Florida, but only sparingly in Bermuda; it does not occur in Brazil. It is most common at 3 to 20 m but reaches at least 40 m.

References: Goreau and Wells 1967, Roos 1971, Squires 1958.

Dichocoenia stellaris Milne-Edwards and Haime

This species resembles *Dichocoenia stokesi* but can be distinguished easily. *Dichocoenia stellaris* has a flattened, plate or pancake-like form and the corallites are less elongate (maximum length about 12 mm) and more rarely y-shaped. *Dichocoenia stokesi* is always rounded or hemispherical with elongate corallites (to 50 mm in length) that are often y-shaped. *Dichocoenia stellaris* generally occurs deeper and is more abundant at 10 to 40 m. It can occur as shallow as 2 m to 72 m. Within the species the corallites are further apart and more protuberant in specimens from deeper water. It is green to brown in color.

The species is presently known from Jamaica, Puerto Rico, Bonaire and the Bahamas. It does not occur in Bermuda.

References: Wells 1973a, Scatterday 1974.

Dichocoenia stellaris has a flattened, pancake-like shape with the polyps elongate or ellipsoidal. Puerto Rico, La Parguera, 24 m depth.

The large septa of *Dichocoenia stokesi* are apparent around the margin of the polyp in this closeup view. The tentacles of the polyps can also be seen. Bahamas, Cat Island, 30 m depth.

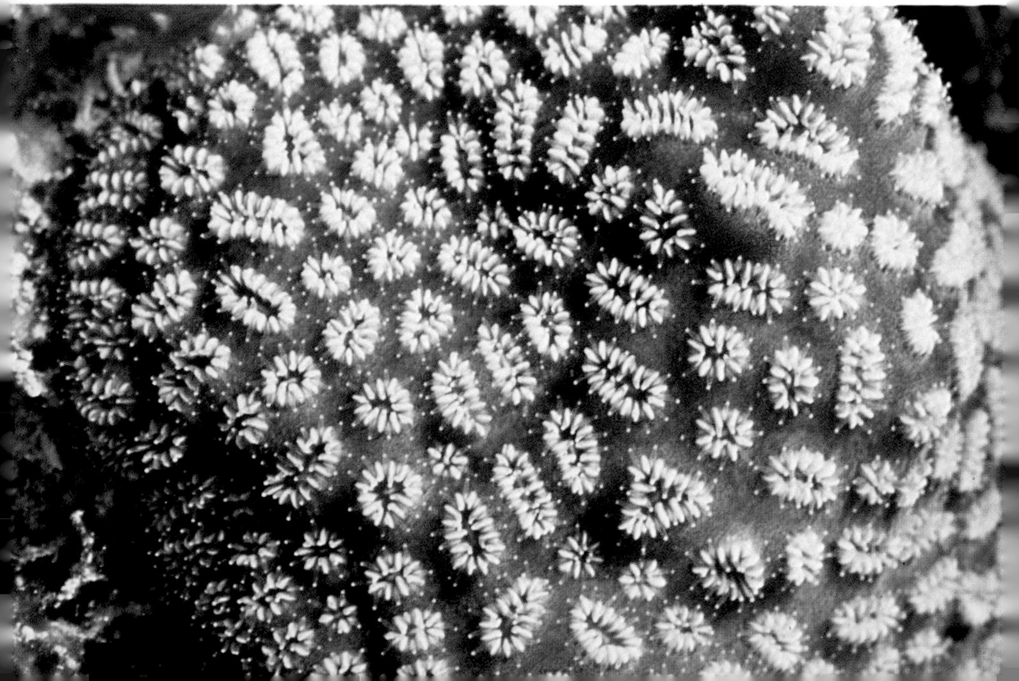

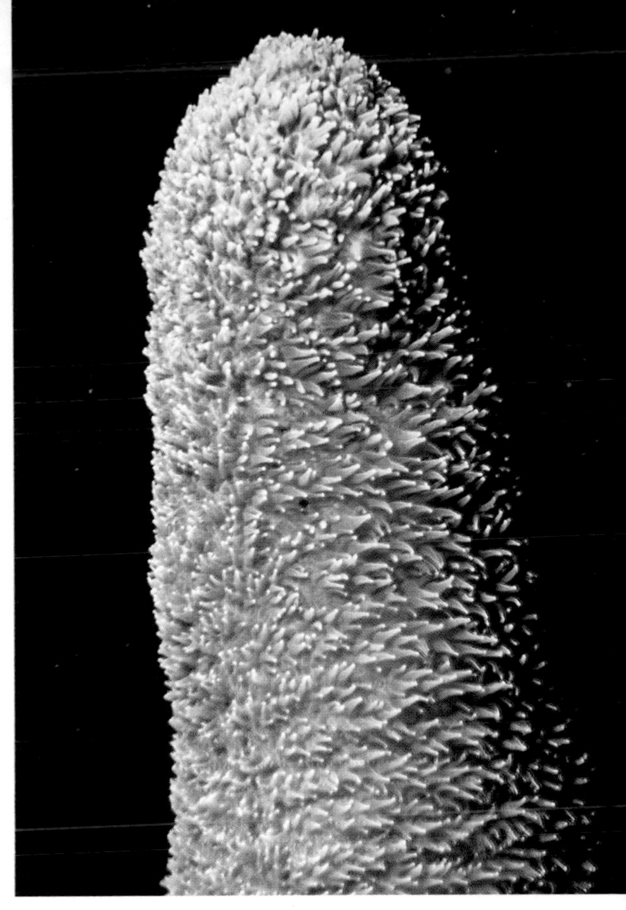

The tentacles of the polyps of *Dendrogyra cylindricus,* pillar coral, are expanded during the day and wave back and forth with the passage of currents. Jamaica, Discovery Bay, 15 m depth.

The large pillars of *Dendrogyra cylindricus* start as a series of knobs rising from a basal plate of the coral as is shown here. Jamaica, Discovery Bay, 15 m depth.

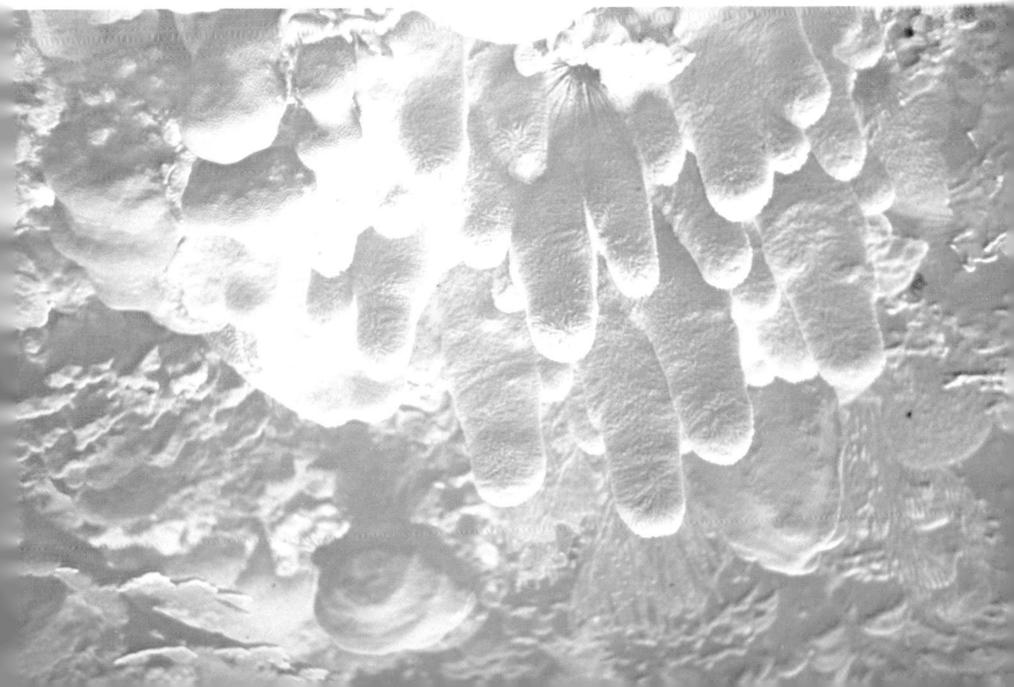

Dendrogyra cylindricus Ehrenberg
Pillar coral

Pillar coral is one of the most spectacular stony corals found on West Indian reefs. Colonies may contain dozens of upright cylindrical branches and reach a total height of nearly 3 m. If a single one of the "pillars" is broken off and comes to rest in a position where it continues to live, the branch will give rise to several new pillars which again grow vertically. Occasionally entire colonies will topple over due to erosion of the basal material by biological activity and will present the appearance of a "fallen giant." Colonies have been observed which have tumbled over at least twice and were growing their third generation of pillars.

Dendrogyra cylindricus is unusual in that the polyps with their tentacles are expanded in the daytime unlike most other stony corals. The tentacles sway with the current and if one portion of the colony is disturbed by touching so that the polyps contract, a wave of contraction of the other polyps can be seen to pass over the entire colony in a period of a few seconds.

Pillar coral varies considerably in abundance throughout its range and is a very minor constructor of reefs. It is found on flat or gently sloping reef bottoms between 1 and 20 m. It is most common on some insular coasts, such as the north coast of Jamaica and some of the Bahamas. It is absent from Bermuda and possibly along the coast of Colombia and Panama. Colonies were once more common along the Florida reef, but commercial collecting has greatly reduced its occurrence.

References: Goreau 1959, Goreau and Wells 1967, Geister 1972, Pfaff 1969.

Family Mussidae

Mussa angulosa (Pallas)

The individual polyps of *Mussa angulosa* are quite large, as much as 12 cm across on the long axis, and generally 3 to 6 cm in diameter in most individuals. The polyps actually occur on the outer end of thick, close set branches forming a hemispherical or rounded clump. The fleshy polyps touch adjacent polyps and make the surface of the skeleton appear

A lovely specimen of *Dendrogyra cylindricus* growing on the side of an *Acropora cervicornis* reef at Discovery Bay, Jamaica, depth 12 m.

Mussa angulosa has very large polyps which tend to form a hemipherical mass and are quite fleshy. Jamaica, Discovery Bay, 30 m depth.

At night the large, fleshy polyps of *Mussa angulosa* expand but the tentacles are relatively small and occur around the edge of the disk. Bahamas, Eleuthera Island, 15 m depth.

Scolymia lacera is the largest solitary polyped coral in the Caribbean. It is also one of the most aggressive on the extracoelentric digestion hierarchy. Puerto Rico, La Parguera, 24 m depth.

Scolymia cubensis occurs in more dimly lit areas than its close relative *S. lacera* and is often more abundant. Jamaica, Spring Gardens, 22 m depth.

Left: *Scolymia lacera* on the side of a steeply sloping rock face at Salt River submarine canyon, St. Croix, Virgin Islands, depth 25 m.

Below: A fallen pillar of *Dendrogyra cylindricus* with a series of new pillars on its length. When the pillars reach a maximum height and weight they topple. New pillars form if the original pillar remains alive. Photographed at Hogsty Reef, Bahamas, 15 m.

as a continuous single head rather than a clump of branches whose outer surfaces inscribe a rounded surface. The polyps, which can be brown, green, pinkish or purplish, act as a clump of individuals rather than as an interconnected colony, and this coral is intermediate between colonial and solitary corals. The polyps are contracted during the day but are expanded at night with a fringe of small tentacles around the margin of the polyps. The coral is evidently able to fluoresce, producing the reddish cast some specimens may have.

The species generally occurs only in areas of abundant coral growth. It is one of the most "aggressive" of Caribbean corals, capable of attacking nearly all other species by extracoelenteric feeding, and being attacked by no others. Its aggressive ability is on an even level with two other species, *Scolymia lacera* and *Isophyllia sinuosa*, neither capable of attacking or being attacked. This ability to eliminate other corals attempting to overgrow and block out the sunlight is important for the relatively small, slow growing species occurring in areas of fast growing foliaceous or branched competitors.

Mussa angulosa occurs in the Caribbean, the Bahamas, southern Florida and the shallow banks off Texas. It does not occur in Bermuda. It reaches from 3 to at least 40 m.

References: Roos 1971, Lang 1973, Hubbard 1973.

Scolymia lacera (Pallas)
Fleshy disk coral

Scolymia lacera is the largest solitary-polyped (monostomodaeal) coral found in the West Indies and reaches over 15 cm in diameter. A second species, *S. cubensis*, which was for many years considered to be identical to *S. lacera*, differs considerably in its aggressive abilities. If the two species occur naturally or are placed immediately adjacent to one another, the *S. lacera* is capable of "aggressing" against the specimen of *S. cubensis*. It does so by exuding its mesenterial filaments to reach the other coral (only over distances of a few centimeters) and literally digests the portion of the other coral within its reach. After this has been accomplished (requiring several hours) the filaments are withdrawn. This is

At night the large, fleshy polyp of *Scolymia cubensis* expands so that it is difficult to recognize the coral relative to its daytime appearance. Puerto Rico, Mona Island, 20 m depth.

Isophyllia sinuosa is one of the most aggressive corals in its ability to "attack" via extracoelentric digestion. Puerto Rico, Mona Island, 15 m depth.

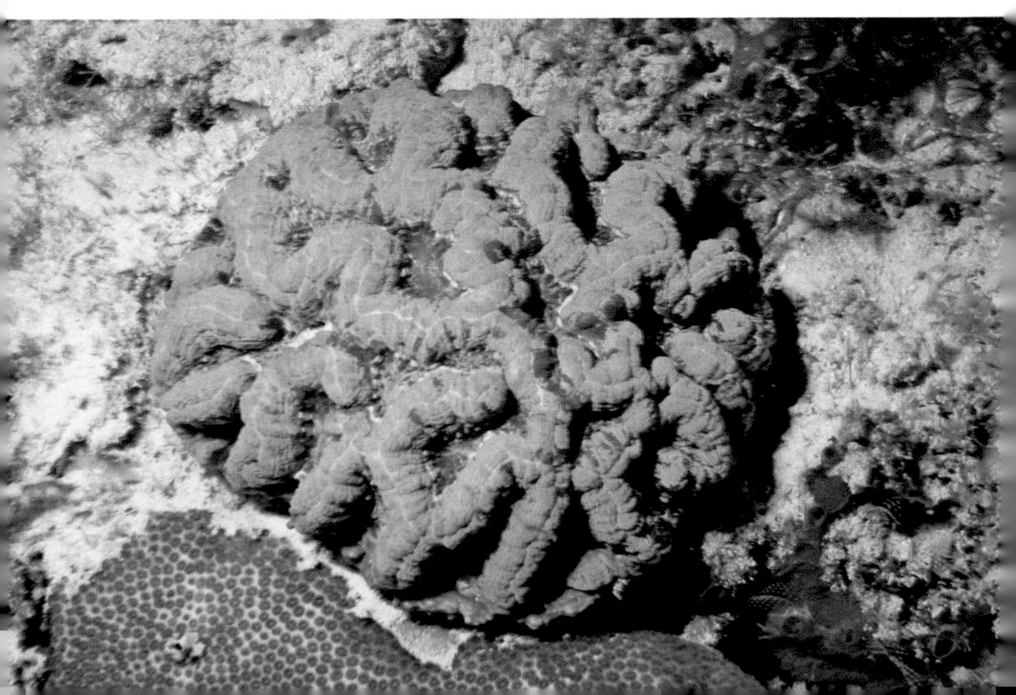

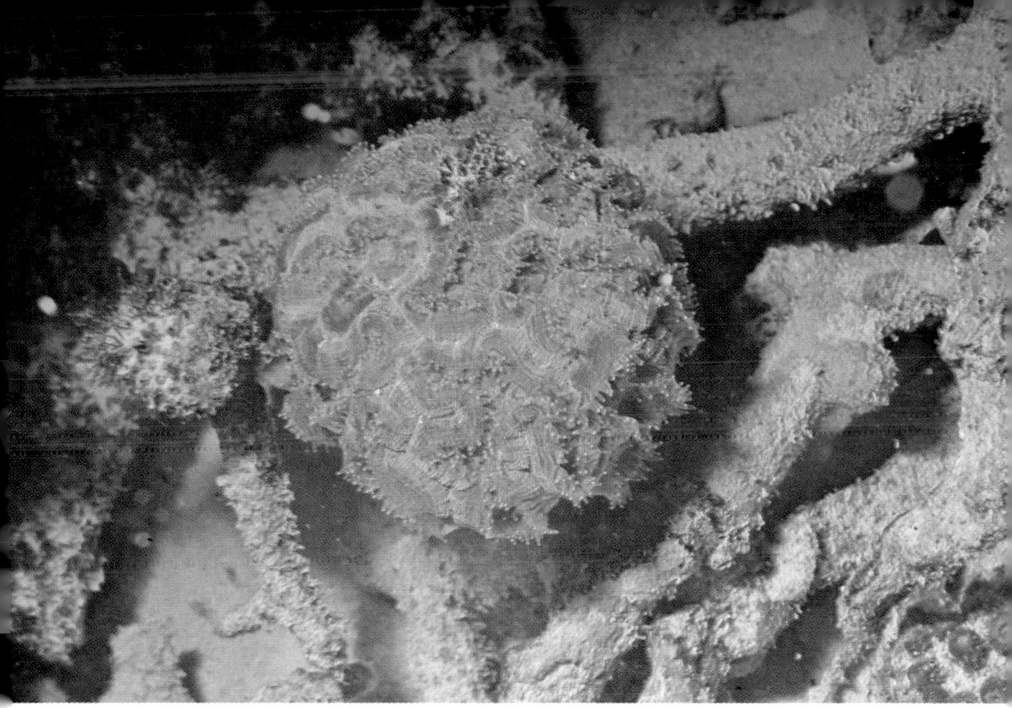

At night *Isophyllia sinuosa* has its polyps expanded and takes on a considerably different appearance than it has during the day. Jamaica, Discovery Bay, 5 m depth.

The valleys of *Isophyllastrea rigida* are closed, unlike the similar appearing *Isophyllia sinuosa*, and colored contrastingly with the ridges. Jamaica, Discovery Bay, 3 m depth.

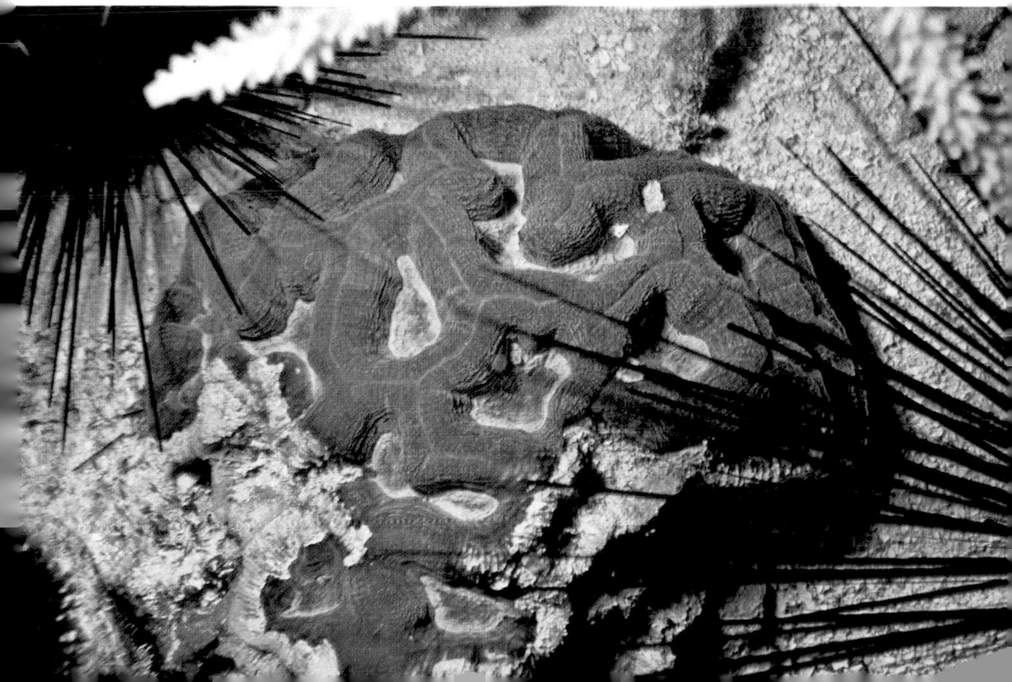

one method in the intense but relatively slow competition for space that exists on reefs whereby a coral can often eliminate another coral species which threatens to grow over it and cut off the light necessary for the zooxanthellae.

The two species of *Scolymia* are found in deeper reef habitats. While neither species is found in shallow areas with high light intensity, *S. lacera* is more restricted to relatively well-lighted areas than *S. cubensis*. If a specimen of *S. lacera* is placed in a reef area only 1 m deep, it will die.

Scolymia lacera grows on rocky substrates, rock outcrops in sand channels and steep slopes of deep reef areas. While *S. lacera* is more "aggressive," *S. cubensis* can live under lower light conditions; where the two species co-occur, individuals of *S. cubensis* are five to fifteen times as abundant. The depth range of *S. lacera* is between 15 and 80 m. It is known from the Caribbean, the Bahamas, southern Florida and Bermuda.

References: Lang 1971, Wells 1971, Wells and Lang 1973.

Scolymia cubensis (Milne-Edwards and Haime)
Smooth disk coral

Scolymia cubensis, like its congener *S. lacera,* is a large (up to 10 cm diameter) solitary-polyped coral. It had been considered the same species as *S. lacera* until recent work revealed differences in aggressive abilities among these two corals.

The polyp of *Scolymia cubensis* is smoother surfaced than that of *S. lacera* and is dark in color. It is found under lower light conditions than *S. lacera* and has a similar depth distribution.

At night the large polyp expands and the small tentacles on the rim of the disk are clearly visible.

A third species of *Scolymia, S. wellsi,* occurs on Brazilian reefs.

Scolymia cubensis is known from the Caribbean on deep reefs at 20 to 80 m and is reported to occur on one of the offshore banks off the Texas coast.

References: Lang 1971, Wells 1971, Rannfeld 1972.

Isophyllia sinuosa (Ellis and Solander)

This species forms small rounded or hemispherical heads of up to 20 cm diameter. It can occur in shallow back reef and fore reef areas. The valleys may have up to three polyps or centra with a narrow ridge between valleys. The polyps are quite fleshy when retracted during the day and the oral portion (in the valley) of the polyp is often colored differently than the ridges. There is variation in color with the following combination of oral surface and ridges recorded: yellow-iridescent green, brown-brown, purplish blue-white and green-bluish green. This species ranks within the top three aggressive (via the extracoelenteric digestion response) Caribbean corals. The abundance of *I. sinuosa* on reefs in shallow water seems quite variable.

It is known widely in the Caribbean and has been reported as rare in Aruba and Curacao but common in Bonaire and very common in the windward Lesser Antilles. It also reaches Bermuda, southern Florida and the Bahamas. It occurs to depths of at least 15 m.

A similar species, *I. multiflora* (Verrill), which is perhaps identical, has been reported from several localities where *I. sinuosa* occurs, but it is of somewhat uncertain status. Reportedly the valleys of *I. multiflora* are narrower (10-20 mm versus 25-35 mm) and shallower (5-10 mm versus 8-10 mm) than in *I. sinuosa*.

References: Almy and Carrion-Torres 1963, Roos 1964, 1971.

Isophyllastrea rigida (Dana)

The small heads of this coral are distinctive in that the valleys are closed, each resembling a rough irregular polygon, with the oral surface of the polyp colored differently than the ridge. The valleys contain only one, two or rarely three polyps. The ridges may be green, brown, pinkish or purple with a white or pale oral disk in the valley. A distinctive golden or greenish line may be seen at the juncture of the polyps on the apex of the ridges. The polyps are fleshy and wrinkled; when expanded at night they reach 4 to 5 cm above the coral skeleton surface.

Isophyllastrea rigida can form lovely heads and is a common member of many reef habitats. Puerto Rico, Mona Island, 15 m depth.

Mycetophyllia lamarckiana is usually found only in fore reef areas and is not particularly important in deep reef areas. Bahamas, Grand Bahama Island, 20 m depth.

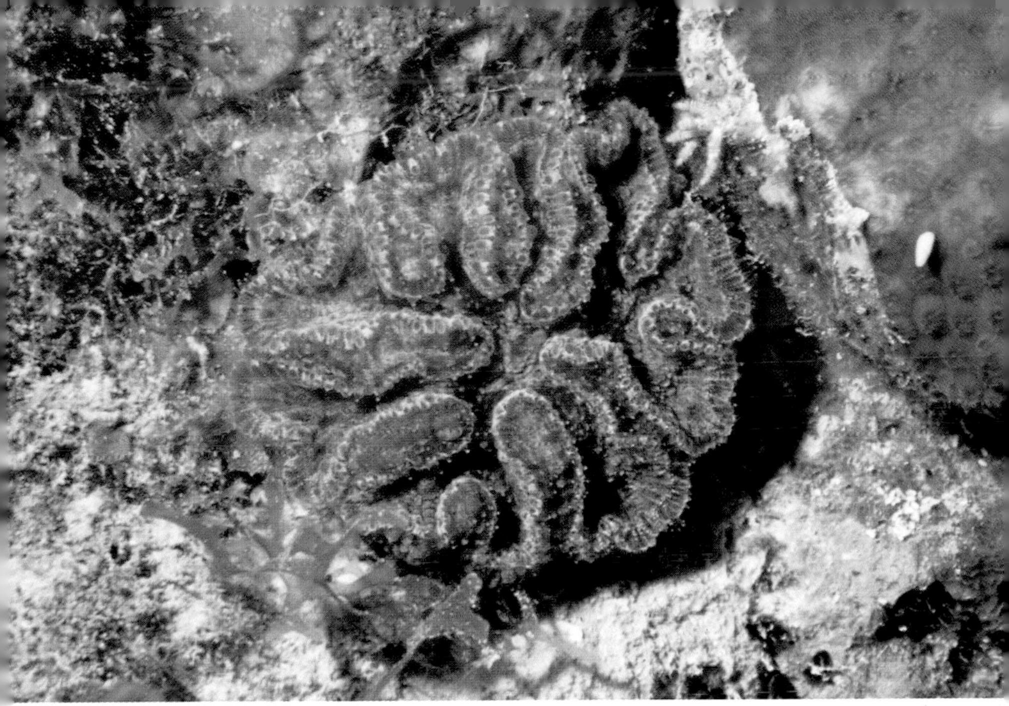

The polyps of *Mycetophyllia lamarckiana* expand at night and the colonies take on quite a different appearance. Bahamas, Crooked Island, 20 m depth.

Mycetophyllia danaana appears quite similar to *M. lamarckiana* and there is some doubt as to whether it is a distinct species. Puerto Rico, Desecheo Island, 6 m depth.

The species occurs in both back and fore reef areas in shallow water. It is moderately aggressive on the extracoelenteric digestion hierarchy. It occurs in the Caribbean, the Bahamas, southern Florida and Bermuda to depths of at least 15 m.

References: Roos 1971, Squires 1958, Almy and Carrion-Torres 1963.

Mycetophyllia Milne-Edwards and Haime

The genus *Mycetophyllia* contains important members of the coral reef facing the open sea at moderate to deep depths. Until recently only a single species was recognized in this genus, but investigations, mostly in Jamaica, indicated that several species existed, and some of these species have now been collected at other localities. At present five species are recognized, and all are discussed below.

Mycetophyllia lamarckiana Milne-Edwards and Haime

This green to brown species occurs as flattened, often nearly circular masses in fore reef areas. It has distinct ridges, rounded on top, with very apparent serrate septa. The valleys are of varying width due to the method of size increase in the colony and are occasionally whiter than the ridges. The colony increases in size after settlement of the planula by budding of new polyps around the initial polyp with no ridges formed between the polyps. Subsequently each polyp on the margin buds outward, forming a progressive chain of budding polyps with a ridge between it and the adjacent chains. Whenever the chain splits into two chains a new ridge is formed. The ridges then appear to radiate somewhat from the center of the colony.

Mycetophyllia lamarckiana occurs most abundantly at moderate depths (13 to 30 m) and is less abundant on the deep reef than other species in the genus. It ranges from only a few meters to over 60 m in the extremes. The geographic range includes the Caribbean, the Bahamas and southern Florida, but not Bermuda.

References: Wells 1973a, Roos 1971.

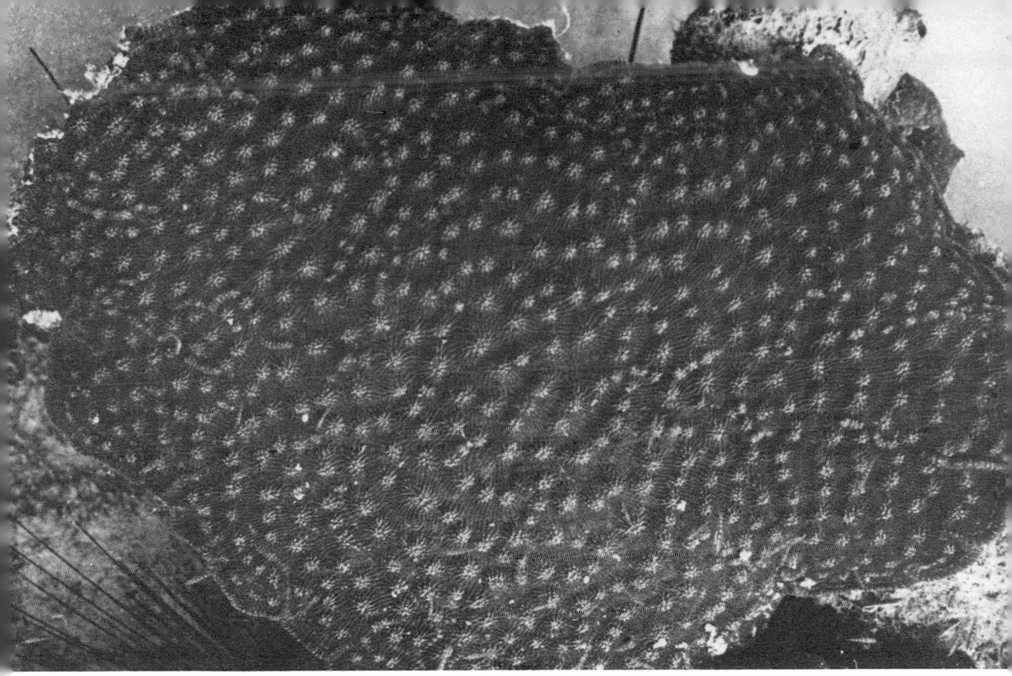

A large flattened plate of *Mycetophyllia aliciae* photographed at a depth of 30 m at Salt River submarine canyon, St. Croix, Virgin Islands. While quite broad this plate is quite thin and fragile. This specimen has few of the ridges developed on the plate and is attached to the substrate only near its central portion.

Another specimen of *Mycetophyllia aliciae* with better development of the ridges. This colony is firmly attached to a near vertical substrate although still thin.

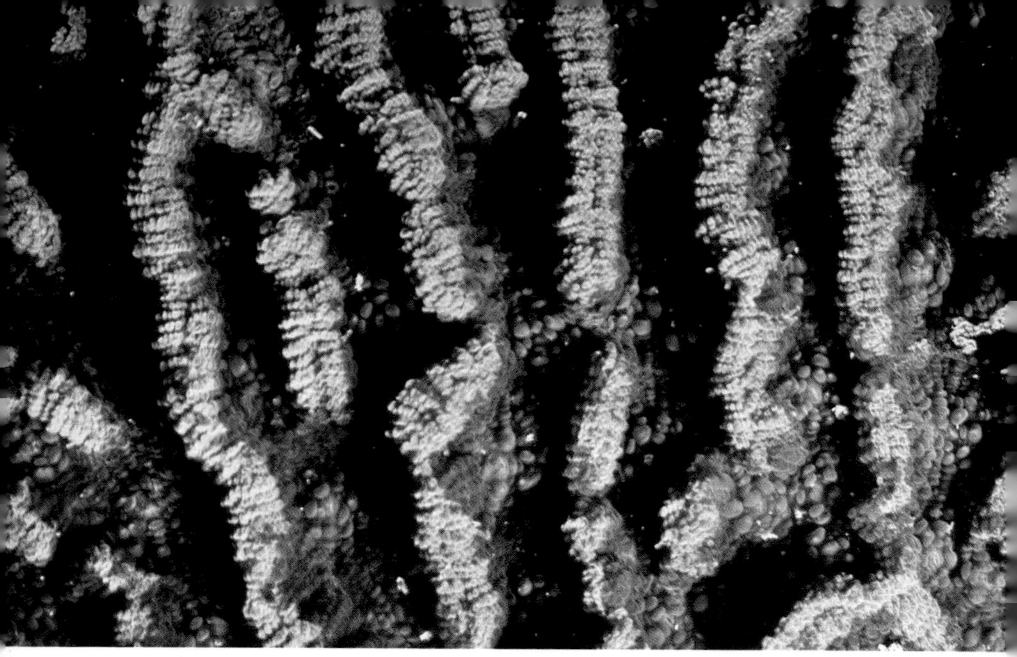

Mycetophyllia ferox is a very distinct member of the genus and its low ridges, almost square in cross section, are characteristic. Puerto Rico, La Parguera, Laurel Reef, 12 m depth.

Mycetophyllia ferox can form sizeable colonies and is often somewhat hidden below corals. It is difficult at times to find a colony sufficiently exposed to photograph completely. Puerto Rico, La Parguera, 30 m depth.

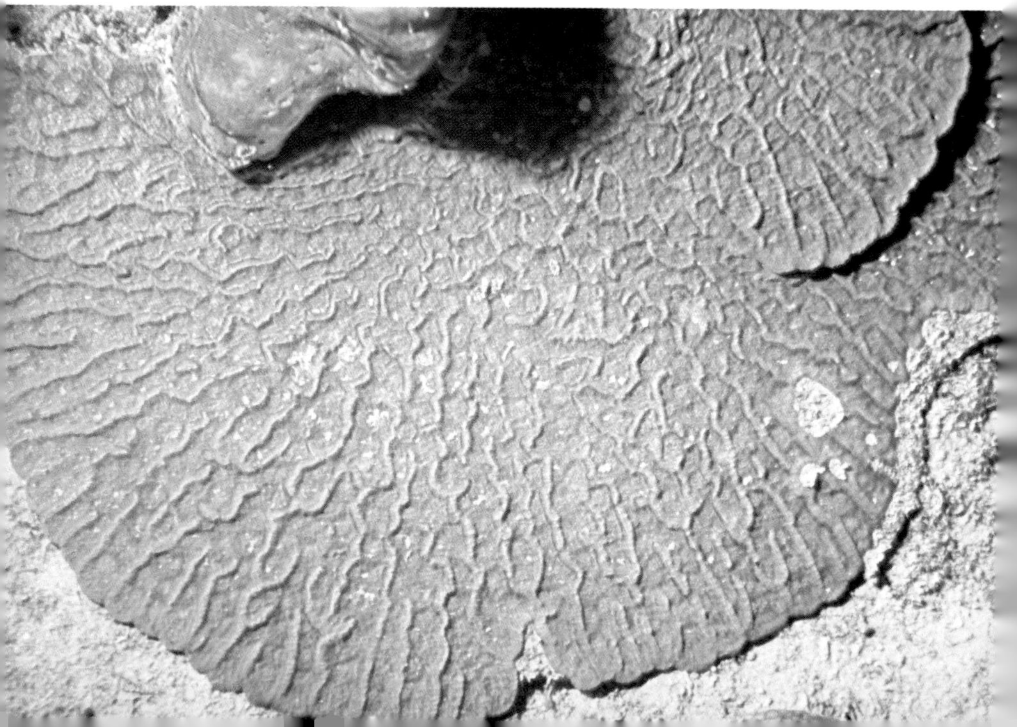

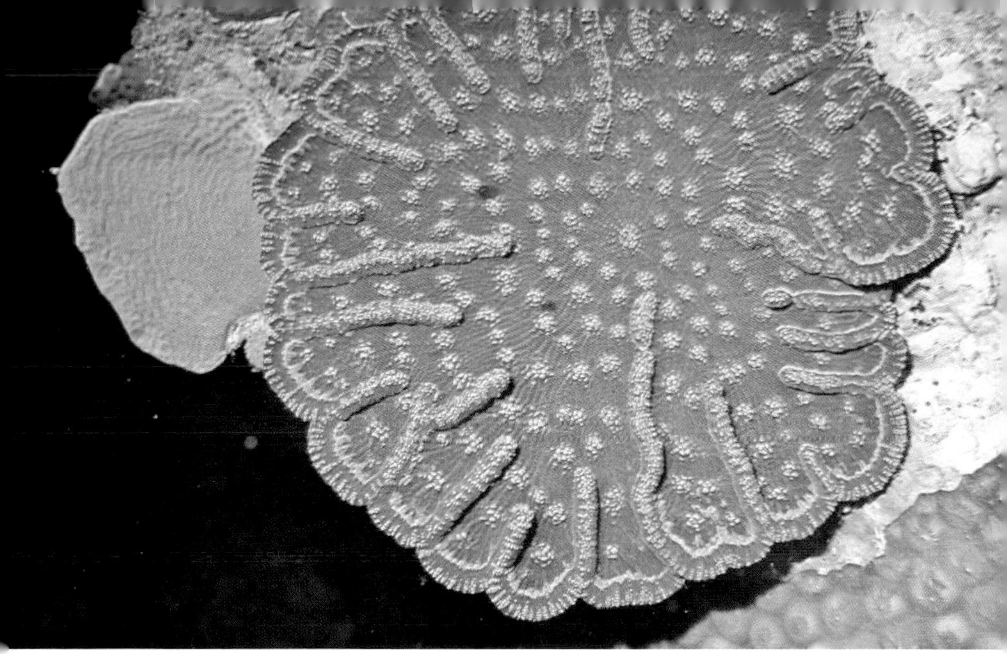

Mycetophyllia aliciae is distinctive with its light colored ridges and raised areas at the center of each polyp. The ridges form a "ruffle" around the outside of the entire colony. Jamaica, Spring Gardens, 18 m depth.

Some specimens of *Mycetophyllia aliciae* lack the contrasting coloration on the ridges and polyp centers, but still possess the "ruffle" around the edge of the colony. Such unusually colored *M. aliciae* could be confused with *M. reesi* but the latter lacks the ridge on its circumference. Jamaica, Spring Gardens, 24 m depth.

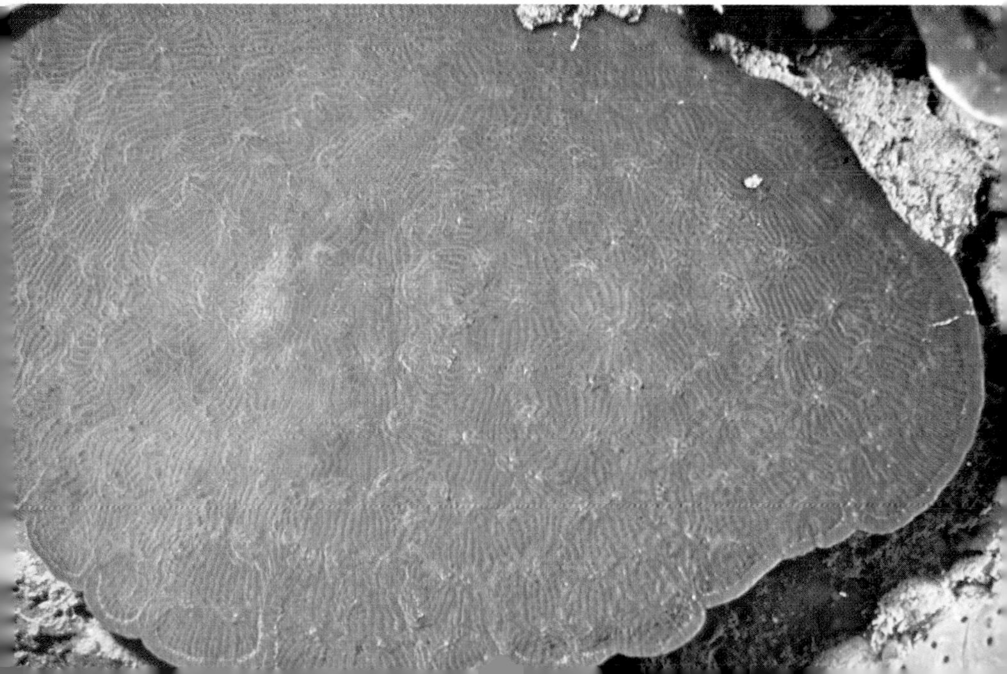

A close-up view of the surface of *Mycetophyllia aliciae.* The development of the ridges in the central portion of the colonies of this species is quite variable with fairly extensive development in this particular specimen.

Large flattened colony of *Mycetophyllia ferox* on the side of a rock outcrop at Discovery Bay, Jamaica, depth 20 m.

Mycetophyllia danaana Milne-Edwards and Haime

There is some doubt as to whether this species is distinct from *M. lamarckiana*. In *M. danaana* the ridges are less radiating than in *M. lamarckiana.* No extracoelenteric digestive reaction has been observed between these two species. While the extremes of each form are quite distinctive, intergrades do exist. The species is generally greenish brown.

Mycetophyllia danaana occurs in both fore and back reef habitats and occurs in abundance shallower than any other species of this genus. Its depth range is from 3 to at least 30 m. It is known from the Caribbean at several locations.

References: Wells 1973a, Scatterday 1974.

Mycetophyllia ferox Wells

This species forms flattened plates, often irregularly circular, with roughly radiating ridges. The valleys are discontinuous, being completely closed by ridges, and *M. ferox* is the only member of the genus in which this occurs. The ridges are low, appearing somewhat flattened on top, and sometimes are lighter in color than the valleys. The species is brownish to greenish in color.

This is one of the most "aggressive" of the Caribbean corals, being attacked successfully by only three species and able to attack nearly all others. It often occurs in or near thickets of *Acropora cervicornis*, a species it can easily digest extracoelenterically if within a few centimeters.

Mycetophyllia ferox generally occurs at 10 to 20 m, but reaches at least 35 m. Presently it is known from Jamaica, Bonaire and Puerto Rico, but it certainly has a wider distribution.

References: Wells 1973a, Scatterday 1974.

Mycetophyllia aliciae Wells

This and the following species are strictly deep reef dwellers. *Mycetophyllia aliciae* is a plate-like, often circular, coral usually with radiating ridges near the outer margin. These ridges are less distinct the deeper the coral occurs, and

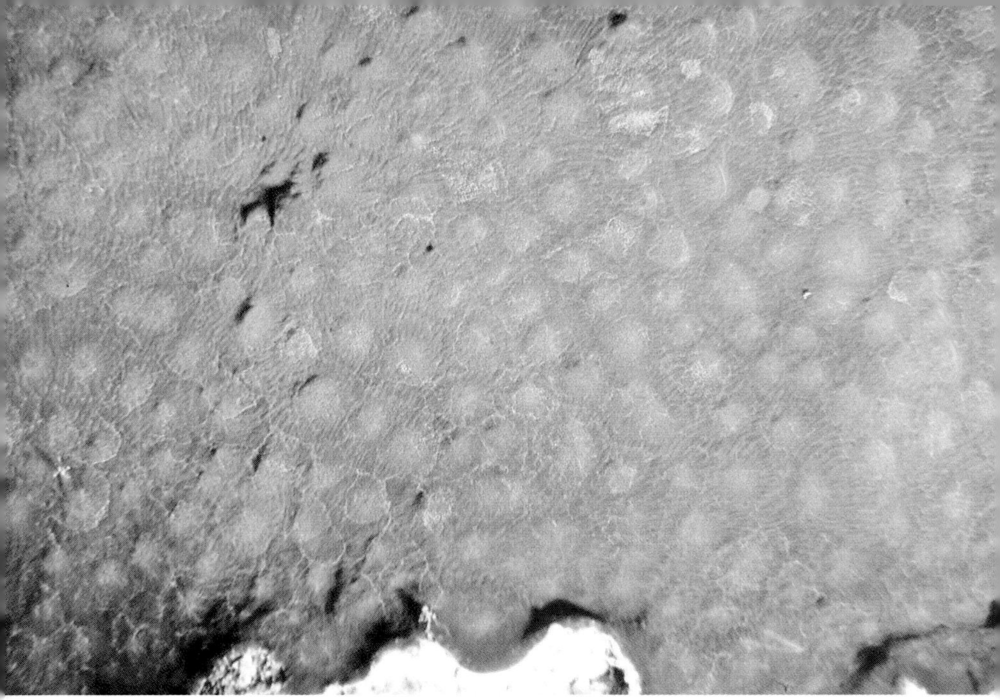

Mycetophyllia reesi is strictly a deep-reef inhabitant and is seldom seen or photographed. It lacks ridges and valleys and has only small "lumps" on the surface of the colony. Jamaica, Discovery Bay, 54 m depth.

Closeup the living surface of *Mycetophyllia reesi* resembles eroded hills rather than a scleractinian coral. The polyps are unusual in that they lack tentacles. Jamaica, Rio Bueno, 50 m depth.

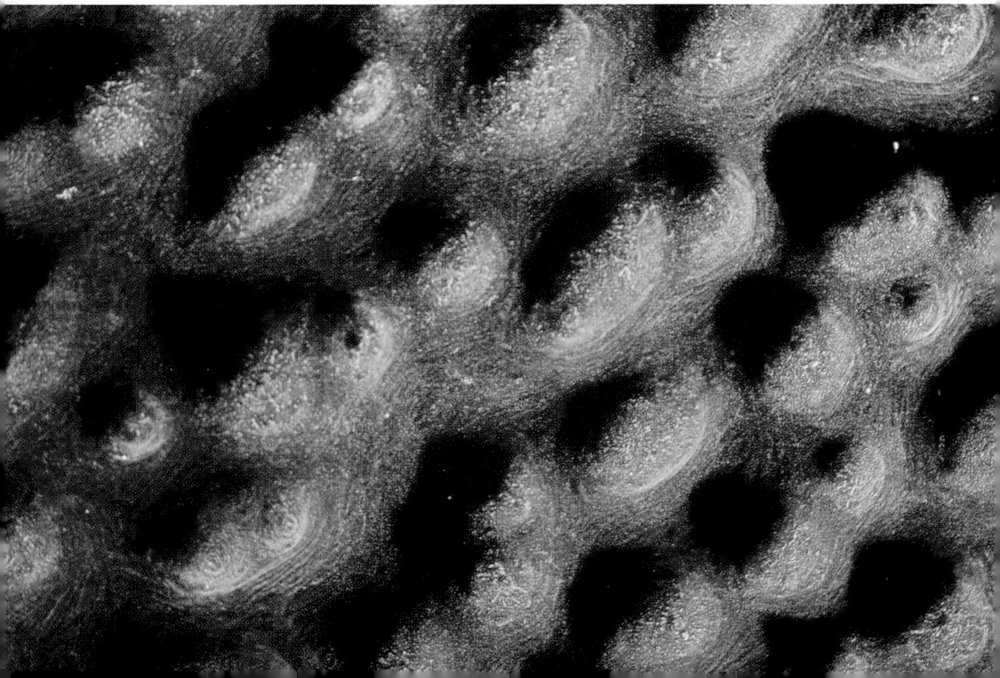

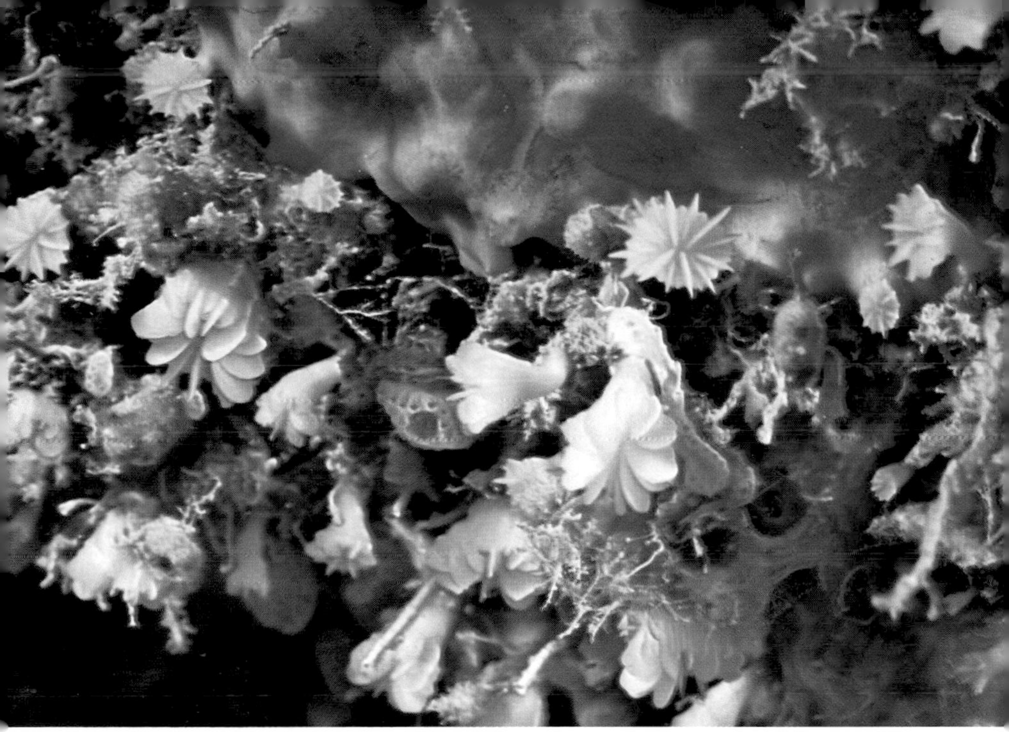

Desmophyllum riisei is a pale, tiny ahermatypic coral which lacks zooxanthellae. It can be found in deep-reef caves but reaches to depths far below the coral reefs. Jamaica, Rio Bueno, 30 m depth.

Flower coral, *Eusmilia fastigiata,* produces a "bunch" of polyps, each on a single branch. Puerto Rico, Desecheo Island, 30 m depth.

in extreme specimens they are barely raised but are still visible as a series of septa. The colony may be gray-brown, brown, green-brown, or green with the teeth on the septa and around the mouth of the polyps light in color.

The species occurs on deep reefs, often on vertical faces, and is thin and often loosely attached to the substrate. The shallowest record of this species is 18 m, and *M. aliciae* reaches 75 m in depth. The species presently is known from several locations in the Caribbean and the Bahamas. This coral is named for H.R.H. Princess Alice, at one time Chancellor of the University of the West Indies.

References: Wells 1973a, Scatterday 1974.

Mycetophyllia reesi Wells

Like *M. aliciae, Mycetophyllia reesi* is strictly a deep reef inhabitant. It forms thin, often convoluted plates, conforming in many instances to the substrate. It lacks both ridges and valleys, the only species of *Mycetophyllia* to do so, and its surface consists of small lumps where the centers of the polyps are located. The species is also the only West Indian *Mycetophyllia* to lack tentacles on the polyps. It varies in color from green or brown to a dark bluish.

Like all *Mycetophyllia, M. reesi* is moderately aggressive and able to keep a number of species of coral from overgrowing it. In the deep reef habitat light is of critical importance to corals in forming the skeleton and any deprivation of the weak ambient light present on the deep reef can be fatal for a coral.

Montastrea cavernosa can assume a thin, flattened growth form at depths around 60 m which resembles that of *M. reesi*. Close examination of the corallites easily distinguishes these two species.

This species is known from Jamaica and the Bahamas. It undoubtedly has a wide distribution but is seldom collected due to the depth at which it occurs.

Reference: Wells 1973a.

Family Caryophyllidae

Desmophyllum riisei Duchassaing and Michelotti

This is an example of one of the small ahermatypic

corals which may be encountered in some reef environments. It occurs in dark caves, beneath overhanging ledges in fairly deep water and below the reef on open rock surfaces. *Desmophyllum riisei* lacks zooxanthellae, its white skeleton clearly visible beneath the colorless polyp; occasionally it may be purple in color. While solitary, it can occur in "groves," groups of individuals close to one another. The species occurs as shallow as 18 m and reaches at least to 300 m. It is known from a number of Caribbean localities.

Eusmilia fastigiata (Pallas)
Flower coral

The polyps of flower coral, appearing to be on stalks arising from a central core, occur in a single, often hemispherical plane, producing a colony up to 50 cm in diameter. Each polyp is usually round or oval, but one form has larger polyps that are elongate and sinuous. This elongate form is characteristic of moderate depths. Even though the polyps are widely spaced, they expand greatly at night so that the expanded tentacles form a continuous cover. The coral can vary from yellow to brown in color.

Eusmilia fastigiata has a wide depth range from 1 to 65 m but is most common at 3 to 30 m. It can occur in a variety of habitats from back reefs to fore reefs and under overhanging sides of larger corals. The species is found widely in the Caribbean, the Bahamas and southern Florida but apparently is absent from Bermuda.

References: Roos 1971, Scatterday 1974, Wells 1973a.

Tubastrea aurea (Quoy and Gaimard)

Even though this coral is non-reef building (ahermatypic), it is on occasion abundant on reefs in the proper habitat. The species is not solitary, with clumps containing a few to hundreds of polyps occurring on undercut waveswept rocks, on overhanging faces in deeper water and in fairly dimly lit caves. One pier off western Puerto Rico has all the area available on the inside of the pilings, beneath a platform providing shade, completely covered by this coral to a depth of 1.5 m.

The individual polyps of *Eusmilia fastigiata* have large septa; the mouth is visible in this photograph. Puerto Rico, La Parguera, 5 m depth.

At night the polyps of *Eusmilia fastigiata* expand and "fill in" the spaces between the branches with their tentacles. Bahamas, Crooked Island, 15 m depth.

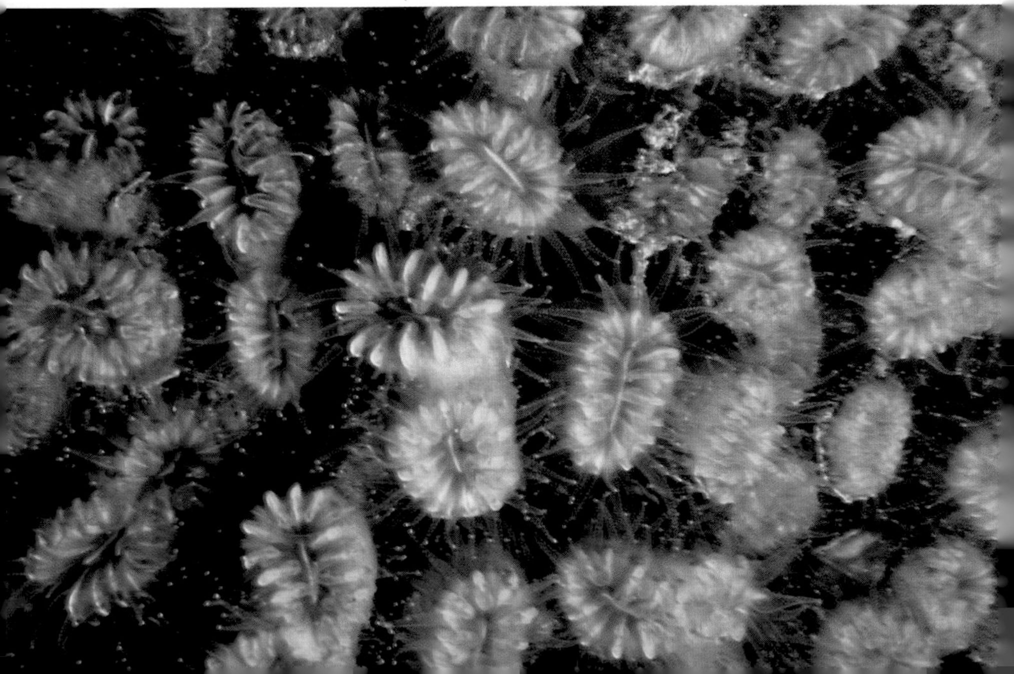

When contracted the polyps of *Tubastrea aurea* are red-orange and the skeleton of each polyp is visible. These clumps of the coral are found on a rock surface along with a variety of sponges. Puerto Rico, Desecheo Island, 5 m depth.

When expanded (at night or in dark situations) the polyps of *Tubastrea aurea* are a striking yellow-orange. When expanded it is easy to mistake these stony corals for anemones as the skeleton is completely hidden. Puerto Rico, Aguadilla, 2 m depth.

The coenosarc (the tissue between the corallites) is bright red, although the species lacks zooxanthellae. The great beauty of this coral comes from its polyps, which are orange-yellow when expanded at night or during the day in dimly lighted areas of the reef.

This species has been recorded widely in the West Indies and the Bahamas from near the surface to 35 m. It also occurs in western Africa, but is not known from Bermuda.

References: Roos 1971, Boschma 1953, Goreau 1959.

ORDER ANTIPATHARIA: BLACK CORALS

Antipatharians, commonly known as "black corals," are one of the most difficult groups of Atlantic reef organisms to identify to species. They are often quite variable and the characters used for distinguishing species are few in number.

At present over thirty species are reported from the western Atlantic. Many of these occur only at depths far below those normally reached by divers. Other species are of such uncertain taxonomic status that their validity is questionable. Finally, to cap this taxonomist's nightmare, there are at least several (including one shallow water species) black corals which are presently undescribed. Relatively few contemporary works have been published regarding antipatharians, most previous work having been carried out in the 19th century.

Antipatharians have a black, spiny keratin-like axis, the basal portion of which is used in the large species for jewelry. The skeleton is secreted in concentric layers generally around a hollow core and the colony is attached by a holdfast to a solid, rocky substrate. The skeleton can be easily cut, sanded and polished, resulting in semi-precious material which has an appearance not unlike shiny black plastic. Aside from the artful way the material can be worked and the settings associated with the black coral, the lure of black coral as a substance for jewelry comes more from the aura of deep-diving bravado associated with it, rather than from some innate, unique property of the material itself. The significance of going to depths of 60 m or more, as a feat of diving, to wrench an antipatharian tree, which may be over a hundred

A colony of *Antipathes pennacea,* one of the most easily identified of the black corals. This specimen was photographed at 25 m depth at the Salt River submarine canyon, St. Croix, Virgin Islands.

Antipathes pennacea is one of the most distinctive black corals with its feather-like pinnate branching and large size. This specimen has a fire worm, *Hermodice carunculata,* on it. Puerto Rico, Monito Island, 27 m depth.

The black coral *Antipathes furcata* has a somewhat fan-shaped colony and does not reach a great size. Bermuda, 53 m depth.

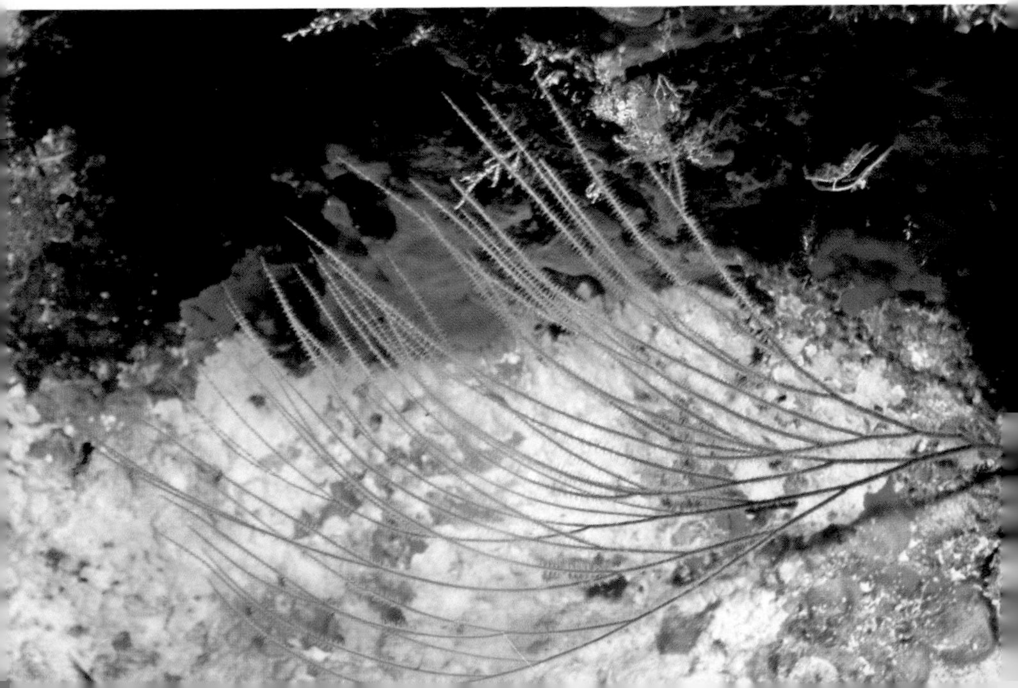

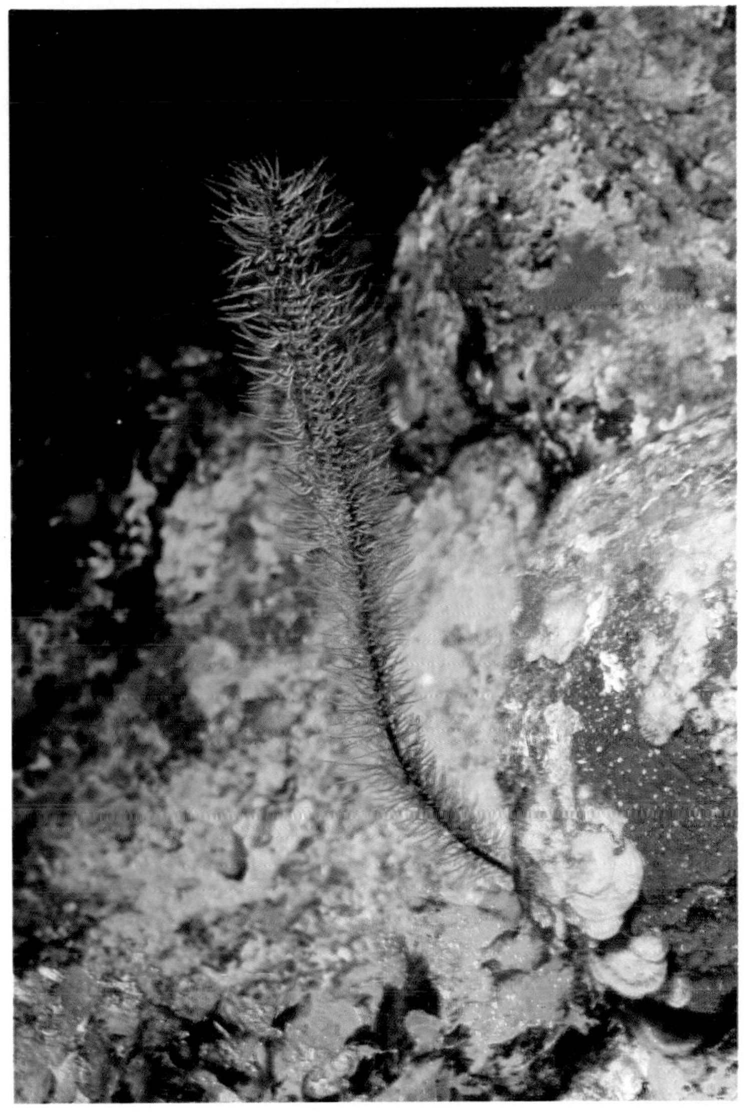

Antipathes tanacetum is a single stalk black coral which has a wide distribution in the tropical Atlantic. Bermuda, 50 m depth.

Small colony of *Antipathes* sp. on a rock promontory at 30 m depth, St. Croix, Virgin Islands. This is the species of commerce in the West Indies although it is still probably undescribed. The white of the polyps on the black skeleton is plainly visible in the photo and a second "tree" can be seen in the left background. A filament of *Stichopathes lutkeni* is visible in the upper part of the photo.

years old, from its home only to discard 90% of the material as useless, is somewhat muted by the fact that black corals capable of being worked into jewelry can be found in depths as shallow as 7 m in certain locations and if desired, could be taken by snorkelers. Excessive collection of black corals by commercial collectors and visiting divers has caused various island governments to restrict the export, either as jewelry or unworked material, of black corals.

The polyps, which have six non-retractile unbranched tentacles, occur on the surface of the skeleton. The size and shape of the polyps can vary considerably in a single colony. Generally the polyps become more elongate the further they are from the holdfast of the colony.

Antipathes pennacea Pallas

The branching and form of this species make it the most distinctive of the western Atlantic black corals. Colonies can be sizable, reaching over 1.5 m in both height and width, but are branched in nearly a single plane. The stems and branches bear fine, simple pinnately branched extensions (pinnules). These pinnules are short (2-6 cm in length) and are alternately arranged in two rows on each branch with only 1-3 mm between pinnules. Adjacent pinnules are of nearly equal lengths and the arrangement of branches and pinnules resembles that of a bird feather. Like feathers, the pinnules are always inclined at an angle toward the outer portion of the branch.

Although the skeleton is black, the polyps often impart a dark red tinge to the entire colony. Patches of white sediment may also be found on portions of colonies, resulting in a blotched appearance.

The species occurs abundantly on deeper portions of some reefs, but has been found as shallow as 3 m in reef caves or under ledges. Below about 20 m *A. pennacea* is found on vertical faces, but it can also grow vertically from a nearly level substrate. It is most abundant at depths of 25 to 60 m and has been taken as deep as 329 m. It is known from the tropical Atlantic, Pacific and Indian Oceans. Atlantic records indicate that it is found throughout the Caribbean,

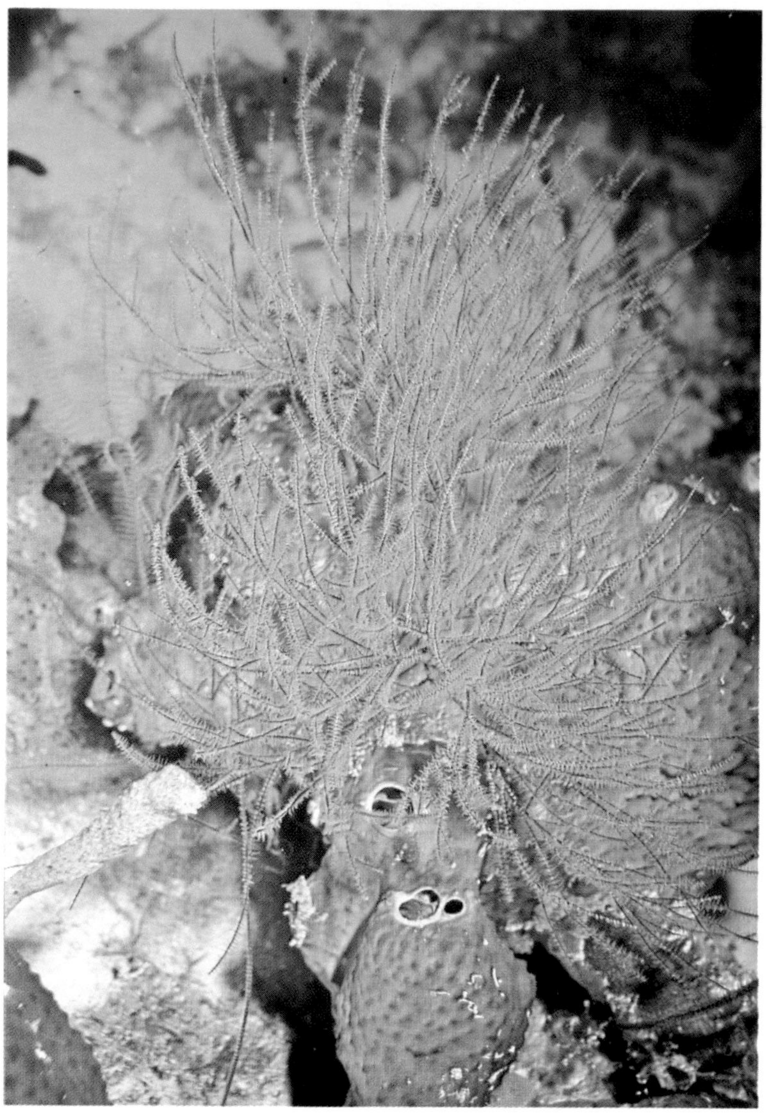

This is a small example of the black coral of commerce in the Caribbean. This species, which may be undescribed, may reach over 3-4 m in height and is harvested occasionally for use as jewelry. Puerto Rico, La Parguera, 45 m depth.

This unidentified species of *Antipathes* called the "scraggly" antipatharian, is found on the undersurfaces of overhanging ledges of dropoffs. It may be an undescribed species. Bahamas, Eleuthera Island, 24 m depth.

This is a closeup view of the commercial black coral in which the pale polyps are visible around the darker skeleton. Jamaica, Discovery Bay, 35 m depth.

in the Bahamas and even in southern Florida. The species has not yet been taken in Bermuda, however.

A number of other organisms occur with *A. pennacea*, including several species of shrimps, molluscs and a fish. Portions of colonies are often overgrown with fire coral (*Millepora*), hydroids and algae.

Reference: Opresko 1974.

Antipathes tanacetum Pourtales

This species always consists of a single stalk up to 30 cm in height with pinnules radiating out from it in four to six rows. There are secondary pinnules growing from the primary pinnules.

It occurs on limestone surfaces, growing either on a vertical to horizontal surface or from beneath small overhangs. It is known throughout the West Indies, Brazil, the Straits of Florida and Bermuda. Its known depth range is from 45 to 900 m, and it occurs commonly around Bermuda at depths below 50 m.

References: Opresko 1972, 1974.

Antipathes furcata Gray

This black coral forms a fan-shaped colony with branching often occurring nearly in one plane. Some specimens are somewhat elliptical in cross section. *Antipathes furcata* is small, reaching a maximum height of only about 40 cm. The individual branches are long for the colony size (up to 20 cm in length) and straight, usually extending to the top of the colony. The portion of the stem from the holdfast to the first branching is short, only a few centimeters long in a sizable colony.

The species occurs at 30 to 72 m and can be locally abundant. In Bermuda *A. furcata* was observed within crevices and exposed on both vertical and horizontal surfaces of rugged limestone bottoms. It is distributed widely in the western Atlantic (Barbados, Curacao, Puerto Rico, Jamaica, Bahamas, Bermuda) and is known from Madiera in the eastern Atlantic.

Reference: Opresko 1974.

View inside (looking out) a typical buttress cave at 20 m depth. While the passage is quite tall, it is relatively narrow, often so narrow that a diver cannot pass. Gorgonians and black corals occur in these areas and sclerosponges are found in the darker recesses off the main passage.

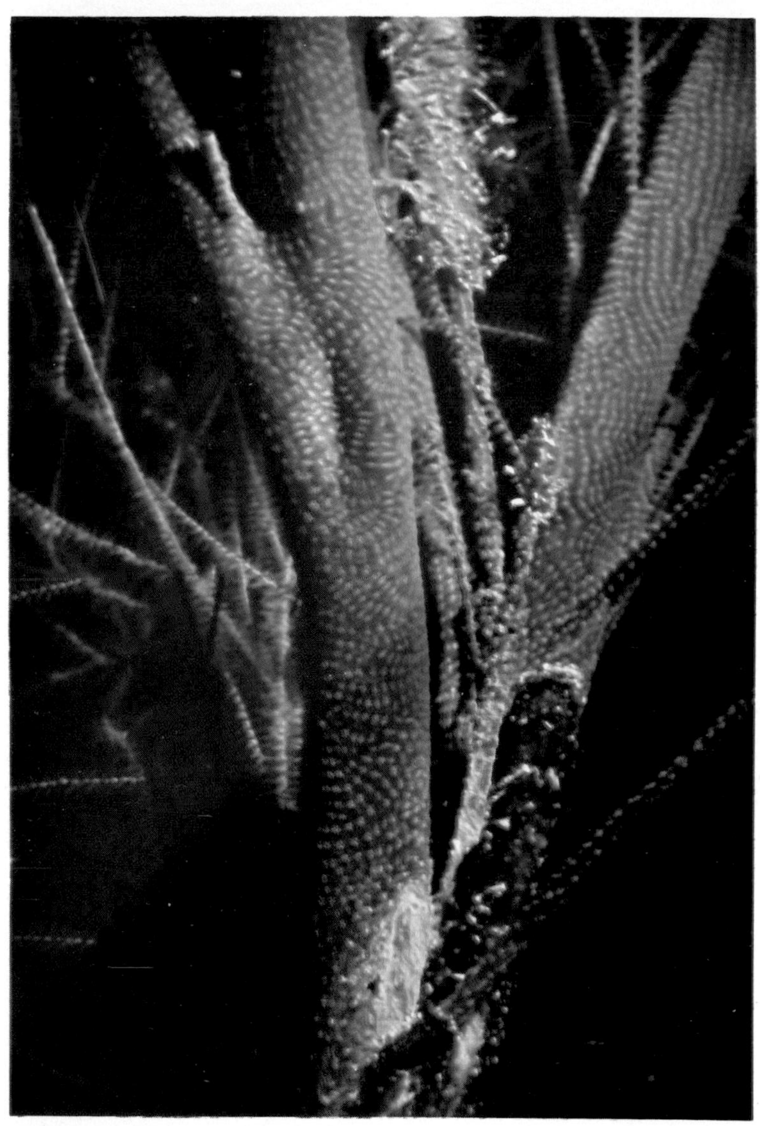

This photograph shows the base of a large colony of commercial black coral. Even in this portion of the colony the polyps still cover the skeleton. Jamaica, Discovery Bay, 40 m depth.

This species of black coral, possibly *Antipathes hirta,* has a very "bushy" skeleton which grows upward from the substrate. Bermuda, 50 m depth.

This delicate fan-like black coral may be *Antipathes atlantica*. Each polyp is visible as a small white dot. Jamaica, Spring Gardens, 24 m depth.

A black coral, *Antipathes* sp., which forms a fan-like corallum. Photographed at 25 m depth at Salt River, St. Croix, Virgin Islands.

Antipathes spp.

Various other Atlantic antipatharians cannot be assigned at present to a definite species. Figures of some of these types are included, but practically nothing is known of their bathymetric and geographic distributions.

One exception to this is a species, probably undescribed, termed here the "scraggly antipatharian." It is uncertain whether this antipatharian even belongs in the genus *Antipathes*. It does, however, occur in abundance in the proper habitat and is an important element of the West Indian black coral fauna. It occurs on the undersurfaces of ledges at moderate depths (about 20 to 50 m), often being found on completely inverted surfaces. It forms an anastomosing network with multiple holdfasts and, while it does not protrude far from the substrate, can cover an area of a meter square with a single interconnected colony. It is known from Jamaica, the Caicos Islands and the Bahamas.

Stichopathes lutkeni Brook

All western Atlantic antipatharians consisting of a single whip-like filament are considered, at present, to be a single species, *Stichopathes lutkeni*. At least two color varieties exist (one is brownish black and the other cream in color) and may represent separate species. The polyps occur on the filaments in a single file, usually on only one side of the colony. They may reach over 4 m in length and grow relatively rapidly for antipatharians. Some sizable specimens have increased their length as much as 25% in five years, but the time required to grow from a newly attached planula to a length of 2-3 m is not known.

The species is most common along vertical rocky faces, growing horizontally outward. *Stichopathes lutkeni* is known from southern Florida, the Bahamas, the Caribbean and Bermuda at 15 to at least 120 m. One small species of shrimp, colored to resemble the antipatharian, often occurs with it.

Reference: Noome and Kristensen 1976.

This unidentified species of *Antipathes* has the polyps very clearly visible. Puerto Rico, Rincon, 24 m depth.

The polyps on this unidentified black coral are fully expanded and form almost a solid filtering network between the branches of the colony. Jamaica, Rio Bueno, 20 m depth.

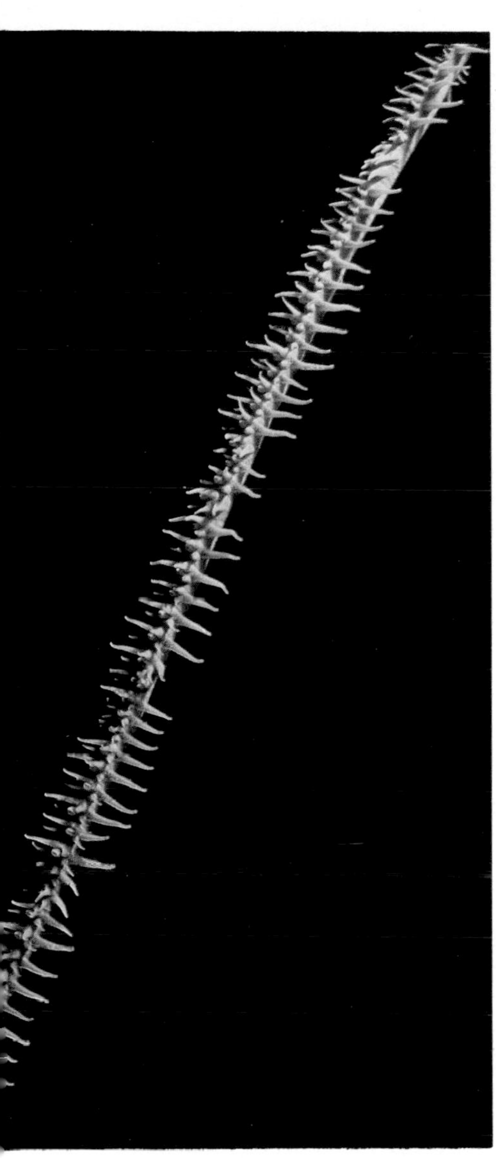

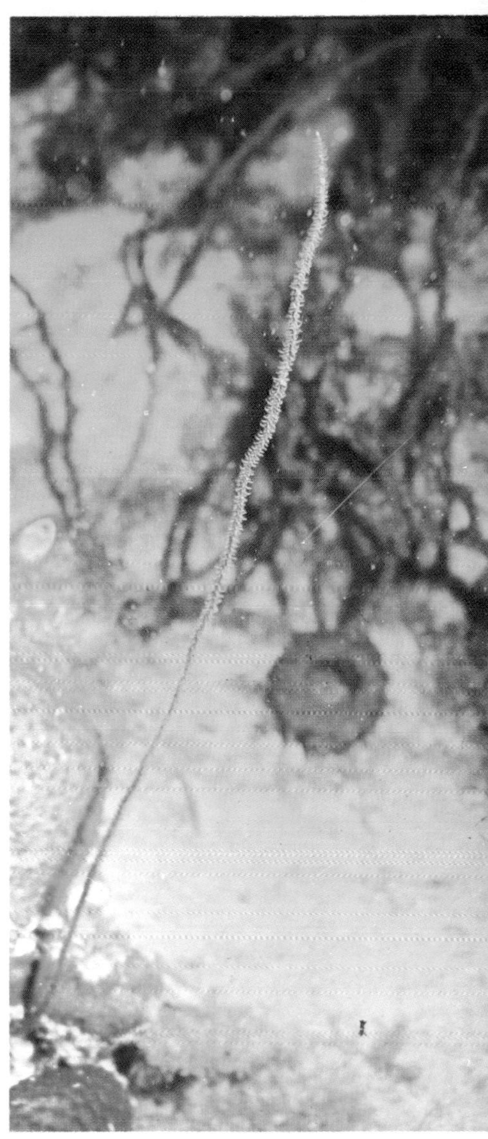

Left: The polyps of *Stichopathes lutkeni* spiral as they extend from their base. This photograph shows two colonies; one spirals while the other extends almost straight through the spirals of the first. Puerto Rico, Rincon, 24 m depth.

Right: The black coral *Stichopathes lutkeni* consists of a single whip-like filament. These colonies can be so dense that they almost seem to form a forest of filaments protruding out from steep or vertical deep-reef faces. Puerto Rico, La Parguera, 30 m depth.

ORDER CERIANTHARIA: TUBE-DWELLING ANEMONES

The cerianthids dwell in a buried tube constructed of mucus, nematocysts and detrital materials. They lack a skeleton completely and, as the lower end of the tube is pointed to rounded, lack a pedal disk. A whorl of oral tentacles surrounds the mouth and larger tentacles occur around the margin of the oral disk.

A few species occur on Caribbean reefs. They are contracted into their tubes during the day and expand greatly at night, with the column and tentacles reaching as much as 30 cm in height. On reefs they occur singly on coarse sandy bottoms. Cerianthids are variously colored; some are pale and one Caribbean species has brown and white banded tentacles. They are quite sensitive to light at night and begin contracting when in the beam of a diving light.

Phylum Ctenophora: Comb Jellies

The comb jellies are nearly transparent, generally planktonic creatures that resemble coelenterates. They possess certain distinctive features that set them apart from the Coelenterata. These include eight rows of comb-like plates on the body, the ctenes. They are used in limited locomotion, although the animals are truly planktonic.

They are carnivorous, generally capturing food with two tentacles containing adhesive cells called colloblasts and depositing it at the mouth. Comb jellies are extremely delicate creatures and descend from the ocean surface during rough seas to the calm water below; otherwise they can be severely injured by the water motion. Some species are also bioluminescent.

While twelve species are known from the tropical coast of North America, the members of the genus *Mnemiopsis* are most common. No ctenophores are exclusively associated with reefs, but they often drift over reefs with water movement and can occur in surprising numbers under the proper conditions.

Reference: Mayer 1912.

Cerianthids are quite spectacular when they are expanded at night, but they are quite sensitive to light and start to withdraw when illuminated by the beam of a flashlight. Jamaica, Discovery Bay, 15 m depth.

The oral tentacles of this cerianthid are visible around the mouth. It is protruding at night from its tube buried in a bottom of coral sand and rubble. Jamaica, Discovery Bay, 15 m depth.

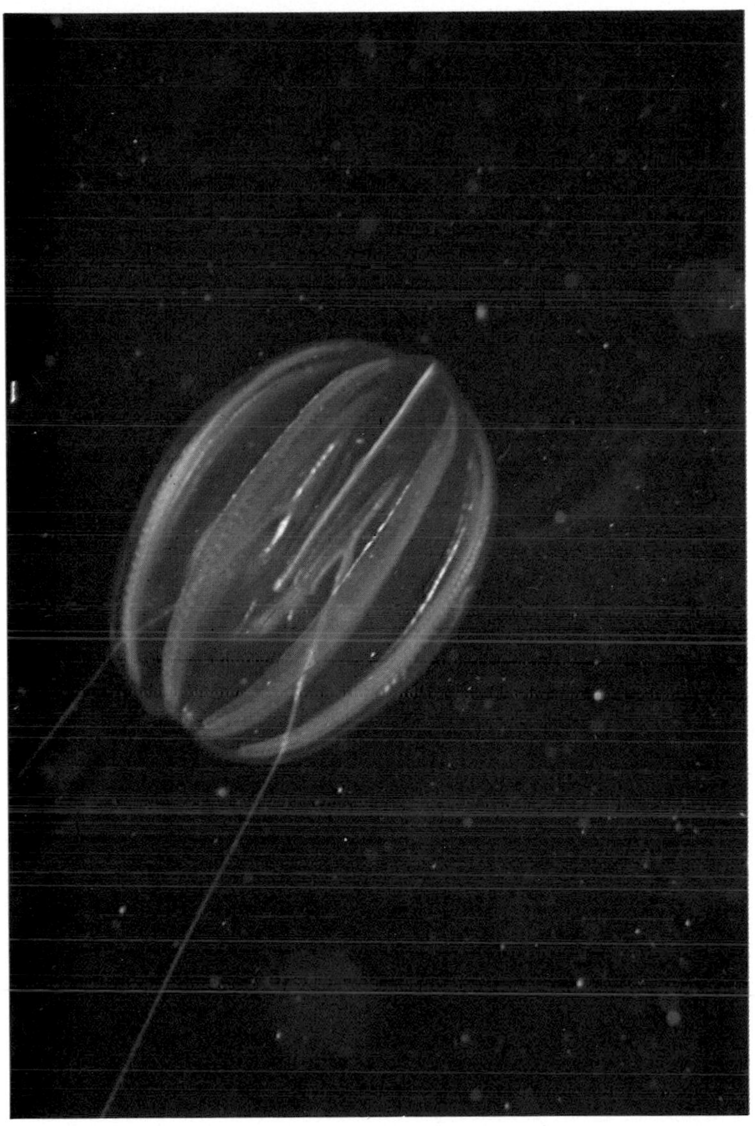

The single pair of tentacles and four of the eight comb-like plates are clearly visible in this photograph of a comb jelly or ctenophore. Photo by Alex Kerstitch.

Phylum Nemertea: Ribbon Worms

The ribbon worms are a phylum of primarily marine worms of which a small percentage occur in tropical waters. Some are planktonic, but most are benthic. Those of coral reefs hide beneath rocks or occur in clumps of algae such as *Halimeda.* The unidentified species figured was found on an open coral rock surface at night.

Most nemerteans are believed to prey on small invertebrates and do not affect any commercially valuable species. They are not used for food and are of no economic significance.

Ribbon worms have an anterior mouth and posterior anus, unlike previously discussed groups. The tropical members of this group are often brightly colored with longitudinal stripes.

References: Bayer and Owre 1968, Correa 1961, 1963, Coe 1951.

Phylum Ectoprocta: Bryozoans

The bryozoans are colonial, largely marine animals numbering around 1000 species which occur attached to a substrate. The individuals of colonies are termed zooids, and each zooid lives within a structure it secretes called the zooecium. The arrangement of the zooids, each with their zooecium, varies considerably between species. Some species form encrusting plates; others form small delicate fans. The zooecium may also contain calcium carbonate, and in some species the colony is sufficiently calcified so that it may resemble a small coral.

On reefs bryozoans are most important in the community occurring on the undersurfaces of plate-like corals. In the Caribbean many species occur beneath *Agaricia* plates along with a multitude of sponges and other organisms. The competition for space and interactions of these many organisms are fascinating problems in marine ecology.

Bryozoans are important in paleontology since their colonies are quite well represented in the fossil record.

Examples of some Caribbean bryozoans are figured, but the small size of the zooids (average 1 mm or less) and the similarity of many species make identification in the field or from photographs nearly impossible.

References: Osburn 1940, Hyman 1959, Schopf 1974, Ryland 1974.

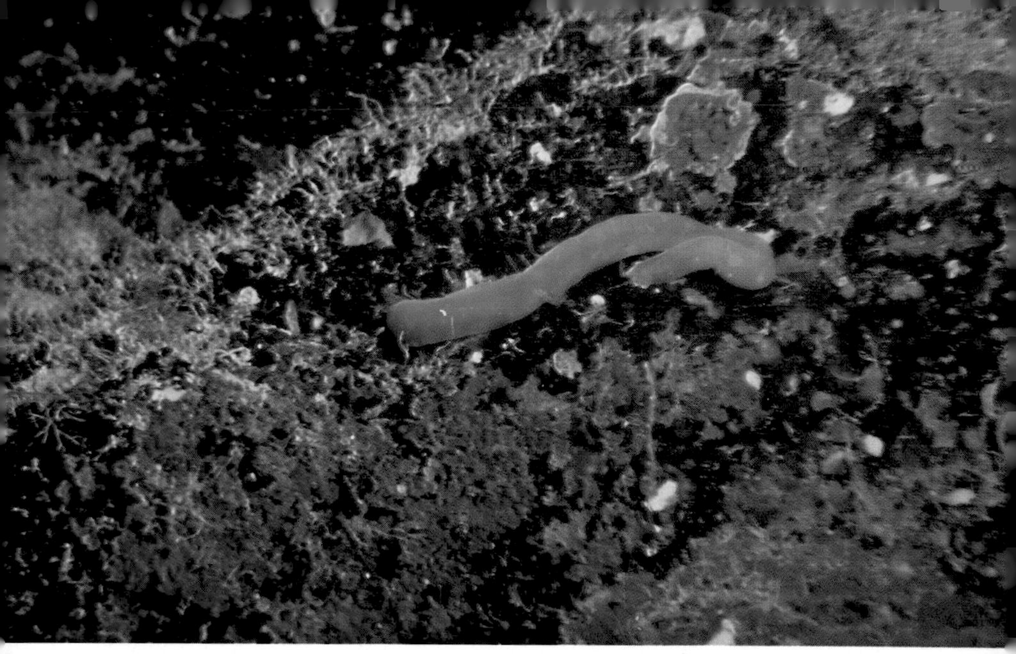

Ribbon worms of the phylum Nemertea are not often seen on reefs due to their secretive habitats, but they are there. This individual was photographed at night on an algae-rock surface. Bahamas, Exuma Islands, 15 m depth.

Unidentified bryozoan. This species forms calcerous plates reaching 20 cm across along vertical reef faces. In form they resemble the plates of *Agaricia* corals. Puerto Rico, Desecheo Island, 6 m depth.

Unidentified bryozoan. This species has digitate calcareous fingers which reach as much as 5-10 cm in height and could easily be confused with a coral on first sight. It is found along steep reef faces in moderately deep water. Puerto Rico, Monito Island, 20 m depth.

Unidentified bryozoan. The individual zooids of this species can be seen. It is growing on the undersurface of a ledge adjacent to some colorful, unidentified sponges. Jamaica, Rio Bueno, 18 m depth.

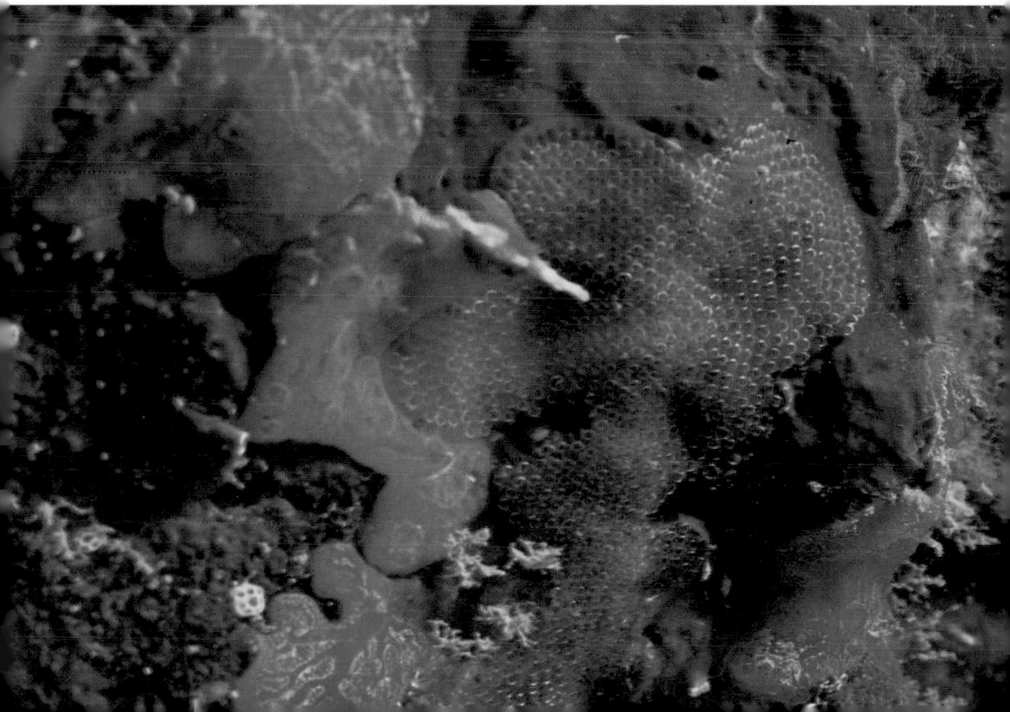

Reef scene with the coral *Montastrea cavernosa* and *Mycetophyllia aliciae* visible among others. The umbrella-like structures seen on top of the coral are the tentacles of the serpulid worm *Spirobranchus.* Photo at Cane Bay, St. Croix, Virgin Islands, 10 m depth.

Phylum Annelida

CLASS POLYCHAETA: POLYCHAETE WORMS

This class is a large group of segmented marine worms numbering over 10,000 species. They are easily divided into the sedentary tube dwellers (Subclass Sedentaria) and the free-moving species (Subclass Errantia). The Errantia include burrowing or crawling forms and some tube-dwelling species which leave the tube for various reasons. The Sedentaria remain in their tubes and rarely expose more than the head from the tube.

SUBCLASS SEDENTARIA

Family Sabellidae

Sabellastarte magnifica (Shaw)
Feather-duster worm

Like other sabellid polychaetes, *Sabellastarte magnifica* dwells in a soft non-calcareous tube. It has brown and white plumose tentacles or radioles projecting in two whorls from the tube. The crown of the expanded radioles can reach 10 cm in this species. These radioles can be quickly retracted into the tube if the worm is disturbed. They are used for feeding, a ciliary current being produced and particles removed from the water by pinnules on the radioles.

This species may be abundant on pilings and on reefs among corals where there is a fair amount of suspended material in the water. The darkness of the radioles varies among individuals. It is found in the Bahamas and the Caribbean at depths of 1 to at least 15 m.

Other Sabellidae on reefs may occur in groups of dozens of individuals. The crowns of radioles sway together with the current or wave surge. Usually when one is stimulated to contract suddenly, all the other individuals will follow.

References: Jones 1962, Marsden 1960.

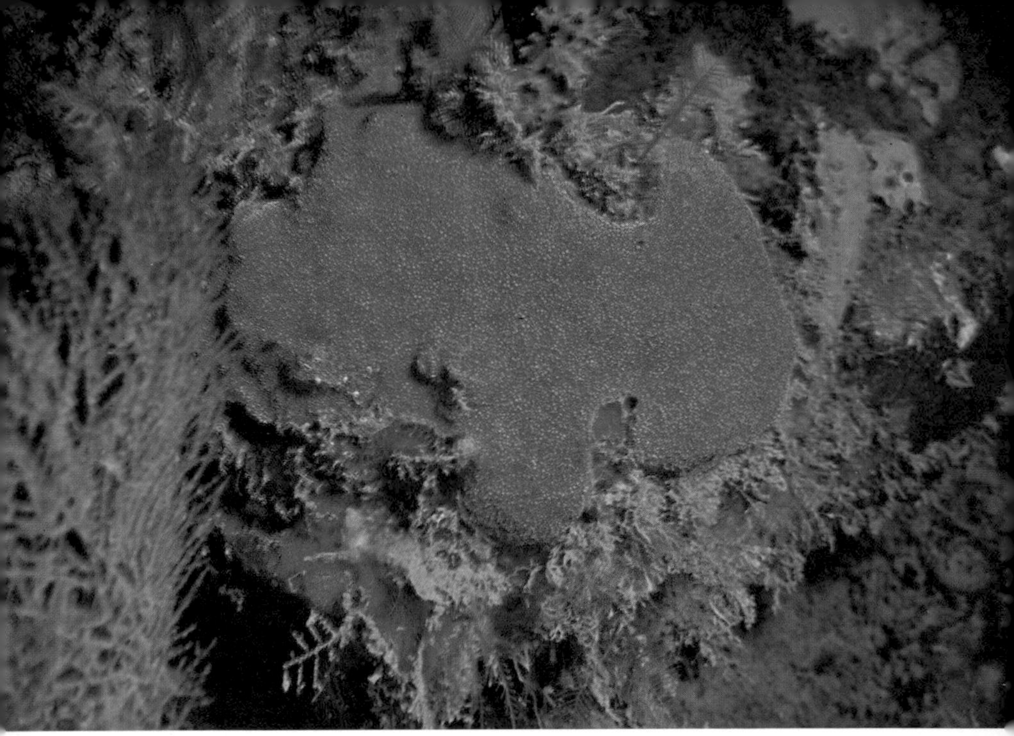

Unidentified bryozoan. The species is growing on an openly exposed surface and appears to be a calcareous encrustation without any vertical or horizontal projections. Puerto Rico, Rincon, 24 m depth.

Unidentified bryozoan. This species has a growth form seen in a number of species of bryozoans. It forms a fan of outward projecting fingers and is not heavily calcified as are some of the larger species. Such fans are found projecting from ledges, as the one shown, and on the undersurface of rocks and ledges. Bahamas, Eleuthera Island, 21 m depth.

Unidentified bryozoan. This small fan-like species is growing on the surface of another unidentified bryozoan. Jamaica, Rio Bueno, 24 m depth.

The feather-duster worm, *Sabellastarte magnifica,* can be found growing from coral heads such as this *Montastrea annularis.* The radioles contract quickly at the slightest touch. Puerto Rico, La Parguera, Mario Reef, 1 m depth.

Family Serpulidae

Spirobranchus giganteus (Pallas)
Christmas tree worm

Unlike the Sabellidae, the Serpulidae possess a calcareous tube. The radioles form two colored spirals in *Spirobranchus giganteus,* with a hardened operculum between them. The operculum bears two or three antler-like horns on its surface which are exposed when the worm is contracted, and the tube itself may bear a large spine at its opening.

Several whorls of radioles exist in each crown, and the color may vary from red or orange to white or a combination of these. The speed with which *Spirobranchus giganteus* can retract the radioles is amazing; often just approaching the location of a worm will cause it to retract. This serpulid can be found on coral heads, rocky substrates or even sponges at a wide range of depths. In the Atlantic S. *giganteus* is known from the Gulf of Mexico to Brazil. It also occurs on the tropical Pacific coast of America, and a subspecies is found from the central Pacific Ocean to the Red Sea.

Two other species of *Spirobranchus* are common on Atlantic reefs. Compared to S. *gigantea,* S. *tetraceros* (Schmarda) is smaller, more gregarious, has more horns on the operculum and lacks the spiraled radioles. *Spirobranchus polycerus* (Schmarda) is colonial and possesses spiral radioles.

Reference: Ten Hove 1970.

Pomatostegus stellatus (Abildgaard)

The radioles of *Pomatostegus stellatus* are usually red with yellow tips. Rather than being spiraled, they form a single U-shaped fan from their calcareous tubes. They occur on coral heads or rocky substrates. This or other similar species may also have the radioles orange, yellow or dark brown.

Reference: Treadwell 1939.

Filograna implexa Berkeley

This colonial serpulid forms masses of fragile, bone-white calcareous tubes. The basal portions of the mass are thickest, while the ends with the actively growing worms are elongate and narrow in diameter. The pale worms have radiating radioles bearing pinnules. The calcareous mass is quite scratchy and prickly to handle, and the finer portions of the skeleton may break under the weight of the basal mass.

This species occurs widely in the Caribbean and has been encountered in dark reef caves at 20 m depth.

Reference: Hartman 1959.

Family Terebellidae

Eupolymnia nebulosa (Montagu)

Only the long white tentacular filaments of this polychaete are easily visible, stretched out in all directions over the bottom from its buried tube. The tentacles may be over 25 cm in length, but when touched they are retracted rapidly into the thin mucous tube where the body occurs. *Eupolymnia nebulosa* evidently feeds on material adhering to the tentacles.

This polychaete typically occurs in back reef environments, often in areas of high sedimentation, from near the surface to at least 10 m depth. It occurs in southern Florida, the Bahamas and throughout the Caribbean.

SUBCLASS ERRANTIA

Family Amphinomidae

Hermodice carunculata (Pallas)

Fire worm

This flattened segmented worm, reaching 30 cm in length, has groups of white bristles along each side. The bristles are hollow, venom-filled setae which easily penetrate the flesh and break off if this worm is handled. They produce an intense irritation in the area of contact, hence the common name of the species. When disturbed the worm flares out the bristles so they are more exposed.

Hermodice carunculata can feed on living corals. It will engulf the last few centimeters of the tip of a branching coral, such as *Acropora cervicornis*, in its inflated pharynx

Sabellastarte magnifica can also be found among seagrasses such as the *Halophila baillonis* shown here. Puerto Rico, Aguadilla, 9 m depth.

The tubes of these unidentified fan worms are clearly seen in this photograph. The tubes are soft and not calcified. Jamaica, Discovery Bay, 15 m depth.

Fan worms, such as this unidentified species shown here, can form colorful clusters which sway with currents. Caicos Islands, Providenciales, 10 m depth.

The christmas tree worm, *Spirobranchus giganteus,* is very common growing on and about coral heads. The worm illustrated is on the edge of a colony of *Montastrea annularis.* Puerto Rico, La Parguera, Laurel Reef, 12 m depth.

and remove the coral tissue from that portion of the skeleton. The worm will remain 5-10 minutes at each branch tip, visiting several, and the branches attacked are apparent by their white ends.

The fire worm is abundant on reefs, beneath stones in rocky or seagrass areas and on some muddy bottoms. It has also been found at or near the surface in flotsam and occurs to at least 60 m. It is found throughout the tropical western Atlantic and at Ascension Island in mid-Atlantic.

References: Marsden 1960, 1962, Glynn 1962, Ebbs 1966, Lizama and Blanquet 1975.

Family Syllidae

Syllis spongicola Grube

This small white polychaete occurs by the tens of thousands as parasites of a number of sponges. It feeds by inserting the proboscis into individual cells of the sponge and may be so abundant in some specimens that the worms comprise 5% of the weight of the sponge. In some dark sponge species, such as *Neofibularia nolitangere*, the worms are easily visible on the inner walls of the lumen. Some sponge-dwelling fishes of the genus *Gobiosoma* (Gobiidae), which reach only about 5 cm in length, feed on *Syllis spongicola* almost exclusively.

The species is found circumtropically at depths where their host sponges occur.

Reference: Reiswig 1973c.

Phylum Arthropoda

CLASS CRUSTACEA: CRUSTACEANS

The crustaceans are a large, very diverse class which encompasses about 30,000 species. They are mostly aquatic, with the relatively few terrestrial members usually retaining ties (reproductive and others) to the aquatic environment. They have segmented bodies with a chitinous exoskeleton, compound eyes and respire by gills or the body surface. The sexes are usually separate and the young undergo some sort of larval development.

In those crustaceans occurring on coral reefs, a separate head, thorax and abdomen rarely occur. The head and thorax are almost always fused into a cephalothorax which is covered by a carapace. There are generally five pairs of head appendages; two are antennae and three are masticating and feeding devices. Chelae (claws) may be present on the latter. Walking legs and swimmerets (pleopods) may be variously developed.

ORDER MYSIDACEA: MYSID OR OPOSSUM SHRIMPS

Although small (generally 10 to 31 mm long), the mysid shrimps occur in such numbers in various marine and fresh water environments that they are quite important in the ecology of these areas. The mysids serve as food for a wide variety of animals and are an important component of the coral reef plankton.

The split thoracic legs (schizopods) of the female are modified to form a brood pouch where the eggs hatch and the young are retained for a time. Generally they are regarded as scavengers.

These specimens of *Spirobranchus giganteus* are growing on a sponge, probably *Neofibularia nolitangere.* There are also a few small unidentified serpulids on the sponge. Puerto Rico, Aguadilla, 5 m depth.

When contracted the tube and the operculum with its horn-like sculpturing of *Spirobranchus giganteus* are visible. The speed with which the worm can retract its radioles into the tube is amazing. Puerto Rico, La Parguera, Laurel Reef, 12 m depth.

Pomastegus stellatus is a serpulid worm which has the radioles in a U-shaped pattern rather than in a spiral. It can occur on rocks or coral heads. Jamaica, Discovery Bay, 15 m depth.

The colonial serpulid *Filograna implexa* has fragile white calcareous tubes which become larger in diameter towards their bases. This specimen was photographed in a dark reef cave and a small black coral is present within the mass of tubes. Little Cayman Island, Bloody Bay, 18 m depth.

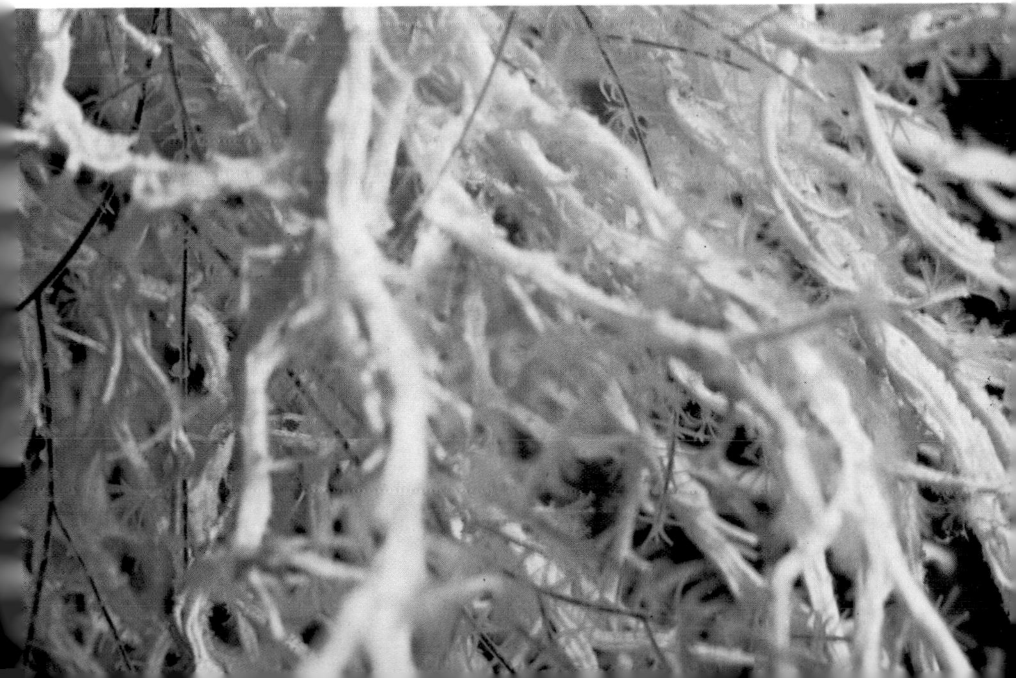

Family Mysidae

Mysidium spp.

The three Atlantic species of *Mysidium, M. gracile* (Dana), *M. integrum* Tattersall and *M. columbiae* (Zimmer), are important members of the coral reef plankton. They occur as shoals of evenly-sized specimens which can be easily mistaken for the young of fishes. The members of the shoal face the same direction and maintain position near overhangs on reefs. When disturbed they retreat into the area between the spines of the sea urchin *Diadema antillarum* or to the territories of pomacentrid fishes which drive off any large intruders.

The Atlantic species are quite similar and impossible to distinguish in the field. Their distributions are not well known, but encompass at least most of the Caribbean, the Bahamas and the southern Florida area. All three species may occur in a single locality such as the Florida Keys.

References: Brattegard 1969, 1970, 1973, Emery 1968, Randall *et al.* 1964, Steven 1961, Davis, C. 1966.

Heteromysis actinae Clark
Anemone mysid

Swarms of bright red-striped *Heteromysis actinae* occur only with the anemone *Bartholomea annulata*. A broad crimson red stripe runs along the dorsal surface of the body from the tail (it is divided on the paried uropods) to the eyes where it branches and continues out onto the antennae. The species is small, reaching less than 15 mm, generally with fewer than 100 individuals occurring with any one anemone.

The anemone mysids are immune to the stinging nematocysts of *B. annulata*, swimming up and down the tentacles and never straying beyond a few centimeters from the anemone. The anemones with *H. actinae* may also harbor *Alpheus armatus, Periclimenes pedersoni* and other crustaceans. *Heteromysis actinae* has been observed in aquaria to feed on wastes egested by the anemone.

The genus *Heteromysis* is found in inshore waters and is conspicuously absent from oceanic plankton. The similar species *H. bermudensis* occurs with *B. annulata* in Bermuda. Other members of the genus are known to occur with hermit crabs.

Heteromysis actinae is known from the Bahamas and the Caribbean at depths above about 15 m.

References: Clark 1955, Brattegard 1969.

ORDER ISOPODA

Family Cymothoidae

Anilocra spp.

The isopods of the genus *Anilocra* occur as parasites on a number of species of reef fishes but are sizable enough that they are often observed while diving. The isopod is located on the outer surface of the fish, plainly visible, and may be on the top of the head between the eyes (interocular) or slightly below the eye (subocular). These crustaceans are protandrous hermaphrodites, initially males but turning into females. Consequently, the large individuals observed are invariably females although small males may often be found on the same fish.

It is not known how detrimental these isopods are to the host fish. The tissue and bone adjacent to the area of attachment are often deformed, but the parasites do not seem to interfere greatly with the host's ability to feed.

Interestingly, the species of fish parasitized vary considerably between islands. Why a fish is parasitized in one area while in another the same isopod occurs only on other species of fish, is not understood. This problem certainly deserves further attention.

Reference: Hochberg and Ellis 1972.

ORDER STOMATOPODA: MANTIS SHRIMP

The mantis shrimp that occur on reefs are usually burrow dwellers, with the opening of these burrows occurring on sand bottoms. They may occasionally be seen moving across the bottom but usually retreat into their burrow or under rocks if disturbed. Mantis shrimps vary from several centimeters to more than 30 cm in length and are somewhat

The polychaete *Eupolymnia nebulosa* has white, spaghetti-like tentacles which spread out over the bottom. At the slightest touch the tentacles are withdrawn into the buried tube where the major portion of the worm is located Puerto Rico, La Parguera, Laurel Reef, 6 m depth.

The fire worm, *Hermodice carunculata,* is an important predator on benthic coelenterates on the reef. They attack corals and are capable of preying on sea anemones, black corals, gorgonians and possibly other groups of coelenterates. Puerto Rico, La Parguera, Laurel Reef, 12 m depth.

Individuals of the parasitic polychaete *Syllis spongicola* are seen here distributed on the inner surface of the oscular cavity of the sponge *Neofibularia nolitangere.* Jamaica, Discovery Bay, 12 m depth.

Most mysid shrimps, such as this unidentified species of *Mysidium,* occur as dense schools which stay close to some sort of shelter to protect them from predators. This school, with each individual mysid barely visible, has taken shelter within the long spines of *Diadema antillarum*, a sea urchin. Puerto Rico, La Parguera, Mario Reef, 6 m depth.

dorsoventrally flattened. They possess a pair of raptorial claws in which the end segments fold upward and back into grooves on the next to last (penultimate) segment. These claws are used in capturing a variety of prey animals and can deliver a deep cut in the skin of humans handling these animals.

Over sixty species of stomatopods occur in the western Atlantic, and a number of these may be found on reefs. A typical mantis shrimp is figured, but an identification to species usually requires having a specimen in hand.

References: Manning 1968, 1969.

ORDER DECAPODA: SHRIMP, LOBSTERS, CRABS

The order Decapoda comprises the largest order of crustaceans, with over 8,000 species. Its members are mostly marine and have the head and thorax fused into a cephalothorax (carapace). They are divided into two suborders, the Natantia (the shrimp and prawns) and the Reptantia (which includes the lobsters and crabs). The Natantia are modified for swimming using the pleopods, appendages of the well-developed abdomen, and usually have a serrated rostrum on the carapace. The Reptantia are modified for crawling or swimming, generally using the legs and not the pleopods, and the abdomen may be reduced.

Within the Reptantia three subgroups (sections) are recognized: the Macrura (lobsters), which have a well-developed abdomen with a tail fan; the Anomura (hermit crabs and others), which have the abdomen bent ventrally without lateral plates; and the Brachyura (crabs), which have the abdomen reduced, without a tail fan and bent under the thorax.

SUBORDER NATANTIA

Family Palaemonidae

Periclimenes pedersoni Chace
Pederson's cleaning shrimp

This colorful cleaning shrimp, with its violet and white markings which are distinctive in its range, is almost exclusively associated with one of a variety of sea anemones. While the anemone *Bartholomea annulata* is most commonly inhabited, others include *Condylactis gigantea, Heteractis lucida* and, in deeper water, *Lebrunia danae*. The shrimp, usually occurring singly or in pairs, have numbered as many as 26 associated with a single anemone. They are immune to the stinging nematocysts of the anemone and can perch directly on the tentacles without ill effect. This immunity is an acquired phenomenon, and if the shrimp is deprived of its anemone for a period of days the immunity is lost. The shrimp must undergo an acclimation process of progressively touching portions of its appendages and body to the anemone for a period of hours before it is not stung by the nematocysts.

Periclimenes pedersoni is known widely in the West Indies, the Bahamas and southern Florida. It does not occur in Bermuda, but its similar appearing relative *P. anthophilus* Holthuis and Eibl-Eibesfeldt occurs there. It has been found at depths between 1 and 52 m.

The shrimp station themselves on or near their anemone and sway the body while lashing their antennae to attract fishes for cleaning. They swim onto the body of fish which pose for cleaning and can enter the mouth and gills with impunity. They will even clean a diver's hand presented to them fully extended and motionless. The species does not have complete immunity to predation though. Wrasses and other omnipresent small carnivorous fish have been observed to eat the species if the shrimp strays too far from its host anemone.

The shrimp evidently remain with one or a group of anemones for some time. If cleaning stations of other organisms such as the neon gobies are nearby, the shrimp will clean the same fish at the same time as the cleaning goby. Although perhaps not strictly territorial, a hierachical system is present within the shrimp on a single anemone. Egg-bearing females have been observed in the Virgin Islands from February through August.

References: Limbaugh *et al.* 1961, Mahnken 1972, Chace 1972, Holthuis and Eibl-Eibesfeldt 1964, Sargent 1975.

The anemone mysid, *Heteromysis actinae*, occurs only around the tentacles of the anemone *Bartholomea annulata*. Many of the red striped *H. actinae* are visible in the photograph. The red and white banded antennae of the alpheid shrimp *Alpheus armatus* are also visible protruding from the anemone. Bahamas, Eleuthera Island, 10 m depth.

A black bar soldierfish, *Myripristis jacobus,* is shown with a large *Anilocra* sp. isopod on the top of the head between the eyes. The isopod is a female and small males could probably be found elsewhere on the fish. Puerto Rico, Mona Island, 15 m depth.

This unidentified species of mantis shrimp was on a back reef area. On occasion mantis shrimp can be observed with only the forward portion of their body protruding from a burrow in sandy bottoms. When approached they back down out of sight into the burrow which invariably is smooth and not lined with stones. Puerto Rico, La Parguera, Laurel Reef, 1 m depth.

The lower fish, a brown *Chromis,* has a large subocular Isopod of the genus *Anilocra*. Usually, among the affected species of fishes in an area, only a small percentage of the individuals will actually have the isopods. Puerto Rico, Aguadilla, 10 m depth.

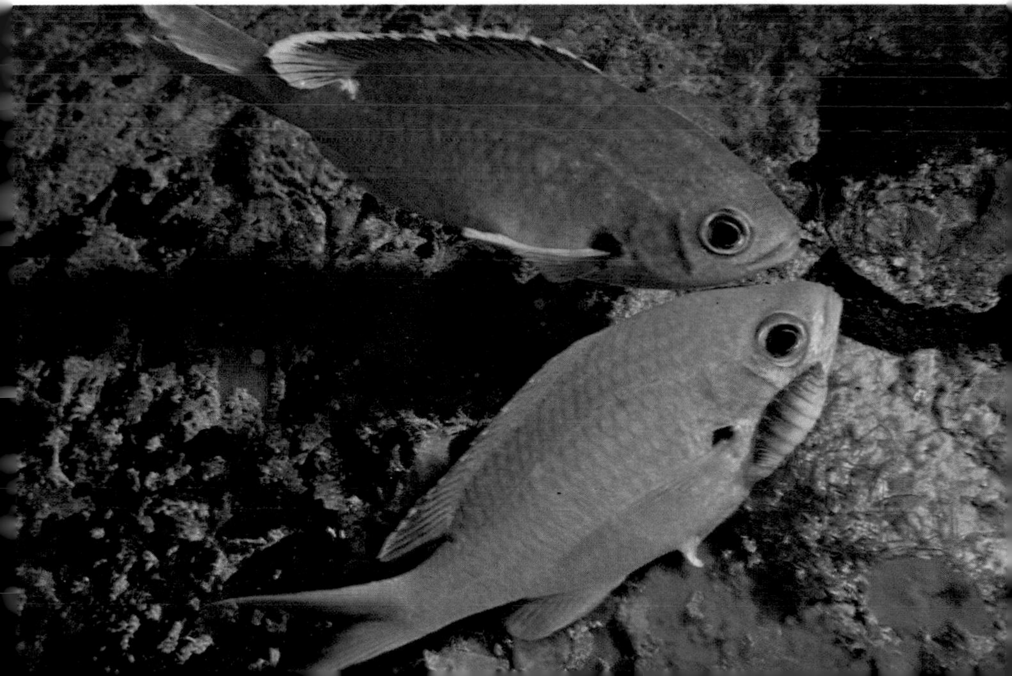

Periclimenes yucatanicus (Ives)

This species, like *Periclimenes pedersoni*, usually occurs associated with sea anemones. As many as six individuals have been observed with *Bartholomea annulata*, *Condylactis gigantea* and occasionally *Lebrunia danae*. In one instance it was encountered on the tentacles of the medusa *Cassiopea* in the Virgin Islands.

The tan and white saddle markings of *P. yucatanicus* are distinctive. While it may possibly be a cleaner, definite evidence is lacking. The swaying of the body and lashing of the antennae are similar to *P. pedersoni*, but *P. yucatanicus* does not leave the substrate if a fish approaches and poses for cleaning. This "attraction" behavior would be inconsistent with a non-cleaning habit.

The species is known from southern Florida to Colombia at depths of up to 24 m. Females carrying eggs under the abdomen have been observed in at least July and August.

References: Limbaugh *et al.* 1961, Mahnken 1972, Chace 1972.

Periclimenes sp.

An unidentified, possibly undescribed, species of *Periclimenes* occurs with the anemone *Stoichactis helianthus*, living on the oral surface of the anemone among the tentacles. The chelae are banded red-brown and white; the abdomen, thorax and legs have red-brown spots and white lines and spots. The antennae are banded also. Small individuals are nearly transparent with only a few markings, and larger shrimp are much more conspicuous on the anemone. It occurs in the West Indies in shallow back reef areas with *S. helianthus*.

Family Rhynchocinetidae

Rhynchocinetes rigens Gordon
Red coral shrimp

This shrimp was believed to be rare until the advent of night diving. It was subsequently found to be quite common, emerging from the deep crevices and caves of the reef after

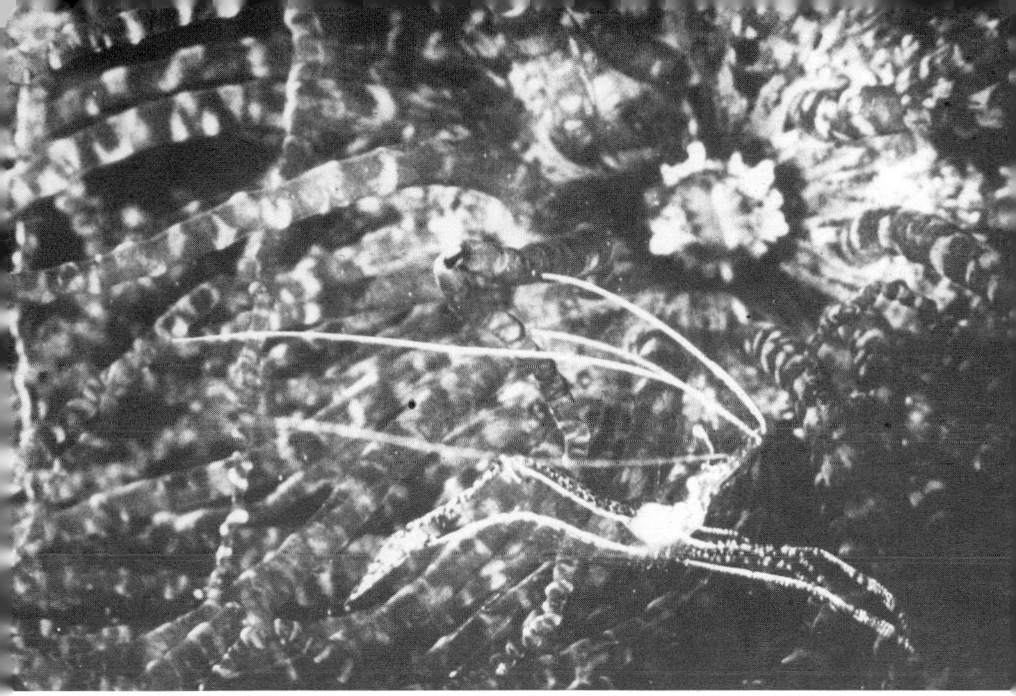

The cleaning shrimp *Periclimenes pedersoni* sitting on the tentacles of the anemone *Bartholomea annulata*.

Close-up view of the corallimorpharian *Ricordea florida*. The tiny mouth is visible at the center of the radiating rows of small tentacles. A small shrimp of the genus *Periclimenes* is hiding behind the upper edge of the disk with only its white antennae visible. Photographed at Discovery Bay, Jamaica, depth 15 m.

Periclimenes pedersoni is usually found associated with sea anemones where it sets up cleaning stations for fishes. This individual is stationed in front of the anemone *Bartholomea annulata.* Jamaica, Discovery Bay, 2 m depth.

Periciimenes yucatanicus is also associated with anemones, *Bartholomea annulata* in this case, and may engage in cleaning behavior. Jamaica, Discovery Bay, 4 m depth.

This example of *Periclimenes pedersoni* has fertilized eggs attached to her abdomen. The white antennae are lashed back and forth by the shrimp in order to attract fishes to its cleaning station. Jamaica, Rio Bueno, 5 m depth.

The shrimp *Stenopus hispidus* resting on the substrate. This individual has lost one of its large chelae but will be able to regenerate another given sufficient time.

dark. *Rhynchocinetes rigens* is readily observable at night, the eyes glowing red when illuminated by a diving light, but is still quite shy, springing away if approached too closely.

The red color of *R. rigens* is distinctive among Caribbean reef shrimps; the males (and some females) have a large red spot laterally on the front portion of the abdomen.

The species has a wide geographic range, being known widely in the West Indies, from Bermuda, the Florida Keys, the Bahamas, Madeira and the Azores. It occurs on outer reefs at depths to at least 30 m. There is some evidence that it feeds on small algae and polychaetes and may ingest sand in search of the organic material it contains.

Reference: Manning 1961.

Family Stenopodidae

Stenopus hispidus (Oliver)
Red-banded coral shrimp

A cluster of gently curved white "whiskers" a few inches in length, sprouting from underneath a coral ledge, usually indicates the presence of one or more *Stenopus hispidus*. Long a favorite with aquarists, the red-banded coral shrimp, with its white antennae and its body and legs boldly banded red and white, is the largest of the cleaning shrimp.

It is often encountered in pairs, and individuals have been known to remain in the same area for lengthy periods of time. The females, when in season, are easily detected by the greenish mass of eggs underneath the abdomen which contrasts sharply with the white and red of the body.

The oversized claws are evidently lost naturally on occasion and tend to break off when a specimen is collected for the aquarium. The cleaning behavior of this species is not well known and it may be most active at night.

Stenopus hispidus is found in a variety of reef habitats, including under coral ledges, rocky ledges and crevices or any other area giving shelter.

It is known from 1 to 210 m and from the Atlantic, Pacific and Indian Oceans plus the Red Sea. In the Atlantic it is widely distributed from southern Florida to the Guianas and Bermuda.

This unidentified species of *Periclimenes* occurs on the anemone *Stoichactis helianthus.* Several individuals can occur on a single anemone. Puerto Rico, Aguadilla, 12 m depth.

The red coral shrimp *Rhynchocinetes rigens* comes out only at night, but even at that time it is still shy and easily disturbed. Jamaica, Discovery Bay, 15 m depth.

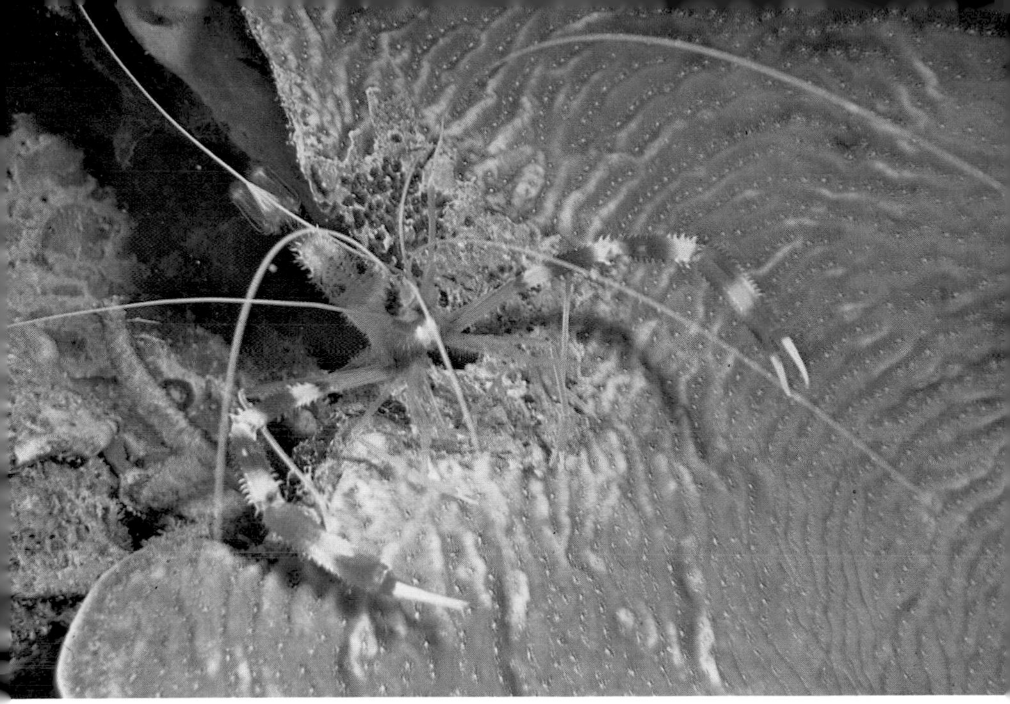

The red-banded coral shrimp, *Stenopus hispidus*, cleans parasites from fishes by using its long chelae and it is often found in pairs. Jamaica, Discovery Bay, 15 m depth, night.

The shrimp *Stenopus scutellatus* does not seem as common as *S. hispidus* and is generally smaller in size. It is also believed to engage in cleaning behavior. Puerto Rico, Aguadilla, 15 m depth.

References: Limbaugh *et al.* 1961, Chace 1972.

Stenopus scutellatus Rankin

While basically similar in appearance to *Stenopus hispidus, S. scutellatus* can be easily distinguished by its yellow and orange-red colors where *S. hispidus* is white and red. It is a smaller species and seems more shy about approaching its host for cleaning. It often simply reaches out from its concealing hole to pick at parasites on the sides of the fish.

It occurs in smaller, darker caves than *S. hispidus* and is not as commonly observed. Still, it can be found in a variety of habitats, such as coral boulders on an algal plain, on solid substrate near *Thalassia* beds and on submerged debris.

Stenopus scutellatus is known to depths of 113 m and can occur as shallow as 10 m. It is distributed in the western Atlantic from Bermuda and the Gulf of Mexico through the West Indies to the islands off Brazil.

References: Limbaugh *et al.* 1961, Chace 1972.

Family Hippolytidae

Lysmata grabhami (Gordon)
Red-backed cleaning shrimp

Perhaps the boldest of the cleaning shrimp is *Lysmata grabhami*, which does not hesitate to attempt to clean fish although divers may be only inches away. With its longitudinal broad red band with a white stripe down the center, it would be difficult to confuse this species with any other shrimp. Its antennae, which are longer than the body, are white, like those of most other cleaning shrimp, and the chelae also have a white stripe on their dorsal surface.

This shrimp normally occurs in shaded crevices where only the white antennae, claws and dorsal stripe of the shrimp, swaying back and forth, may be visible. It may leave its shelter to clean a host but does not usually climb onto the body of the fish. It has been observed to climb onto moray eels, however.

Pairs or groups of a few individuals are often encountered, and in one instance occurrence with the anemone *Stoi-*

chactis helianthus has been reported. It has often been observed occurring with other cleaning shrimp (*Stenopus hispidus* and *S. scutellatus*) and occurs from shallow depths to at least 55 m. It is known widely in the tropical Atlantic (Gulf of Mexico, Florida, Bahamas, West Indies, Madeira) and the Pacific (Hawaii, Society Islands, Fiji Islands).

References: Limbaugh *et al.* 1961, Chace 1972, Bruce 1974.

Lysmata wurdemanni (Gibbes)

This shrimp has been observed several times within the lumens of tubular sponges on Caribbean reefs. More than one may occur per sponge tube, and other inquilines such as sponge-dwelling fishes may share the same tube. Little is known of the biology of this shrimp. It occurs from Virginia to Brazil to depths of 30 m.

Reference: Chace 1972.

Thor amboinensis (DeMan)

This small shrimp is always associated with sea anemones and can occur on the same anemone as other shrimp such as *Periclimenes pedersoni*, *P. yucatanicus* and *Alpheus armatus* and the mysid *Heteromysis actinae*. *Thor amboinensis* seems particularly common with the anemone *Stoichactis helianthus* in shallow back reef areas, and at least one shrimp occurs with each anemone.

This shrimp is distinctive, with the rostral area and particularly the tail pointing upward, white saddles on a brown ground color and short banded antennae. It may occur directly on the tentacles of, beneath the expanded disk of or on rock surfaces adjacent to the anemone.

A second species, *T. manningi*, has also been reported to occur associated with the anemone *Bartholomea annulata* but is definitely not restricted to anemones. *Thor amboinensis* is known from the tropical Atlantic, Pacific and Indian Oceans; in the Atlantic it occurs from the Florida Keys to Tobago and the western Caribbean in shallow water.

Reference: Chace 1972.

Lysmata grabhami is another cleaning shrimp, but sways back and forth to attract fishes to be cleaned. The antennae and the stripe on the back are brilliant white. Puerto Rico, Aguadilla, 12 m depth.

The shrimp *Lysmata wurdemanni* is usually found within the lumen of tubular sponges such as the species of *Verongia* pictured here. It is shy and does not come out onto the outer surface of the sponge. Puerto Rico, Mona Island, 30 m depth, night.

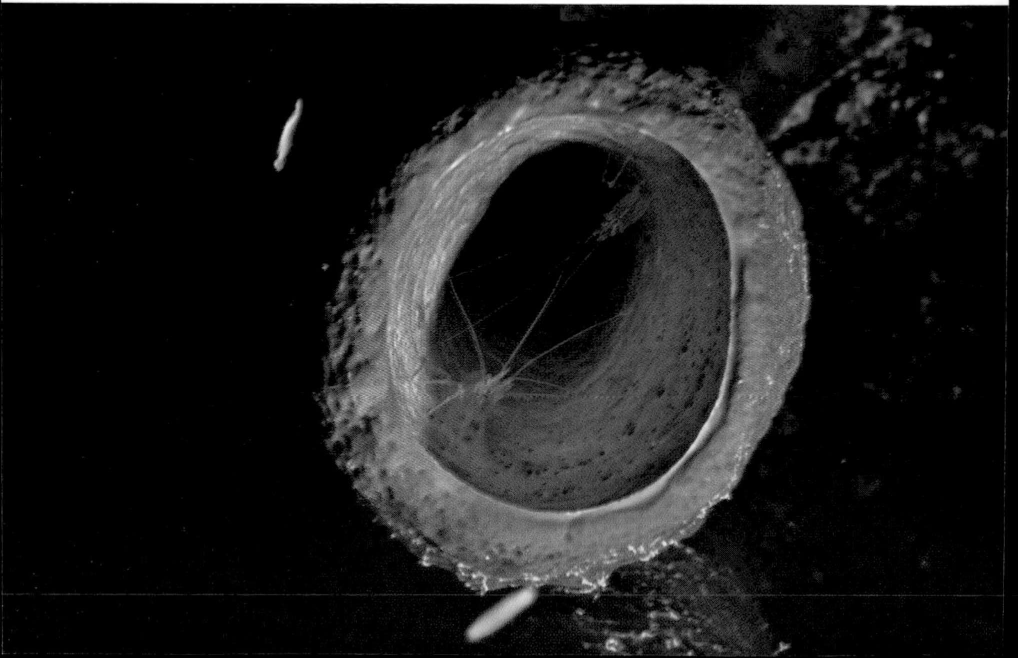

The tiny shrimp *Thor amboinensis* occurs on and around sea anemones, often with other commensal shrimps. Often there are considerable numbers of *T. amboinensis* where they do occur and they spring rapidly from place to place when disturbed so that their movements are difficult to follow. Jamaica, Discovery Bay, 3 m depth.

Alpheus armatus, snapping shrimp, occurs with the anemone *Bartholomea annulata* most often. Usually a pair of shrimp are present and they defend the anemone. Puerto Rico, La Parguera, Laurel Reef, 12 m depth.

Family Alpheidae

Alpheus armatus Rathbun

This alpheid shrimp associates with sea anemones, particularly *Bartholomea annulata*, and is usually first observed by its red-brown and white banded antennae protruding out among the anemone tentacles. It hides adjacent to the column of the anemone and may excavate a chamber next to the column. The other species of shrimp and mysids associated with anemones may occur with it. Often male and female pairs are found with a single anemone; this relationship seems to endure for some time.

Like all alpheids, *Alpheus armatus* possesses an enlarged chela used for defensive purposes; the chela, like the body, is red-brown speckled with white and on occasion has golden highlights. A mutualistic relationship may exist with the anemone, the shrimp defending the anemone against predators and the anemone providing a secure refuge for the shrimp from its enemies. *Bartholomea annulata* from which all individuals of *A. armatus* have been removed often disappear shortly thereafter, their exact fate unknown, but perhaps eliminated by predators undeterred by the alpheid shrimps.

Alpheus armatus occurs throughout the West Indies, the Bahamas and southern Florida in shallow water.

References: Chace 1972, Limbaugh *et al.* 1961.

SUBORDER REPTANTIA
SECTION MACRURA

Family Palinuridae

Panulirus argus (Latreille)
Spiny lobster

The spiny lobster is a common member of the Caribbean reef community and forms the basis for one of the most important fisheries in the area. The carapace bears a number of strong sharp spines, including the large "horns" above the eyes, and spines are also found on the lengthy antennae and at the lower margins of the plates covering the tail. Spiny lobsters shelter under crevices, caves, ledges or any debris available during the day and scavenge about the bottom at night. Often during the day only the antennae are visible protruding from the hiding place.

In Florida *Panulirus argus* spawns from March to June with April being the peak month. Females carry a spermatophore, appearing as a gray patch of material, on the ventral surface of the carapace between the walking legs. This spermatophore was deposited there by the male and is used for externally fertilizing the eggs when ready; the eggs are then attached to the underside of the tail and are easily visible as an orange to blackish mass. They hatch in three weeks or less and become phyllosoma larvae. These larvae have a life in the plankton of about six months, metamorphosing and settling into a juvenile spiny lobster. At this stage it still takes several years for the young *P. argus* to reach commercial size.

Migrations of large numbers of *P. argus*, marching in lines or small groups head to tail, have been observed on numerous occasions. The purpose for these "marches" is not known but may be involved with spawning. In the Bahamas the marches occur in the fall, usually a few days after a storm, and the spiny lobsters migrate along the edge of the bank and then at a certain point go into deep water.

Panulirus argus is known from North Carolina to Brazil and is common throughout the Caribbean. It seems outstandingly abundant in certain areas such as the Turks and Caicos Islands and the Bahamas. It is known to depths of 90 m.

References: Williams 1965, Opresko *et al.* 1973, Herrnkind 1969, 1970, Herrnkind and Cummings 1964, Herrnkind and McLean 1971, Berrill 1975.

Panulirus guttatus (Latreille)
Spotted spiny lobster

While similar to *Panulirus argus* in shape, the coloration of this smaller species immediately sets it apart. The body, tail and legs are dark with white to cream spots. Some red spines and spots occur on the body, and the last digits of the legs have brown stripes.

Besides being smaller than *P. argus*, *P. guttatus* is much more rarely observed while diving and is only occasionally

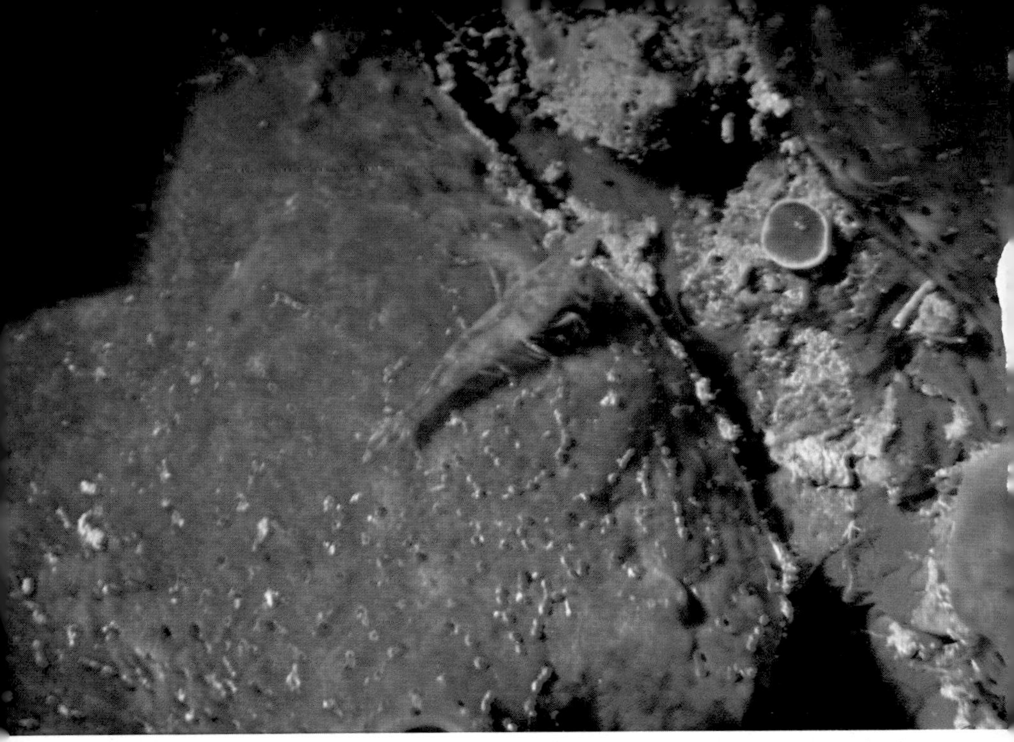

A few species of penaeid shrimps, the family of great commerical importance, can be found on the reef on occasion. This specimen was photographed at night on the surface of a sponge possessing zoanthids. Puerto Rico, Mona Island, 30 m depth.

The spiny lobster, *Panulirus argus,* emerges into the open more at night than during the day. Their long antennae are directed at any object, such as an approaching diver, which might pose a threat to them. Florida Keys, Alligator Reef, 10 m depth, night.

Panulirus argus has "horns" above the eyes which are quite strong and sharp. Much of the body and antennae have many small spines which give this lobster its common name.

Panulirus guttatus, the spotted spiny lobster, is much more attractively colored than *P. argus* and is also a more secretive species. It is most easily seen at night when it emerges from the reef caves and crevices it occupies during the day. Jamaica, Discovery Bay, 18 m depth.

taken in commercial catches. Surprisingly, sizable populations can occur around rocky jetties in southern Florida, and the reason why *P. guttatus* is more common in this habitat than *P. argus* is not known.

The spotted spiny lobster occurs in Bermuda, Florida, the Bahamas, West Indies and Brazil; it has also been taken at St. Pauls Rocks in the tropical mid-Atlantic. Its depth range is from 2 to at least 24 m. This is a secretive species occurring deep in reef caves, often on the roof, during the day and remaining near the entrance of such caves at night. It does not range as far from cover at night as does *P. argus* and is not known to engage in group migrations like *P. argus*. In Florida maximum spawning occurs in June, but in more southerly waters some spawning may occur year-round.

References: Opresko *et al.* 1973, Calliouet *et al.* 1971, Munro 1974.

Justitia longimanus (Milne-Edwards)
Long-armed spiny lobster

This strange lobster is occasionally observed on deeper reefs. The first pair of legs have wide dark red crossbands, are large and end in down-curved claw-like tips. The tail is brick red in color with yellow spots and stripes. *Justitia longimanus* is smaller than the members of the genus *Panulirus*.

The species emerges to the openings of reef caves at night but usually remains well hidden during the day. It occurs as deep as 300 m but has been observed as shallow as 15 m. It is most often seen at depths near 30 m by divers.

It is known from the Atlantic (Bermuda, southern Florida, the Bahamas and the West Indies), the Pacific (Hawaii) and the Indian Oceans (Mauritius and Reunion).

References: Opresko *et al.* 1973, Robertson 1969b, Baisre 1969.

Palinurellus gundlachi Von Martens
Copper lobster

The copper lobster, the most primitive Atlantic member of the spiny lobster family, is unusual in appearance, with a squat body and small chelae. The antennae, legs and chelae are hairy and the lobster lacks the large, numerous spines of the genus *Panulirus*. It is reddish brown with the ventral surface being more yellowish.

Palinurellus gundlachi is quite shy, retreating rapidly when observed on the reef, and is found only in caves at moderate to deep depths. It is uncommon compared to the species of *Panulirus;* even its phyllosoma larva is much less common in the plankton than that of the other lobsters. It is rarely taken in commercial fishing operations. The copper lobster has been observed at night feeding on a portion of the tail of *Panulirus guttatus*. It is known from several locations in the West Indies.

References: Sims 1966, Holthuis 1946, 1965.

Family Scyllaridae

Scyllarides aequinoctialis (Lund)
Spanish lobster

The Spanish lobster appears quite unusual, lacking elongate antennae, claws and spiny projections. It has a broadly flattened carapace, generally reddish brown or orange-brown in color, with a coarse but not spiny texture. The species hides in caves during the day and forages on the open reef at night.

Some other members of the family Scyllaridae might be encountered on reefs, particularly at night. The slipper lobster, *Parribacus antarcticus* (Lund), and the ridged slipper lobster, *Scyllarides nodifer* (Stimpson), are among these species. *Scyllarides aequinoctialis* is taken occasionally in commercial trapping and is sold along with more common spiny lobsters. Its edibility is good and it is sometimes eaten by large reef fish such as the jewfish, *Epinephelus itajara*.

The species probably spawns in the spring. Fertilization may be internal since external spermatophores are not common in this species. The phyllosoma larva is known and has a larval life of about eight or nine months in duration.

Scyllarides aequinoctialis is known from Bermuda and southern Florida through the Caribbean to Brazil at depths of 10 to 180 m.

References: Lyons 1970, Opresko *et al.* 1973, Robertson 1969a.

Justitia longimanus is a strange little lobster found in some deeper reef areas. The pictured individual does not have the downcurved chelae, but otherwise is similar in coloration to typical *J. longimanus.* Jamaica, Discovery Bay, 20 m depth.

The copper lobster, *Palinurellus gundlachi*, is seldom seen but is quite distinctive in appearance. The illustrated individual was in a deep-reef cave and had a piece of the carapace of *Panulirus guttatus* it was apparently feeding on in its grasp. Jamaica, Spring Gardens, 27 m depth, night.

The spanish lobster, *Scyllarides aequinoctialis,* is one of the "slipper" lobsters which lack elongate antennae and hardly resemble other lobsters. This individual was photographed among plume-like gorgonians of the genus *Pseudopterogorgia* and an unidentified black sponge or tunicate. Florida, Alligator Reef, 10 m depth, night.

Petrochirus diogenes is the largest of Caribbean hermit crabs and often utilizes the shell of the queen conch, *Strombus gigas,* as its shelter. Bahamas, Eleuthera Island, 12 m depth, night.

SECTION ANOMURA

Family Diogeniidae

Petrochirus diogenes (Linnaeus)
Conch hermit crab

This species is the largest hermit crab in the West Indies, reaching a size that allows it to utilize nearly any size shell of the queen conch, *Strombus gigas*. Its use of *S. gigas* shells may not simply be a matter of chance. It has been observed attacking live queen conchs and feeding on recently dead individuals also. The entire process (if it truly exists) of attacking, killing and eating the conch plus using its shell has not yet been observed.

Petrochirus diogenes has distinctive red and white banded antennae and generally reddish claws with the right slightly larger than the left. The eyes are green or blue-green and areas of the body are often covered with algae.

This hermit crab occurs from North Carolina to Brazil at depths to 35 m. Two other species occur in the genus *Petrochirus*, one on the tropical western coast of America and the other in the eastern Atlantic. Several individuals of the small crab *Porcellana sayana* may occur with *P. diogenes* and are found on the hermit crab and on the inner and outer surfaces of the shell.

References: Provenzano 1959, 1968, Randall 1964.

Dardanus venosus (Milne-Edwards)

This medium-sized hermit crab has the chelae (right larger than the left) with many purple tuberculations on the yellowish surface and black tips to the fingers. Its eyes are greenish blue with a central black spot. A very similar species, *D. fucosus* Biffar and Provenzano, occurring with *D. venosus* over much of its range has bluish or greenish eyes with a broad black bar horizontally when the crab is viewed from the front.

Dardanus venosus occurs in a wide variety of habitats. Often the sea anemone *Calliactis tricolor* occurs on the shells inhabited by *D. venosus*, the crab actually placing the anemone on its shell. In depth *D. venosus* ranges from 1 to 138 m and occurs from Bermuda and southern Florida to Brazil. The closely related *D. fucosus* has a similar depth range and occurs from North Carolina to the Amazon River.

References: Biffar and Provenzano 1972, Cutress and Ross 1969.

Calcinus tibicen (Herbst)

This small hermit crab has the carapace dark brown-red with tiny white spots. The antennae and eyestalks are a pale orange, with the area beneath the dark eyes white in color. The left cheliped (claw) is larger than the right.

Calcinus tibicen occurs on rocky and reef substrates from 0 to 35 m. It may occur in small aggregations during the day and range out as individuals at night for short distances to feed. It is known from Bermuda and southern Florida to Brazil.

References: Provenzano 1959, Hazlett 1966.

Paguristes cadenati Forest
Red hermit crab

The bright red carapace, legs and chelipeds of this reef-dwelling species make it unmistakable. The eyestalks are pale and the eyes greenish. It is a small species with the chelipeds nearly equal in size. There are no setae (bristles or hairs) on the body or appendages.

This species has been observed climbing on corals at night, the time it is most active. It is restricted to reefs at 5 to 30 m and is known from southern Florida, the Bahamas and the West Indies.

References: Hazlett 1966, Provenzano 1960.

Family Paguridae

Pagurus spp.

There are a number of small species of the genus *Pagurus* which may occur on Caribbean reefs. They are difficult to positively identify, particularly in the field. Generally they are pale in color and the right cheliped is much larger than the left.

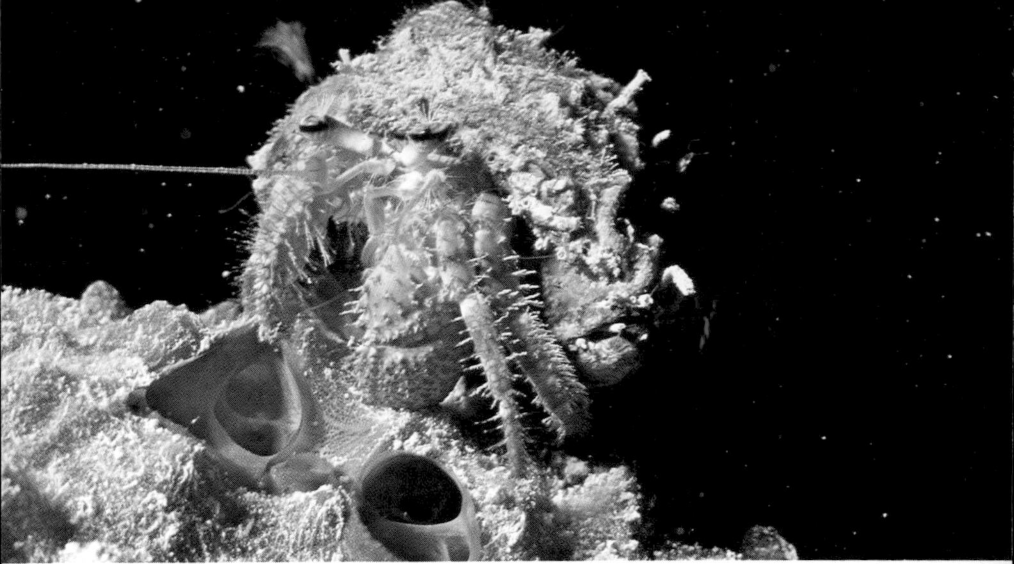

Dardanus venosus is a typical hermit crab utilizing the discarded shell of a dead gastropod mollusc as a shelter which it carries with it. The abdomen of the hermit crab is extensively modified, as compared to other crustaceans, to hold the shell. Puerto Rico, La Parguera, Laurel Reef, 10 m depth.

Calcinus tibicen is a small species of hermit crab which may occur in small aggregations during the day. Jamaica, Discovery Bay, 3 m depth.

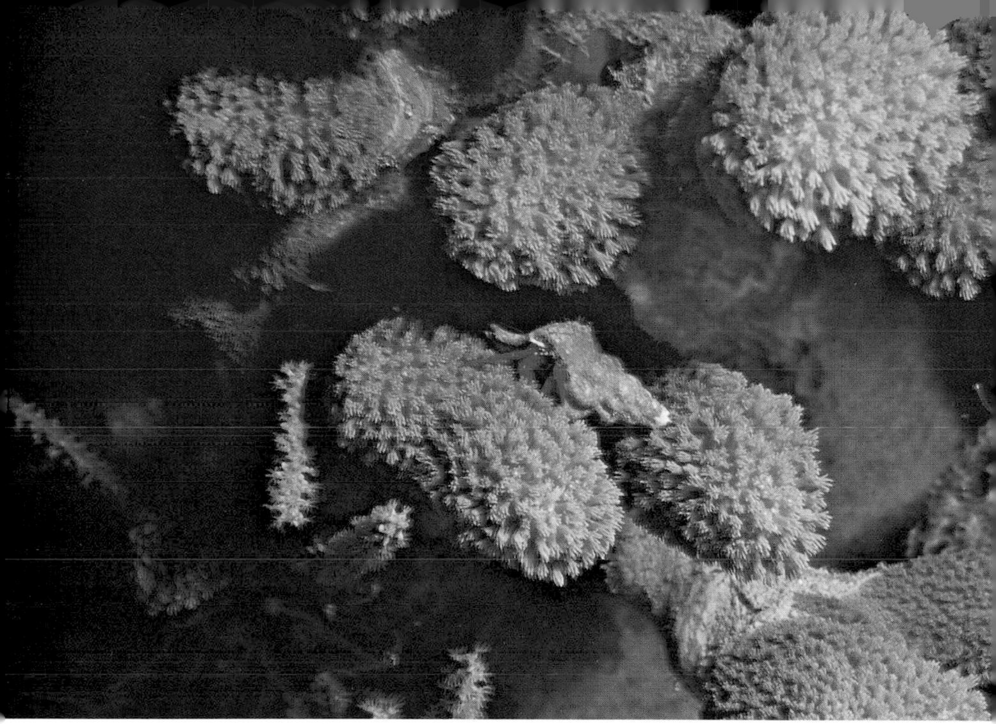

Paguristes cadenati is perhaps the most distinctive Caribbean hermit crab. With its brilliant red legs it would be difficult to confuse with any other species. Jamaica, Discovery Bay, 15 m depth, night.

The small species of the genus *Pagurus,* such as illustrated here, are nearly impossible to distinguish in the field but can be quite common in certain areas. Puerto Rico, La Parguera, Laurel Reef, 3 m depth.

The shell fighting (competition between an aggressor who wants the shell and the crab attacked) and reproductive behavior of various species are known.

References: Provenzano 1959, Hazlett 1966, 1972, McLaughlin 1975.

SECTION BRACHYURA

Family Cancridae

Carpilius corallinus (Herbst)
Coral crab

The attractive coral crab has the surface of the carapace red-brown with a large number of yellow to white markings. The walking legs and arms are reddish and purple and the fingers of the chelae have dark tips. This species is among the largest of West Indian crabs and is quite common in some areas. It is used for food and experiments are underway to commercially rear this species.

On the reef *Carpilius corallinus* occurs in caves or crevices during the day but often does not retreat deeply into cover, remaining near the opening where it can occasionally be seen. At night it moves into the open to forage on the bottom. Its distribution is strictly tropical, from the Bahamas to Brazil with records from many West Indian islands. There are no definite records from Florida or Bermuda.

Reference: Rathbun 1930.

Family Portunidae

Portunus ordwayi (Stimpson)

A typical portunid crab, *Portunus ordwayi* has the last pair of walking legs modified into flattened paddles for swimming. When swimming this species and other portunids move sideways rather than forward and can move at a surprising rate. The upper surface of the carapace is reddish, the lower yellowish and the fingers of the chelae are banded light and dark. The margin of the carapace and segments of the chelae are well armed with spines and the lateral spine of the carapace is considerably larger in males than females. The segment before the chelae also has an iridescent area on its outer surface.

Portunus ordwayi ranges from North Carolina (Massachusetts in summer) to Brazil, including Bermuda, at depths of 5 to 72 m.

References: Rathbun 1930, 1933.

Portunus sebae (Milne-Edwards)

This species is the easiest of the Caribbean portunid crabs to identify. It has a pair of reddish spots on the posterior part of the dorsal surface of the carapace. Otherwise it is similar to most species of *Portunus.*

During mating the male crab (usually larger) mounts the female on her dorsal surface and remains there until the female molts. At that time the actual mating occurs, as the female is without the normal hard exoskeleton for a short period and the sperm, in the form of spermatophores, are stored in seminal receptacles. Later the eggs, fertilized by the sperm transferred during this mating, are deposited beneath the abdomen of the female, where they remain until hatching.

Portunus sebae is known from the Gulf of Mexico, southern Florida and Bermuda to Brazil at 6 to 30 m.

References: Rathbun 1930, 1933.

Family Grapsidae

Percnon gibbesi (Milne-Edwards)
Urchin crab

This shy flattened crab usually occurs with the long-spined sea urchin, *Diadema antillarum* Phillipi, taking shelter in the numerous spines when it feels threatened. The carapace is nearly disk-like with a maximum dimension of only about 25 mm and is brown in color. The long legs are brown with golden yellow banding and splotches of gold at the joints. An iridescent greenish line runs around the front of the carapace beneath the eyes and a thinner greenish line occurs on the undersurface of each eye. The chelae of males are considerably larger than those of females and in both sexes are adapted for snipping at a thin film of filamentous algae.

The coral crab, *Carpilius corallinus,* is a large, delicately colored crab which is secretive during the day. They come out into the open at night and are easily photographed. Puerto Rico, Mona Island, 15 m depth.

Portunus ordwayi is typical of the swimming crabs of the family Portunidae. They have a flattened body with the last two walking legs modified into swimming paddles. Puerto Rico, Aguadilla, 10 m depth.

Portunus sebae is another swimming crab. This individual is preparing to defend itself against the photographer and the modified last legs are clearly visible, being held up posterior to the animal. Puerto Rico, La Parguera, Caracoles Reef, 12 m depth, night.

This pair of *Portunus sebae* is preparing to mate. The female is below, the male above. Mating can only be accomplished when the female molts; the male will hold onto her until such time as this is accomplished. Puerto Rico, La Parguera, Caracoles Reef, 12 m depth, night.

The relationship between the crab and sea urchin is loose, and either species may occur without the other. The crab gains protection from predators by association with *D. antillarum*, but the urchin apparently derives no benefit. *Percnon gibbesi* feeds both day and night, typically on algae grazed from rock surfaces.

The urchin crab is known from near the surface to 23 m but is most common in the 2 to 7 m depth range. It occurs widely in the Atlantic, in the west from North Carolina and Bermuda to Brazil and from the Azores to the Cape of Good Hope in the east. It also occurs in the eastern Pacific from Baja California to Peru and the offshore islands of Clipperton and the Galapagos.

References: Craft 1975, Williams 1965.

Family Majidae

Mithrax spinosissimus (Lamarck)

This crab is the largest species of *Mithrax* and the largest species of Caribbean reef crab, with the length of the carapace alone reaching 17 cm and the whole crab weighing 2400 gm. (over 5 lbs.). It is brownish red in color but is often covered with considerable amounts of encrusting organisms presenting a fuzzy or whitish appearance. The walking legs are long and hairy. In males the chelae are larger and stronger than in females (longer than the legs) and the gape produced when the fingers are closed is considerably larger.

Mithrax spinosissimus emerges onto the open reef only at night, when it is believed to feed on algae. It may also occur among rocks. Females with eggs have been observed every month in the West Indies and there does not seem to be any maxima of spawning. The juveniles are not usually taken in commercial fishing operations (lobster pots) where the adults are captured and may occupy a different habitat. The species occurs from the Carolinas (rare) to the Caribbean and is known from several West Indian islands. It is locally abundant in some areas, such as the cays off Kingston, Jamaica. Its depth range is from 2 to 180 m.

References: Rathbun 1925, Williams 1965, Munro 1974.

Mithrax cinctimanus (Stimpson)
Anemone crab

This small crab usually occurs with sea anemones, particularly *Stoichactis helianthus.* It has an oval carapace with light and dark areas on its surface. The legs are banded and the outer segments have dense short dark brown hair. The fingers of the chelae are gaping, with spoon-like tips.

Another species (*Mithrax commensalis* Manning) has been recently described, but most authorities feel this species is identical with *M. cinctimanus.*

Mithrax cinctimanus may be found on the tentacles or around the margin and column of the anemone *Stoichactis helianthus* in shallow back reef areas where the anemone is common. The crab may also occur on the anemone *Lebrunia danae.* It hides beneath the expanded oral disk of the anemone when disturbed and is immune to the nematocysts of the tentacles. The shrimp *Periclimenes* sp. and *Thor amboinensis* may occur with *M. cinctimanus* on *S. helianthus.*

This species is known from the Florida Keys and the Bahamas to Curacao and widely in the West Indies at depths generally shallower than 6 m.

References: Manning 1970, Rathbun 1925.

Mithrax sculptus (Lamarck)

This small green crab, often partially covered with whitish material, has the tips of the fingers of the chelae spoon-like and white in color. The legs are also tipped in white and often carry a considerable amount of encrusting material.

The species occurs in a wide variety of habitats but is perhaps most abundant on shallow back reefs. It occurs among the branches of *Porites furcata* and feeds on the expanded polyps at night. The chelae are alternately employed, one conveying food to the mouth while the other is attacking another polyp; crabs have been seen to devour 10 polyps in a minute. *Mithrax sculptus* also feeds on organisms encrusting blades of *Thalassia testudinum.* It may also be found near to the anemone *Stoichactis helianthus* and will hide beneath the expanded column of the anemone.

Percnon gibbesi, the urchin crab, is found almost exclusively with the sea urchin *Diadema antillarum*. The crab has very lovely markings in gold, but is difficult to observe as it is quite shy. Puerto Rico, La Parguera, Laurel Reef, 6 m depth.

Mithrax spinosissimus is a large species of spider crab which is only openly exposed at night. Bahamas, Eleuthera Island, 15 m depth, night.

Mithrax cinctimanus usually occurs with sea anemones. It is small and "hairy" and may be found on the anemone itself or around the margin. Puerto Rico, La Parguera, Laurel Reef, 1 m depth.

The anemone crab, *Mithrax cinctimanus,* is illustrated within the tentacles of *Condylactis gigantea.* The nature of its immunity to the stinging nematocysts of the anemone is not known. Jamaica, Discovery Bay, 2 m depth.

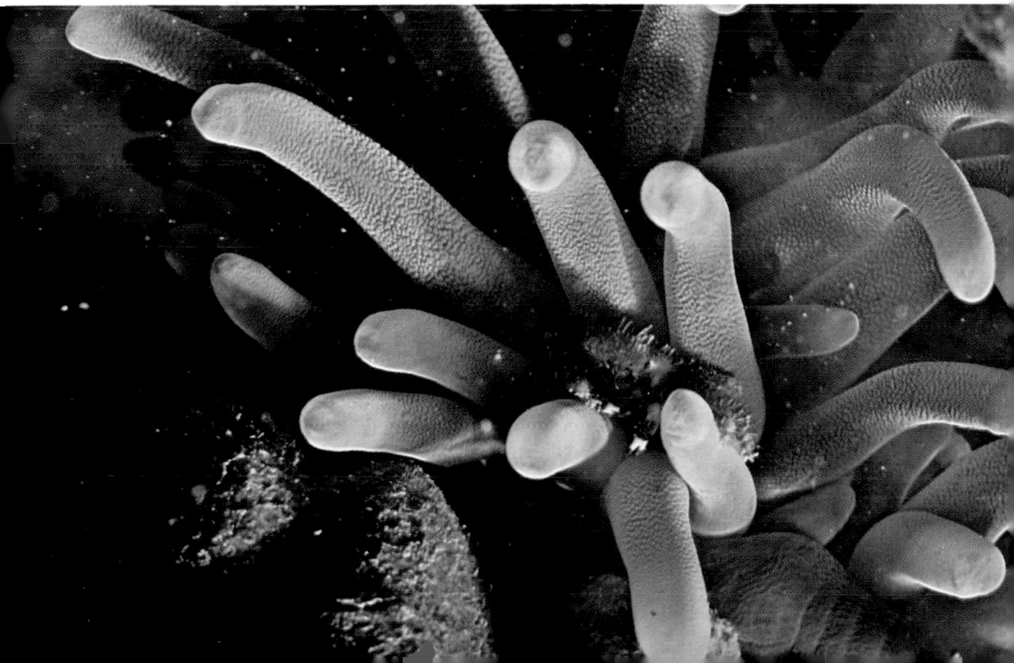

Mithrax sculptus occurs from southern Florida and the Bahamas to Brazil at depths of 0 to 54 m.

References: Glynn 1962, Rathbun 1925.

Stenorhynchus seticornis (Herbst)
Arrow crab

The extremely long legs, pointed cone-like body and spidery appearance easily identify the arrow crab. The body is striped with white, brown and tan, and the eyes protrude from either side of the pointed rostrum. The legs, several times the length of the carapace, are also striped; small blue-tipped chelae occur on the first pair of legs.

Stenorhynchus seticornis occurs in a wide variety of habitats, including reefs, rock, pebbles and sand. On reefs they may occur near crevices, on sponges or close to sea anemones. Often algae may grow upon the tip of the rostrum.

The arrow crab occurs throughout the tropical western Atlantic from North Carolina to Rio de Janeiro, Brazil and in the eastern Atlantic at Madeira, the Canary Islands and Angola. Its known depth range is 0 to 1500 m, but it is rare below 180 m. Females bearing eggs are common in spring and summer.

References: Williams 1965, Rathbun 1925.

DECORATOR CRABS

Various species of spider crabs, particularly of the family Majidae, are "decorator crabs," attaching bits of sponges, algae and other material to their bodies. These items are picked up with the chelae, brought to the mouth and then to a position on the body where they are believed to be held by tiny hooked hairs. Once attached this camouflage seems to grow well by itself. In many cases the crab is nearly indistinguishable from the bottom, with the anterior and lateral portions of the carapace densely covered. Only the chelae may be clearly visible, and if they are hidden from view the crab is completely inconspicuous until it moves. One small species of decorator crab actually becomes more distinctive, with red sponges attached to it, and is clearly visible on the surface of sponges at night.

Family Xanthidae

Domecia acanthophora (Desbonne and Schramm)

This small xanthid crab normally occurs on the branches of corals of the genus *Acropora*. In Puerto Rico it is found on all three species of *Acropora*, but most commonly on elkhorn coral, *A. palmata*. The crabs occur in pit-like structures on the coral surface, usually at the intersection of two branches.

The growth of the coral is modified by the presence of the crab to produce the pit-like deformation. The coral polyps in the center of the pit cease to grow while those on the edge continue to lay down calcium carbonate. The end result is a structural deformity which may be shallow, deep or slit-like, and in which the crab spends much of its time. Since corals of *Acropora* are rapid-growing it would be possible for a sizable pit to be produced in as little as six months time.

Unlike the true gall crabs, *Domecia acanthophora* can leave its pit at any time and move over the coral surface. There is some exchange of pits among crabs and occasionally slightly more crabs than pits on a given coral colony. This crab is possibly a filter feeder although its food habits are poorly known.

While the genus *Domecia* is worldwide in the tropics, *D. acanthophora* is known only from the Atlantic. It is known from South Carolina to Brazil and a second form of the species is known from the eastern Atlantic.

Reference: Patton 1967.

Mithrax sculptus is emerald green often with parts of its carapace covered with whitish material. It can be seen scooping material to its mouth with its spoon-like chelae and has been known to prey on coral polyps. Puerto Rico, La Parguera, Mario Reef, 2 m depth.

The arrow crab, *Stenorhynchus seticornis,* is one of the more unusual crustaceans occurring on coral reefs. Puerto Rico, Aguadilla, 12 m depth.

The legs of the arrow crab, *Stenorhynchus seticornis,* are very long relative to its body size. Puerto Rico, Aguadilla, 12 m depth.

The walking legs of *Stenorhynchus seticornis* are visible in this photograph taken from above. The eyes of the crab on opposite sides of the pointed head are visible and it is resting on an unidentified sponge. Caicos Island, Providenciales, 20 m depth.

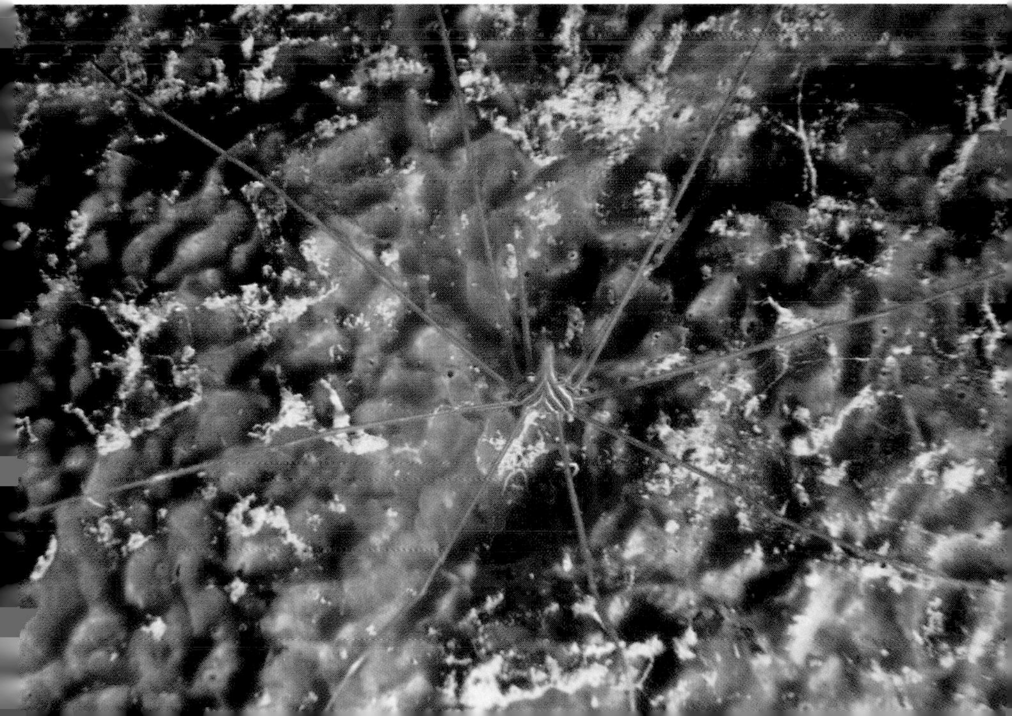

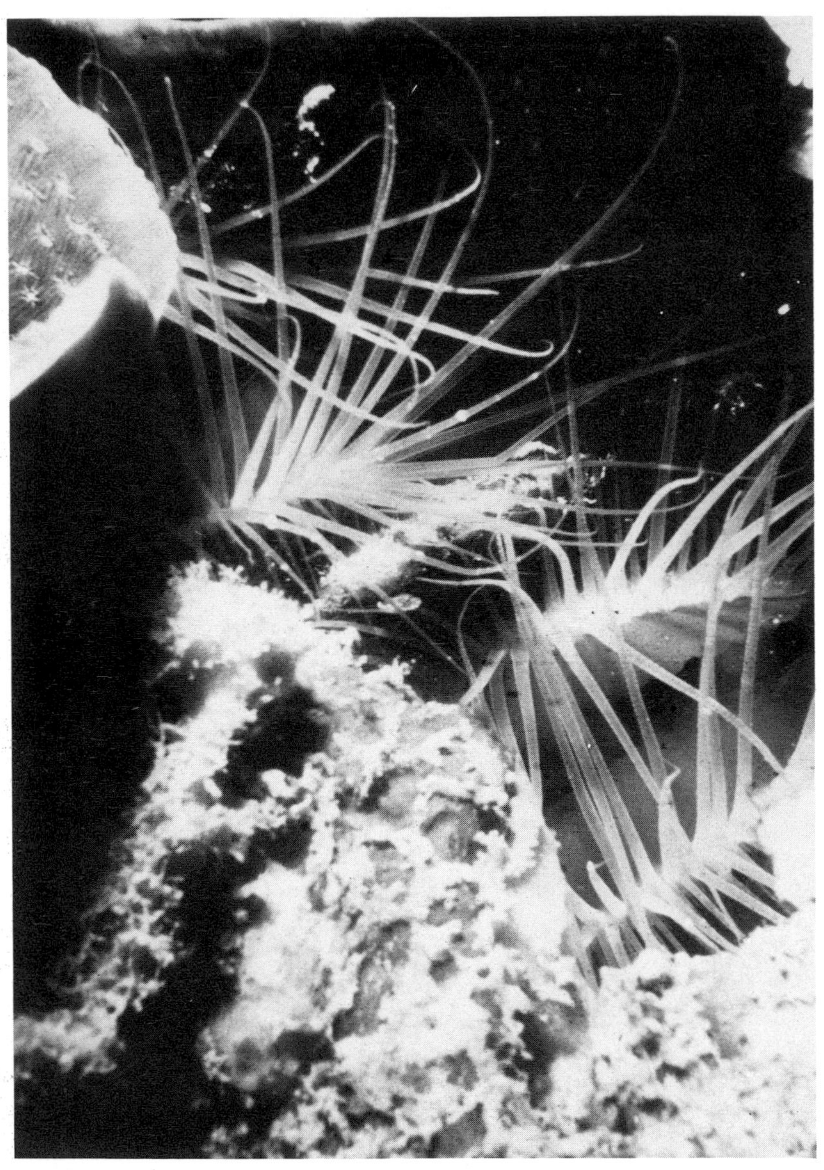

Three scallops, *Lima scabra,* protruding from a crevice in the reef.

Phylum Mollusca: Molluscs

The molluscs are a very familiar group of marine organisms to the average person since nearly all the "sea shells" are skeletal remains of members of this large phylum. There are around 100,000 species of Mollusca world-wide; among phyla their numbers are second only to the arthropods, and their geological importance is very great.

The phylum is usually divided into six classes; however, only four (Amphineura, Gastropoda, Pelecypoda and Cephalopoda) are of interest in discussing reefs.

CLASS AMPHINEURA

The Amphineura are the most primitive of molluscs. Most species are within the subclass Polyplacophora, the chitons. The chitons lack eyes or tentacles and have eight overlapping calcareous plates covering the body. They attach to hard substrates by a large foot covering nearly the entire ventral surface and generally are flattened. The mantle may cover much of the plates externally; the numerous external gills are found between the plates alongside the foot.

The chitons are typically intertidal or very shallow subtidal. Those that occur on Caribbean reefs are found beneath rocks on the reef crest or attached to rocky shores along fringing reefs. They cling tightly to the substrate and feed on algae growing on these surfaces.

CLASS GASTROPODA: SNAILS AND NUDIBRANCHS

The gastropods are the largest class of molluscs. There are two subclasses, the Prosobranchia (generally having a spiral shell and operculum) and the Opisthobranchia (often with a very reduced or absent shell and no operculum). The gills are variously developed in the mantle cavity or as secon-

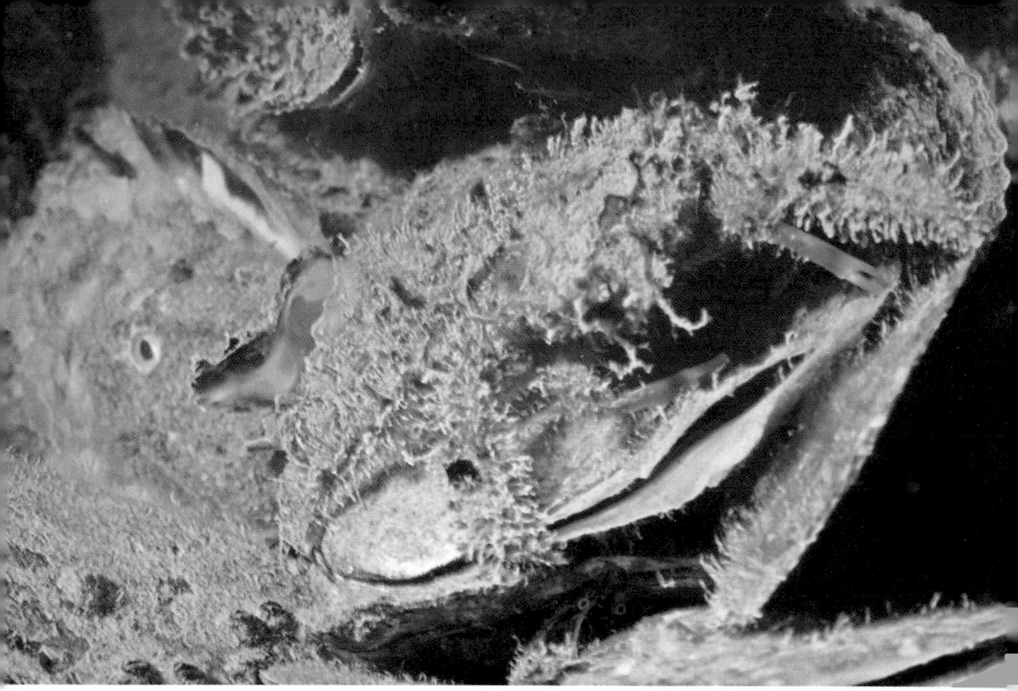

Decorator crabs cover themselves with an assortment of material so that it is very difficult to discern the crab sitting on the bottom. Only when they move can they be easily spotted. Puerto Rico, La Parguera, Laurel Reef, 12 m depth.

Tiny decorator crabs of the family Majidae sometimes cover themselves with brightly colored sponges. Perhaps the crab is relying more on the noxious nature of the sponges to repel predators than on their concealing properties. Puerto Rico, Mona Island, Playa Sardinera, 15 m depth.

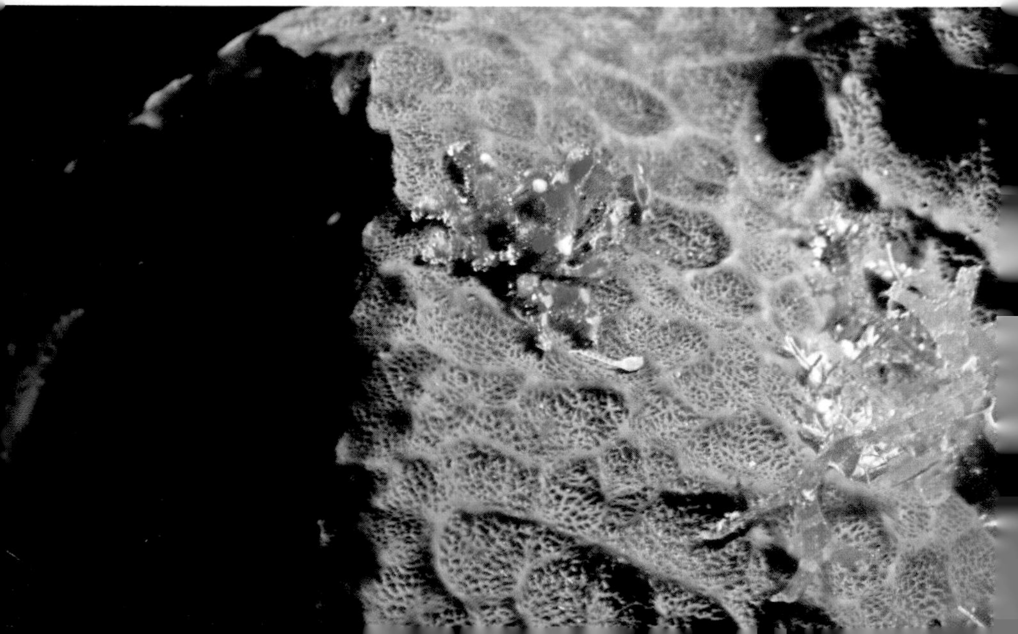

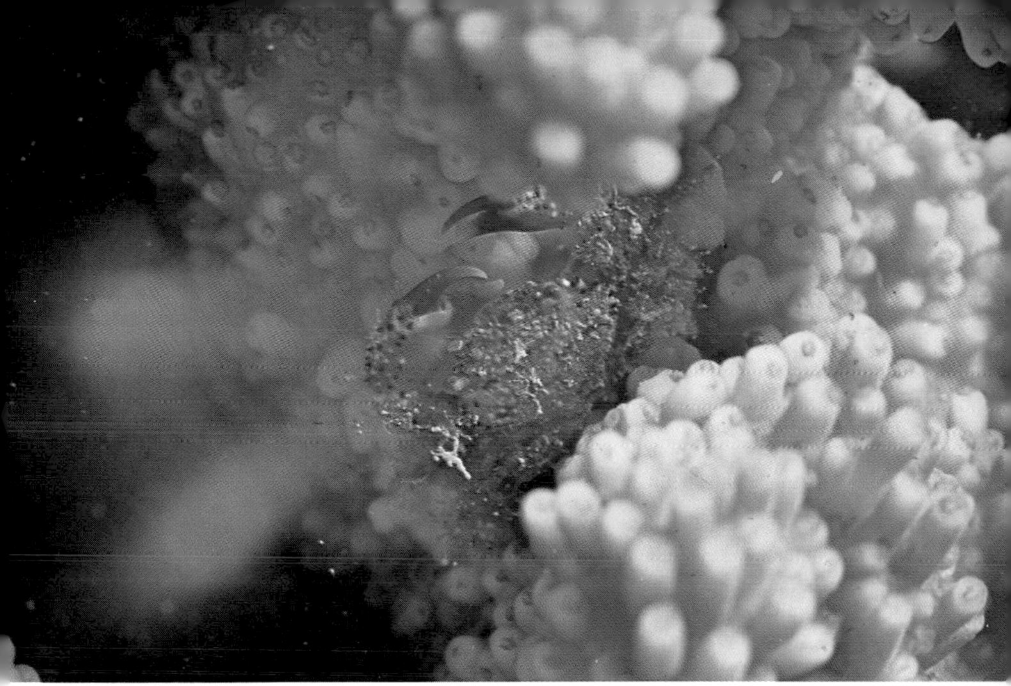

Domecia acanthophora occurs on the branches of the coral of the genus *Acropora.* The crabs form small pits, particularly at the axis of the branches, but roam over the surface of the coral when not disturbed. Puerto Rico, La Parguera, Mario Reef, 6 m depth.

An individual of *Domecia acanthophora* on the coral *Acropora palmata* within the pit produced by the coral growing up around an area where the crab stops the coral from growing. The crab is free to leave the pit when it desires. Puerto Rico, La Parguera, 5 m depth.

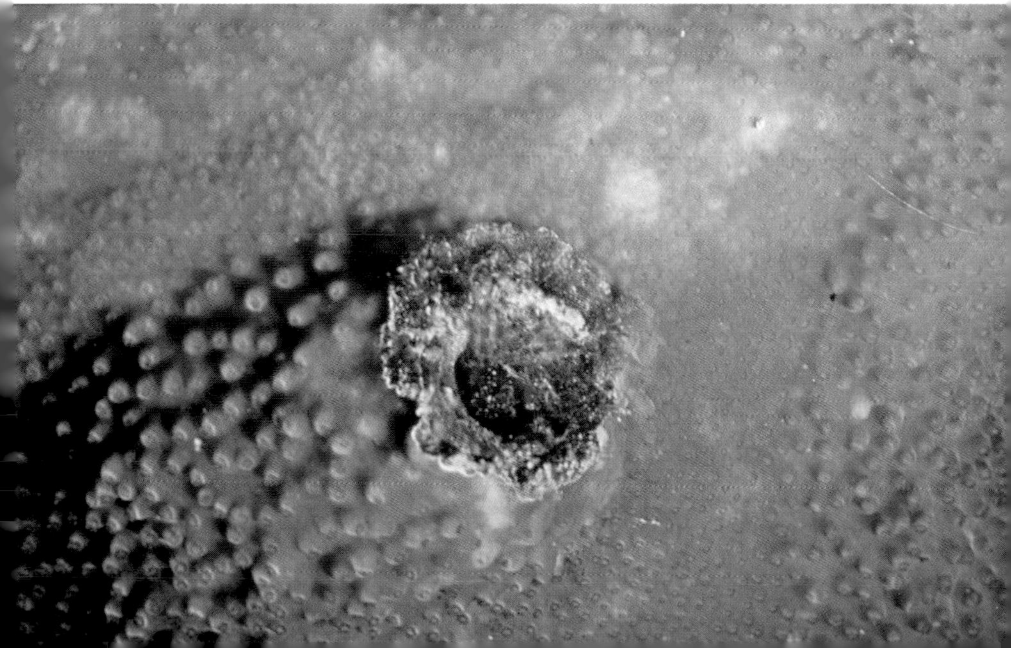

dary structures (cerata) externally or may not be present at all. The Prosobranchia have separate sexes while the Opisthobranchia are hermaphroditic. Gastropods range from herbivores scraping algae from rocks to predators, some with venomous structures, attacking other animals including other molluscs.

Family Turbinidae

Astraea tecta (Lightfoot)
American star-shell

This gastropod occurs in a wide variety of habitats including grass beds, under rocks and in reef crevices. The shell, with its sharp-angled spire, is often heavily coated with growths including coralline algae and reaches about 5 cm in length.

Astraea tecta tecta is the West Indian subspecies and *A. tecta americana* (Gmelin) is the subspecies occurring in southern Florida. This species is found from low water to at least 6 m.

Reference: Abbott 1974.

Family Fissurellidae

Diodora spp.
Keyhole limpets

Keyhole limpets have a conical shell with a small opening at the apex. They attach strongly to rocky surfaces by their foot and are quite difficult to remove. Typically they occur on reefs beneath rocky rubble, particularly in back reef areas. Several species of *Diodora*, basically similar in appearance, occur in the Caribbean, with none exceeding 5 cm in diameter. Most occur from Florida to Brazil and reach from shallow water to depths well over 100 m.

Reference: Abbott 1974.

Family Strombidae

Strombus gigas Linnaeus
Queen conch

The large shell of *Strombus gigas*, with its pink inner

surface and thickened flared lip, is one of the animals most commonly associated with the West Indies. Its shell reaches its maximum length of about 30 cm in two to three years and after that time it is thickened by additional layers of calcium carbonate added on the internal surface.

The queen conch is most typical of beds of turtle grass, *Thalassia testudinum*, but also occurs adjacent to reefs on sandy areas. It feeds largely on microscopic algae often grazed from the surface of the blades of *T. testudinum*. Movement is slow, being accomplished by short "steps" extending the foot with its horny operculum to drive the animal with its heavy shell along.

Young *S. gigas* are open to predation by a variety of gastropod molluscs, cephalopods, crustaceans and fish. The hermit crab *Petrochirus diogenes* is believed not only to devour unfortunate queen conchs but also to expropriate the shell of the *S. gigas* as its new home. The loggerhead turtle, *Caretta caretta*, is capable of crushing the shell of adult queen conchs during feeding, the only marine animal believed able to do so. This gastropod is also highly prized as human food, constituting a staple in some areas, and this has led to great decreases in populations where they are heavily fished.

Some commensal organisms occur with *S. gigas*. The conchfish, *Astrapogon stellatus* (Cope), spends the day within the mantle cavity of this mollusc, emerging at night to feed. A porcellanid crab may also occur with *S. gigas*. A conch pearl is formed when a particle, either animal or mineral, stuck beneath the mantle is covered over with layers of calcium carbonate until it reaches nearly the size of a marble.

Mating is accomplished in open sandy areas, the two partners lined up fore to aft. Packets consisting of gelatinous strings of eggs are laid on sandy bottoms, hatching into planktonic larvae. *Strombus gigas* occurs to a depth of at least 30 m, often in great abundance, and ranges from Bermuda and southern Florida to at least Trinidad in the south.

References: Randall 1964, Work 1969, Abbott 1974, Little 1965, D'Asaro 1965.

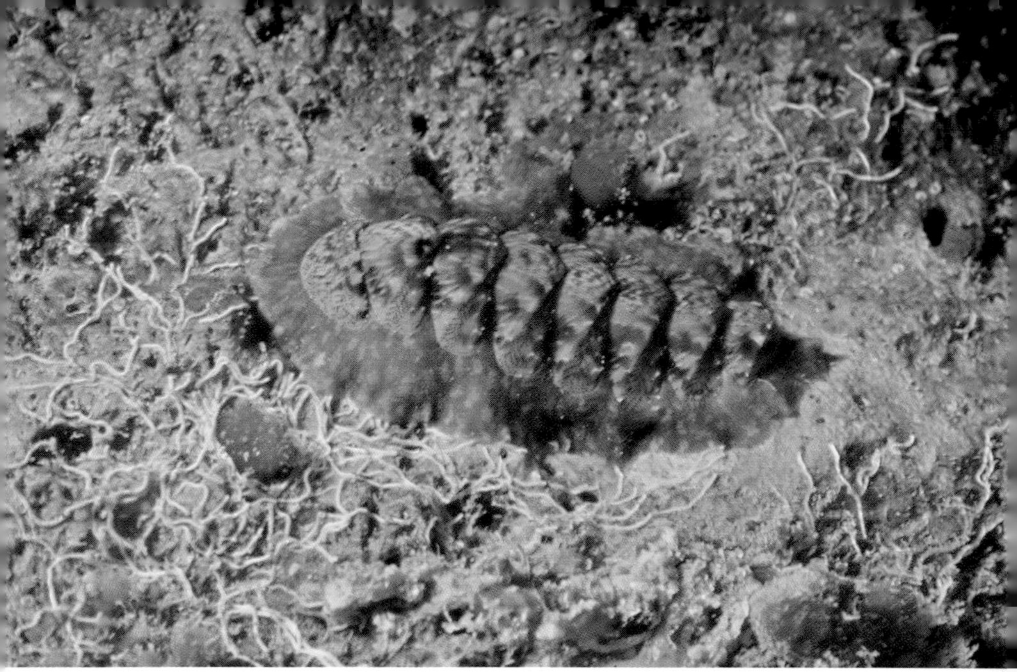

This unidentified species of chiton was on the undersurface of a piece of reef rubble. Also visible are serpulid worm tubes and red tests of the foram *Homotrema rubrum.* Puerto Rico, La Parguera, Laurel Reef, 1 m depth.

Astraea tecta, the American star-shell, is a gastropod mollusc seen occasionally on Caribbean reefs. The outer surface of the shell is coated with a variety of organisms, including coralline algae and other algae. Puerto Rico, Mona Island, Carmelita Reef, 6 m depth.

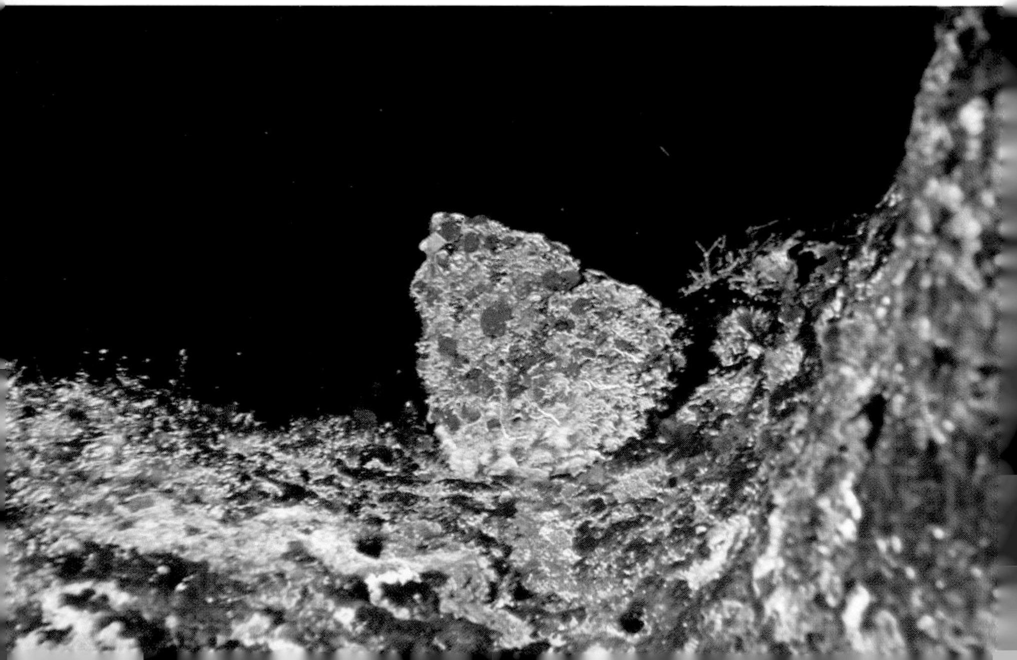

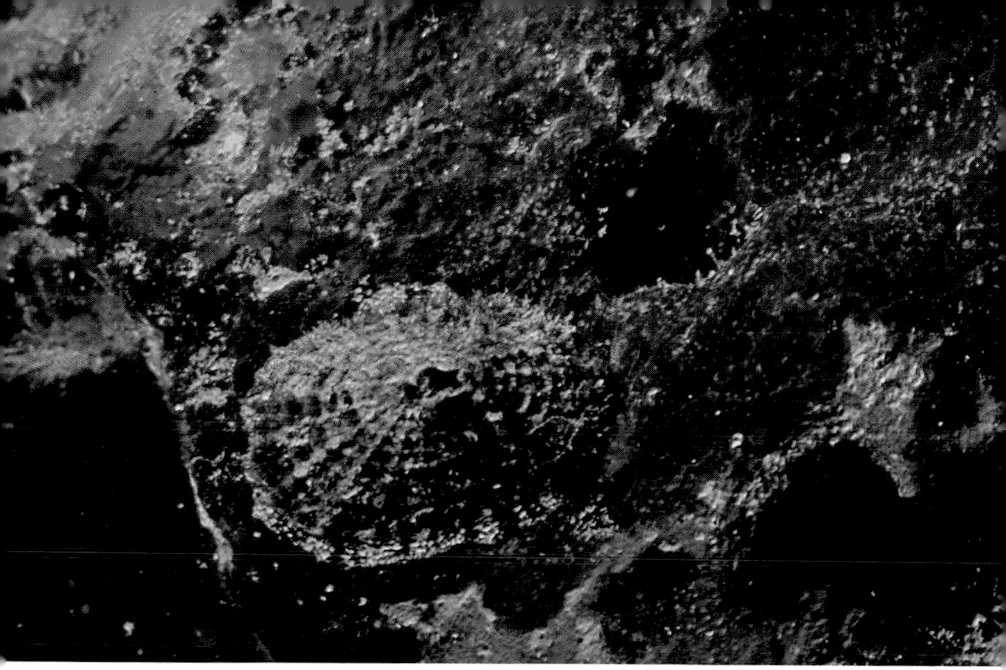

This keyhole limpet has the typical conical shell with a single tiny opening at its apex. This specimen was found beneath a piece of reef rubble. Puerto Rico, La Parguera, Laurel Reef, 1 m depth.

When seen in the field the queen conch, *Strombus gigas,* has the shell covered with a variety of material which serves to make the animal less conspicuous. The spires of the shell are still evident. This individual was photographed in a mixed bed of the seagrasses *Thalassia testudinum* and *Syringodium filiforme.* Bahamas, Cat Island, 5 m depth.

Family Cassidae

Cassis tuberosa (Linnaeus)
King helmet

The king helmet is one of the most spectacular gastropods of Caribbean waters and consequently its shell is prized as a curio. It is most often found in sea grass beds, where it preys on sea urchins, but also occurs on the sandy margins of reefs. Its parietal shield (the flattened part of the shell around the opening) is triangular, which distinguishes it from the similar *Cassis flammea* (Linnaeus), the flame helmet, which has a rounded shield. A third species, *C. madagascarensis* Lamarck, the queen helmet, has much larger blunt spines than the other species, with the topmost largest.

The attack of *C. tuberosa* is an amazing process to observe. The speed and ability of the final leap by *C. tuberosa* onto the prey urchin is startling in light of the size and normally slow movements of this creature and the spines of the urchin, which do not deter this mollusc in the least.

All three species occur from southern Florida throughout the Caribbean at depths of about 20 m.

References: Abbott 1974, Warmke and Abbott 1961, Hughes and Hughes 1971.

Family Cymatiidae

Charonia variegata (Lamarck)
Trumpet triton

The well-known trumpet triton is one of the most spectacular of Caribbean gastropod molluscs. It reaches 35 cm in length and is highly prized for its shell by collectors. It occurs on sandy bottoms near reefs and seems most active at night. It has been observed to attack and eat sea cucumbers, with the holothurians often secreting a toxic substance which occasionally breaks off an attack.

Charonia variegata occurs at depths to about 10 m in the western Atlantic from southern Florida and Bermuda to Brazil. It is also recorded from the eastern Mediterranean Sea, the Cape Verde Islands and St. Helena.

References: Abbott 1972, Parrish 1972.

Family Ovulidae

Cyphoma gibbosum Linnaeus
Flamingo tongue

This unusual gastropod, with its spindle-shaped shell reaching about 30 mm in length, occurs on gorgonians of the reef. The cream-colored mantle with irregular yellow-tan spots edged in black covers the shell in undisturbed individuals. If the animal is bothered, the mantle is retracted, exposing the pinkish orange and white shell.

Cyphoma gibbosum occurs on both fan-shaped and whip-like gorgonians, feeding on the gorgonian as is often evidenced by an area of damaged tissue adjacent to the snail. More than one individual may occur on a single gorgonian colony. Aggressive behavior has been observed between individuals of *C. gibbosum* under unnatural conditions, but the role of this behavior in nature is not known.

The flamingo tongue occurs in shallow water (generally less than 10 m) and is known from southern Florida, the Bahamas and the West Indies. Other members of the family Ovulidae in the genera *Cyphoma* (*C. signatum* Pilsbry and McGinty) and *Neosimnia* (*N. acicularis* Lamarck, *N. uniplicata* Sowerby) also occur on gorgonians.

References: Ghiselin and Wilson 1966, Warmke and Abbott 1961.

Family Olividae

Oliva reticularis Lamarck
Reticulated olive

The reticulated olive is a small carnivorous gastropod which can be found crawling over sandy areas near reefs at night and occasionally during the day. It reaches about 6 cm in length. The snail slides along just below the surface of the substrate on its expanded foot searching for food.

A similar but somewhat larger species, *Oliva sayana* Ravenel, is found from the Carolinas and Florida to Yucatan (and supposedly Brazil) but seems to be absent from the Caribbean. *Oliva reticularis* occurs throughout the western Atlantic tropics.

Cassis madagascarensis, the queen helmet, is seen here on a sediment plane adjacent to a reef. Jamaica, Discovery Bay, 15 m depth, night.

The trumpet triton, *Charonia variegata*, is one of the most spectacular Caribbean gastropods. It is known to attack and eat sea cucumbers, a group of animals which are seldom preyed upon. Puerto Rico, La Parguera, Caracoles Reef, 10 m depth.

The flamingo tongue, *Cyphoma gibbosum,* lives and feeds on gorgonians. The mantle is covered with ocellated spots and is quickly withdrawn into the shell. Puerto Rico, La Parguera, Laurel Reef, 5 m depth.

The olive shell, *Oliva reticularis,* can be found in back reef areas on the substrate searching for food. Puerto Rico, La Parguera, 2 m depth.

Reference: Abbott 1974.

Family Elysiidae

Tridachia crispata Moerch

The dorsal surface of the body is distinctive on this opisthobranch which reaches about 10 cm in length. The margin of the convoluted folding of the dorsal surface is iridescent and the basal portions green or bluish. The cells of the dorsal surface possess chloroplasts, acquired from algal food eaten by the nudibranch, which produce photosynthetically fixed carbon compounds. Mucus, secreted from the pedal gland, over which the animal crawls is partially produced from carbon fixed by the chloroplasts. In addition, oxygen produced by photosynthesis is available to the nudibranch.

Tridachia crispata occurs on reefs and algal covered areas. It has been observed on dead portions of the coral *Acropora cervicornis.* The species occurs from the southern and western coasts of Florida, throughout the Caribbean and the Bahamas to at least Venezuela in water generally less than 20 m deep.

References: Marcus and Marcus 1960, 1967, Trench 1969, 1973a, Taylor 1971.

Family Phyllidiidae

Phyllidiopsis papilligera Bergh

This nudibranch can occur on the surface of living corals, where it moves about slowly. Its coloration, white with gray splotches edged in black, certainly makes it easily observed. The head end of the animal is distinguished by a pair of tentacles with white bases and tips but gray in between; these are thrust up when the animal is not disturbed.

Phyllidiopsis papilligera is known from the southern Gulf of Mexico and the West Indies at moderate depths.

Reference: Abbott 1974.

CLASS PELECYPODA: BIVALVES

The shell of pelecypods or bivalves consists of a pair of

valves which are usually hinged. These mollusks are often burrowing forms and typically are filter-feeders. They are usually sedentary, are quite often attached to the substrate by a mass of threads called the byssus and have a foot with which they can pull themselves down into the sediment.

The gills are within the mantle cavity and water is actively pumped over them, entering and exiting the cavity through siphons, tube-like extensions of the mantle.

Family Pinnidae

Pinna carnea Gmelin
Amber pen shell

This unusual bivalve has a large but thin and fragile shell with several low ribs running the length of each valve. It is triangular, fan or paddle-shaped and reaches a maximum of about 25 cm in length. This mollusc lives in muddy or sandy areas with the shell buried and only the top few centimeters exposed. The apex may also be anchored by byssal threads. It can occur on reefs in areas of high sediment and has been observed growing nearly horizontally from reef faces also overgrown with *Isognomon* sp. and is much more exposed in such situations.

Pinna carnea occurs from southern Florida, Texas and Bermuda to Brazil at a wide range of depths. It is reported to be rare in Florida, however.

References: Turner and Rosewater 1958, Abbott 1974.

Family Pteriidae

Pteria colymbus (Roeding)
Atlantic wing oyster

The Atlantic wing oyster attaches to gorgonian stems on the reef, utilizing both living and dead gorgonians. It has the hinge ends drawn out a considerable distance making it quite distinctive; it is anchored to its substrate by a strong byssus. *Pteria colymbus* reaches only about 7 cm in length. It is known from North Carolina and Florida to Brazil at depths to 20 m.

Reference: Abbott 1974.

Tridachia crispata has a group of ruffled ridges on its back and is seen in a number of reef and algal habitats. It possesses chloroplasts from ingested algal food. Jamaica, Discovery Bay, 6 m depth.

Phyllidiopsis papilligera is a distinctively colored nudibranch which has been found on the surfaces of living corals. Puerto Rico, La Parguera, Laurel Reef, 12 m depth.

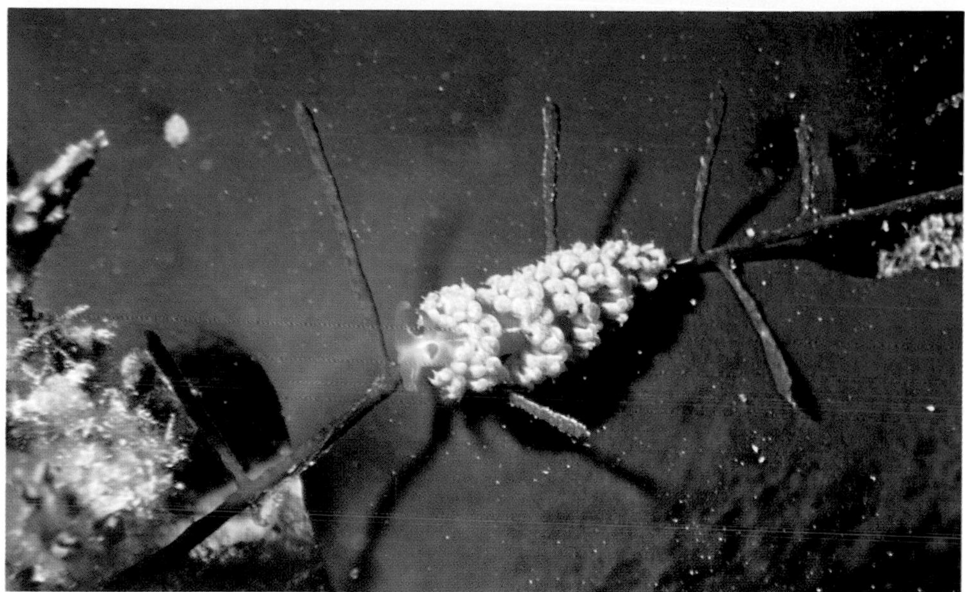

This unidentified nudibranch was found at night on the branch of a gorgonian. The flexible, pointed projections on the back are cerata and many nudibranchs possess them. Bahamas, Eleuthera Island, 20 m depth.

The pen shell *Pinna carnea* is most often found on reefs buried in the sediment. The top of the shell is exposed and the valves slightly opened in the photographed individual. Puerto Rico, La Parguera, Caracoles Reef, 6 m depth.

Family Isognomonidae

Isognomon spp.
Tree oysters

The three Caribbean species of the genus *Isognomon*, *I. alatus* (Gmelin), *I. radiatus* (Anton) and *I. bicolor* (Adams), occur on the roots of mangrove trees or on shallow rocky surfaces. Some also occur under rocky ledges on coral reefs in areas of high sedimentation where they form thick mats of hundreds of individuals. They also occur on the shells of the Atlantic thorny oyster, *Spondylus americanus.* The Caribbean species are small, no more than 7 cm in length.

All three species occur from Florida throughout the Caribbean and two range to Brazil. On reefs they range to at least 15 m.

References: Abbott 1974, Warmke and Abbott 1961.

Family Limidae

Lima scabra (Born)
Rough lima

This strange bivalve, reaching about 8 cm across, hardly resembles most clams. *Lima scabra* occurs in rock crevices where the long white or red tentacle-like extensions of the mantle are seen first. Fringes of mantle hang like curtains on the inside margin of each valve. This bivalve may occur in groups with several piled nearly on top of one another in a narrow crevice. They may be attached by byssal threads or free. Individuals which are not attached can swim jerkily by closing and opening the valves, producing a jet of water.

Two other species of *Lima*, similar in appearance, also occur in the Caribbean. *Lima scabra* ranges from South Carolina and Texas to Brazil at depths to about 130 m.

Reference: Abbott 1974.

Family Spondylidae

Spondylus americanus Hermann
Atlantic thorny oyster

The Atlantic thorny oyster is a bivalve which is almost

totally unrecognizable in its wide variety of natural habitats due to the thick accumulations of sediment and organisms on its shell. The cleaned shell, which may be 10 cm across, has erect spines reaching 5 cm in length on its surface. In nature the spines are seldom visible due to the material on the shell; the easiest way to spot this animal while diving is to look for a small portion of the "bottom" to move, a result of the closing of the valves of S. *americanus.*

The young of S. *americanus* are much less spinose than the adults and resemble the jewel boxes, members of the genus *Chama.* It can be quite common in certain habitats, such as deeper reefs off the Florida Keys and Caribbean reefs in high sediment areas, and often occurs on wrecks and sea walls.

The range of S. *americanus* is from North Carolina and Texas to Brazil, with large specimens generally found at 9 to 45 m.

References: Abbott 1974, Logan 1973.

Family Ostreidae

Lopha frons (Linnaeus)
Frons oyster

This bivalve attaches singly or in clusters to reef gorgonians, some dead corals and pilings. Its deeply folded margins of the valves are characteristic, with one valve firmly attached to the substrate and the opening of the valves oriented vertically. The planktonic larvae settle on calcareous encrustations on the gorgonians rather than on the living surface itself. A number of different species of gorgonians may be utilized as a substrate, but *Lopha frons* seldom occurs on sea fans (genus *Gorgonia*).

The species is limited to clear reef waters, usually on the seaward slope, but has been confused in the literature with similar species occurring on mangrove prop roots. *Lopha frons* is known from North Carolina, Bermuda and the Gulf of Mexico to the West Indies at depths from 7 to 60 m.

References: Forbes 1971, Abbott 1974.

Pinna carnea has been observed on some occasions growing outward from the sides of coral heads or from beneath ledges on reefs. Jamaica, Discovery Bay, 5 m depth.

The Atlantic wing oyster, *Pteria colymbus,* is found on reefs most often attached to gorgonian stalks. Puerto Rico, Rincon, 18 m depth.

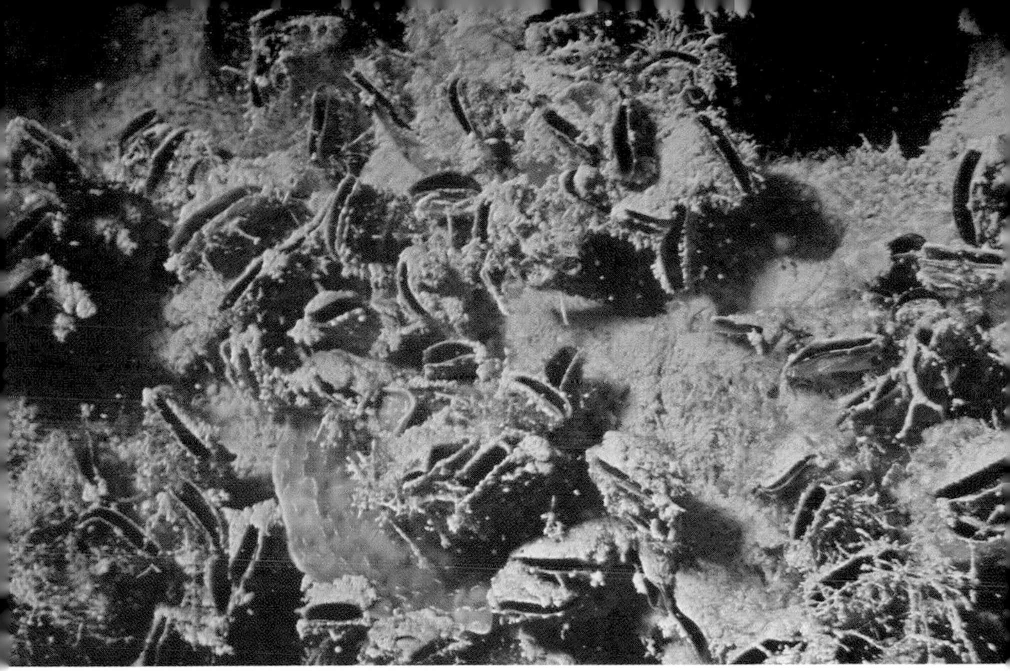

Tree oysters of the genus *Isognomon* can occur on the undersurfaces of reef leges, particularly in areas of high sedimentation. Jamaica, Discovery Bay, 3 m depth.

The rough lima, *Lima scabra,* has tentacle-like extensions of the mantle which are its most outstanding feature and the first seen. They are capable of swimming by clapping the valves together rapidly. Puerto Rico, La Parguera, Laurel Reef, 3 m depth.

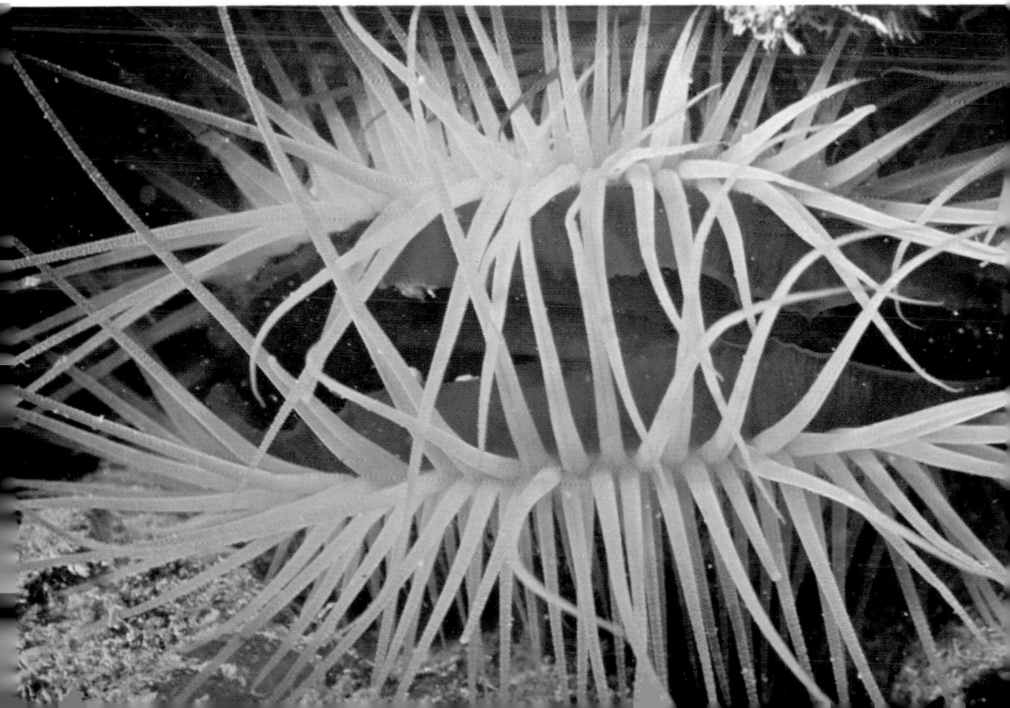

Family Chamidae

Chama macerophylla (Gmelin)
Leafy jewel box

This bivalve, reaching about 8 cm in size, attaches in reef crevices by its left valve where only the colorful foliations along its margin can be easily seen. The foliations are yellow, purple or pink with much of the adjacent shell the same color. The outside is densely covered with other organisms.

Chama macerophylla is the most common species of jewel box in the Atlantic although several others exist. *Chama sinuosa* Broderip typically occurs on reefs as does the tiny species *C. sarda* Reeve. Most species occur from southern Florida to Brazil at moderate depths, although they may be rare in the northern part of the range. Some species also occur in Bermuda.

Reference: Abbott 1974.

CLASS CEPHALOPODA: SQUIDS AND OCTOPUSES

The cephalopods are the most specialized and advanced of the classes of Mollusca. They have a very reduced internal shell (squids) or none at all (many octopuses). They have well-developed brains and sensory systems and surpass other invertebrates in their learning ability. Their eye is surprisingly like that of vertebrate animals but is believed to be similar due to convergent evolution.

Besides the typical squids and octopuses, this class also contains the cuttlefishes and nautiluses, both small groups. The squids possess ten arms bearing sucking disks on their inner surface and are adapted for living above the bottom. The octopuses have eight arms, again with sucking disks, and are more adapted for crawling or walking than active swimming.

Cephalopods are reknowned for their ability to change colors. A wide variety of chromatophores (pigment cells) occur in the skin and are under nervous control. These can be made visible or not almost instantly by the cephalopod, hence its ability to alter its color rapidly.

Family Loliginidae

Sepioteuthis sepioidea (Blainville)
Reef squid

This is the only species of squid normally encountered on Caribbean reefs. By day they are found in small groups of individuals, usually a pale reddish green or reddish brown in color, hovering above the bottom. They are quite shy, not allowing an observer to approach closely. If sufficiently disturbed, they release a cloud of black ink and jet away by contraction of the mantle cavity, forcing water out of the funnel. At night they become darker and, mesmerized by a diving light, often can be closely examined.

The ten arms are short, much less than the length of the body. The fins, sticking out from the mantle, are nearly as long as the body. At night a number of small iridescent blue-green spots are found on the mantle and color changes can be quite rapid. This species has been observed to capture and feed on fishes.

After mating, a number of egg capsules, each containing several large eggs, are attached to hard substrates. They hatch in just over one month and reach sexual maturity in about seven months. The maximum life of the species is about one year.

Sepioteuthis sepioidea lives in shallow water from Bermuda and southern Florida to the southern Caribbean.

References: Voss *et al.* 1973, LaRoe 1971, Arnold 1965.

Family Octopodidae

Octopus briareus Robson
Reef octopus

The length of the four pairs of arms in this medium sized octopus varies. The upper (fourth) pair are shortest, the second and third pairs longest. The animal is colored bluish green to greenish brown, but can change color rapidly (within seconds).

Octopus briareus occurs on coral reefs, rocky areas and *Thalassia* beds. It hides within cover during the day and openly exposes itself only at night when it hunts for bottom-

The Atlantic thorny oyster, *Spondylus americanus,* is one of the best camouflaged molluscs on the reef. Usually the only way it is detected is by a portion of the "bottom" moving when it closes its valves. The individual photographed has had most of its coating organisms removed so its shell with the long spines would be visible. Jamaica, Discovery Bay, 5 m depth.

The frons oyster, *Lopha frons,* has deeply folded margins to its valves. Puerto Rico, Rincon, 22 m depth.

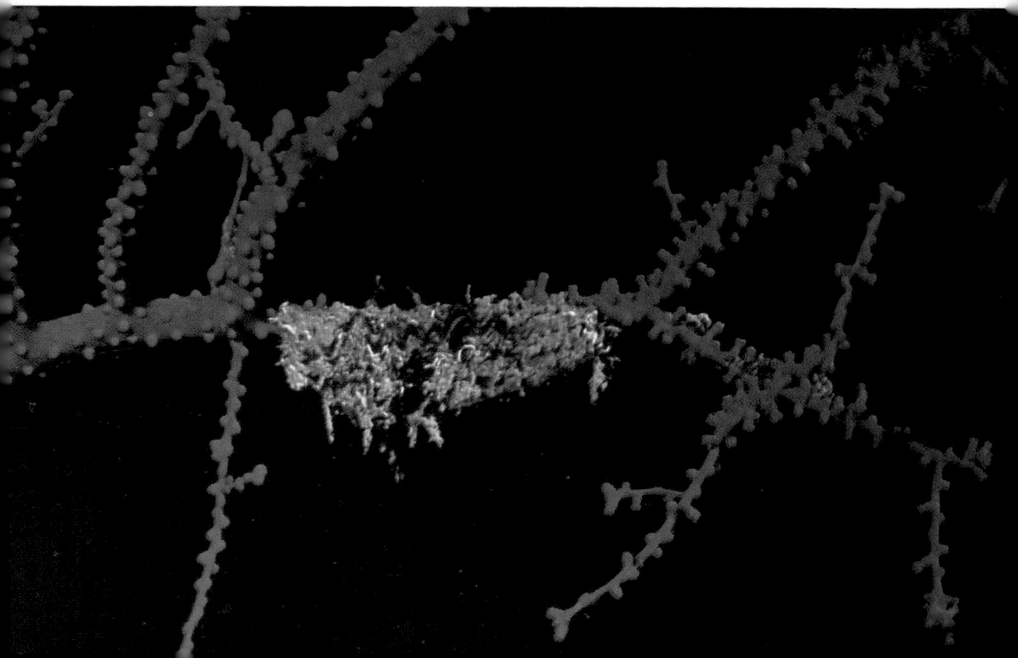

This example of the frons oyster, *Lopha frons,* is nearly hidden among other organisms covering it and a rock surface. The folded margin of the opened valves is the only thing that reveals its presence. Puerto Rico, Aguadilla, 3 m depth.

The leafy jewel box, *Chama macerophylla,* has leafy projections on the margins of the valves but otherwise is nearly indistinguishable from the surrounding organisms. Puerto Rico, La Parguera, Laurel Reef, 10 m depth.

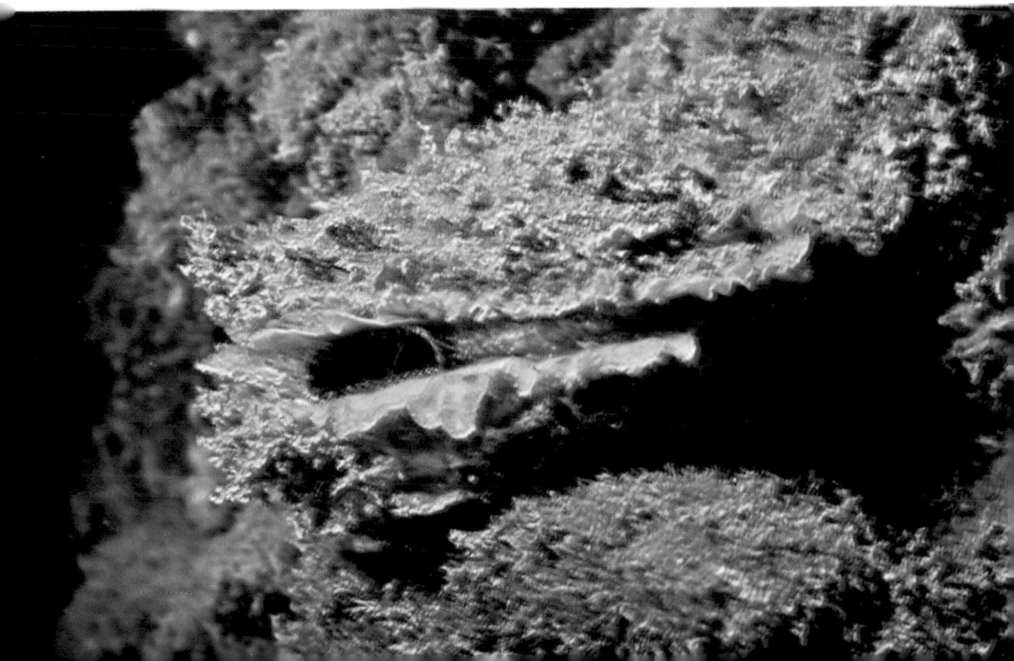

dwelling invertebrates. It reaches a maximum diameter (arm spread) of about 60 cm and a weight of 1 kg.

This octopus spawns only once, the male dying shortly after mating and the female after the eggs, which she guards, have hatched. Mating is accomplished by the transfer of spermatophores by the male's hectocotylized arm (the right or left third arm) into the mantle cavity of the female. The arm bearing the hectocotylus is shed after mating. The eggs are fertilized as they emerge from the oviduct and are attached to the substrate in clusters numbering about 500 eggs total. The eggs hatch in about two months and the young quickly take up a bottom-dwelling habit. The species survives a year or slightly longer.

Octopus briareus occurs in shallow water from the southern Gulf of Mexico and southern Atlantic coast to the Guianas. It is the most common octopus encountered on coral reefs.

Two other large octopods might also be encountered on Caribbean reefs. The common octopus, *O. vulgaris* Cuvier, is larger than *O. briareus* and more brownish in color. Its arms are more nearly equal in length and stouter than those of *O. briareus*. *Octopus macropus* Risso, the white-spotted octopus, is easily distinguished by its white spots on a blue-green to red body. The young of *O. macropus* have been taken near the surface at night. A third species, *O. maya* Voss and Solis, occurs in the Campeche Bank - Yucatan area.

References: Voss *et al.* 1973, Voss 1956, Voss and Phillips 1957, Voss and Solis 1966.

Phylum Echinodermata

The echinoderms are a large group of marine invertebrates possessing an internal skeleton of calcareous plates and a water-vascular system of fluid-filled vessels and appendages. They are really biradially symmetrical but often appear to be radially symmetrical as adults. The body structure often consists of multiples of five in skeletal plates, spines, arms, etc. Tube feet, the tactile extensions of the water-vascular system, occur on the arms and body (test).

A crinoid of the genus *Nemaster* photographed at 60 m depth at Discovery Bay, Jamaica. The central disk is hidden from view in a crevice and much of the substrate Is covered by a varicty of sponges.

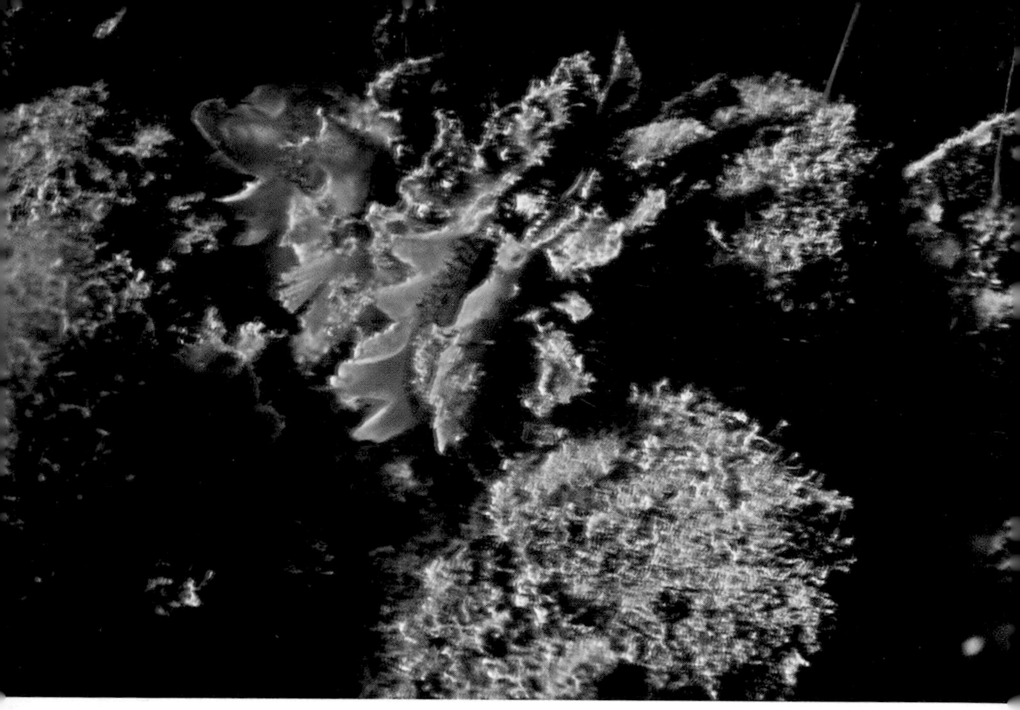

These small specimens of *Chama macerophylla* have the projections like larger individuals. Puerto Rico, La Parguera, Caracoles Reef, 12 m depth.

The reef squid, *Sepioteuthis sepioidea,* is seen both day and night on and around Caribbean reefs. It is difficult, however, to approach the squid during the day but the use of a bright light at night tends to immobilize these shy molluscs. Bahamas, Eleuthera Island, 5 m depth, night.

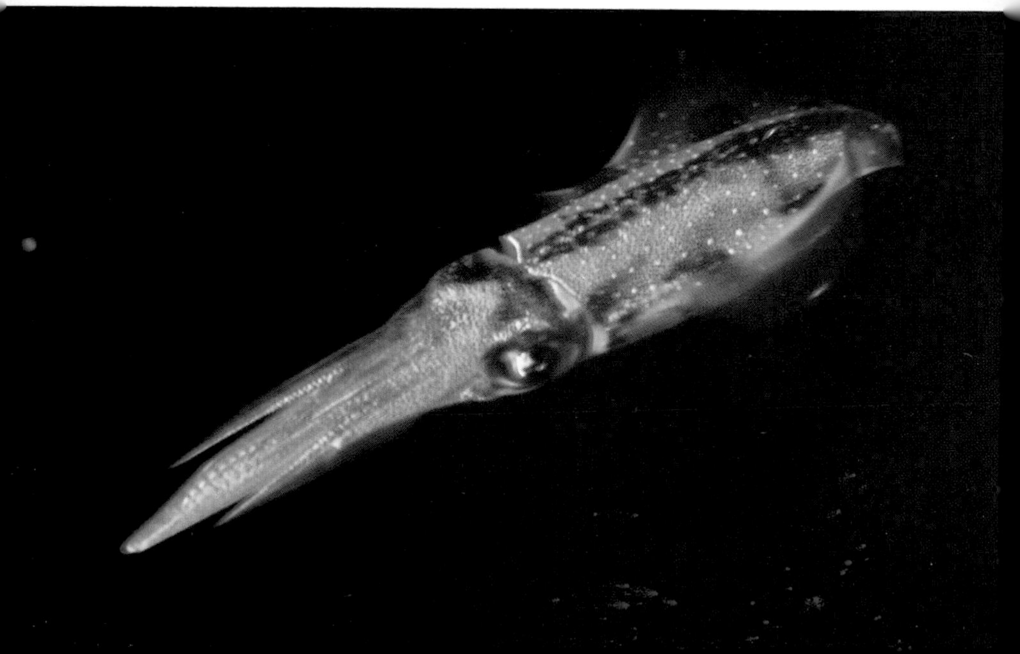

Octopus briareus is the most common octopus seen on reefs. They are secretive during the day and can be observed moving over the bottom at night. Jamaica, Discovery Bay, 15 m depth, night.

The crinoid *Nemaster rubiginosa* keeps its central disk in crevices but extends the arms out into the water. Bahamas, Grand Bahama Island, 12 m depth, night.

Large basket starfish, *Astrophyton muricatum,* on top of a gorgonian at night. Photographed at 10 m depth at Alligator Reef, Florida Keys.

Echinoderms have external fertilization and free-swimming larvae. They are found in three basic forms: star-like (sea stars, brittle stars, crinoids), spherical (sea urchins, heart urchins) and cylindrical (sea cucumbers).

CLASS CRINOIDEA: CRINOIDS

Of the five orders of Echinodermata the life habits of the crinoids are the most poorly known. This situation is rapidly changing, with the leaders in "crinoid watching" being geologists or paleontologists attracted to the living members of this group in order to interpret the habits and occurrence of geologically important crinoids in the fossil record.

Crinoids occur both in shallow reef waters and in deeper areas far below diving depths. Up to seven species have been found in a single shallow reef area in the Atlantic. Water movement seems to be a critical factor in the distribution of crinoids on a reef. Some species are current avoiding (rheophobic), remaining close to reef crevices both day and night, while others seek currents (rheophilic) and climb, usually at night, to locations well exposed to water movement. They are filter-feeding organisms and use the fine pinnules on the arms for straining material from the water.

Family Comasteridae

Nemaster rubiginosa (Pourtales)

This crinoid is common on reefs and is by far the most conspicuous member of this group. While the central disk is generally hidden in a crevice, the protruding orange or yellow arms are easily observed. There is some variation in color, with the pinnules (the small branches of the arms) being orange with yellow tips or whitish with black tips. The ambulacral groove of each of the 20 arms is generally black.

This crinoid can occur under corals, in sponges or in crevices. They move out slightly from cover at night, but generally only half or less of the arms are engaged in filter feeding at any one time. This and the following species extend the pinnules alternately at right angles from the axis of

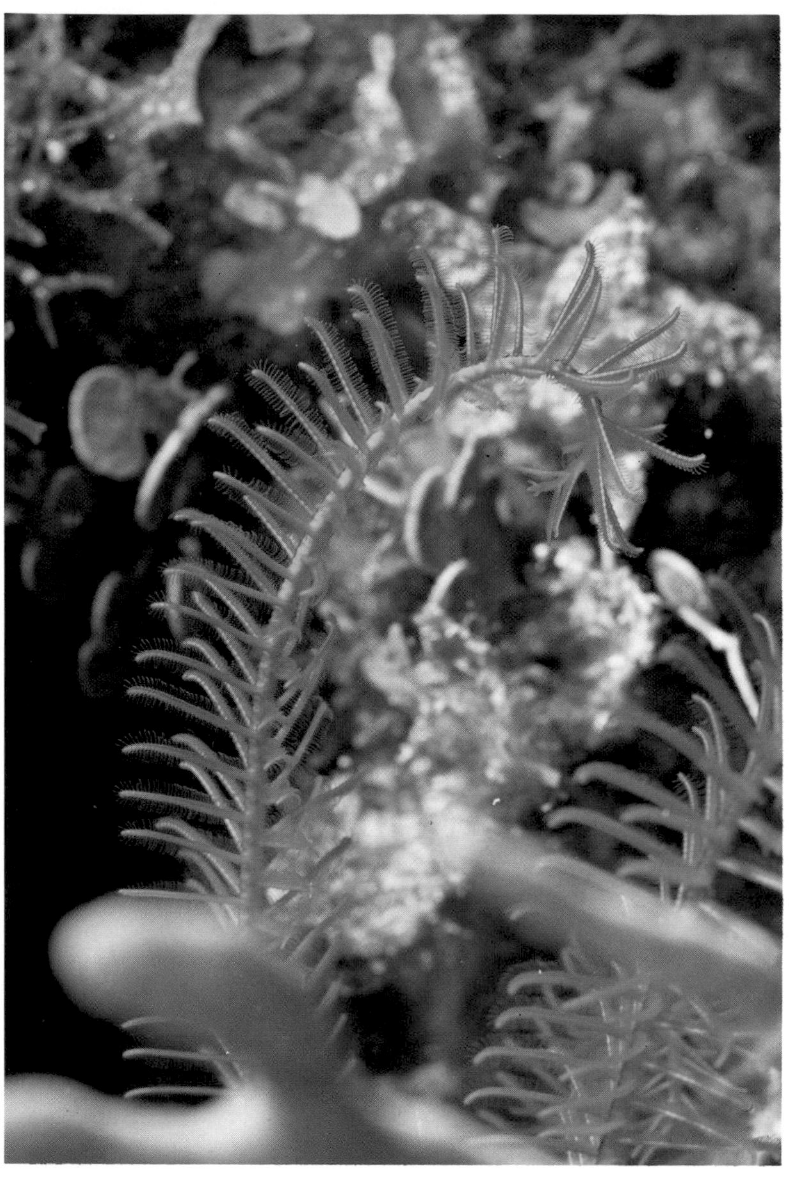

Each of the arms of *Nemaster rubiginosa* has many small pinnules which in turn possess fine "hairs" protruding from them. The pinnules are used to filter material from the water passing by the crinoid. Bahamas, Cat Island, 12 m depth.

Nemaster discoidea is a deeper reef crinoid than *N. rubiginosa.* Here is an example protruding from beneath a sponge. A lovely stylasterine coral, *Stylaster roseus,* and a variety of hydroids are also visible. Jamaica, Discovery Bay, 30 m depth.

Comactinia echinoptera is a crinoid which is active only nocturnally and has a different arrangement of the pinnules than do the species of *Nemaster.* Bahamas, Exuma Islands, 30 m depth.

the arm, an arrangement which is probably more efficient for feeding in this rheophobic crinoid. Under heavy wave and surge conditions feeding by the crinoids ceases and the arms are retracted into cover.

Nemaster rubiginosa occurs at depths of 10 to 30 m and on some reefs is the most common crinoid, with densities of one individual per meter-square having been observed. This species does not swim gracefully like *Analcidometra caribbea* and does not appear to move very much. It occurs in the Caribbean and the Bahamas.

References: Macurda 1973, Meyer 1973a, 1973b.

Nemaster discoidea (Carpenter)

This crinoid is superficially similar to the previous species, *N. rubiginosa*. There is variation in its coloration, with the arms black, grayish green, yellowish or even white and the pinnules beaded black and white, whitish with black tips, black or gray.

Nemaster discoidea is rheophobic and, like *N. rubiginosa*, attached underneath corals and crevices with its cirri. Rough weather conditions also cause it to cease feeding. While it is slightly smaller than *N. rubiginosa*, it is probably more abundant in depths below 15 m and reaches at least 45 m in depth. On reefs where the two species are found together and considerable bottom relief occurs, *N. discoidea* is generally on the lower portions while *N. rubiginosa* is on the higher areas of the reef. The species does not swim gracefully in open water but has been seen on rare occasions to move across open sand by walking with its arms on the bottom.

References: Macurda 1973, Meyer 1973a, 1973b.

Comactinia echinoptera (Muller)

This crinoid differs dramatically from the members of *Nemaster* in its life habits. It is active only nocturnally, when it extends its arms from crevices to feed. By day the arms are coiled and hidden from view.

This species has only ten arms. These have a stouter

appearance than those of *Nemaster* and the pinnules are arranged in a plane on either side of the arms, not at right angles to each other as in *Nemaster*. The arms are reddish, darker toward the disk, with the pinnules red, orange, yellow or beaded red and white.

On high profile reefs *Comactinia echinoptera* generally occurs high upon the reef, often with *N. rubiginosa*. It does not swim and seems limited in its movements.

References: Macurda 1973, Meyer 1973b, Kier 1966.

Family Colobometridae

Analcidometra caribbea (Clark)
Swimming crinoid

The ability of this colorful species to swim with amazing agility and coordination is its most esthetically pleasing attribute. It has ten arms with broad red and white bands and pinnules also banded red and white or orange-yellow.

It is rheophilic, typically being found attached by cirri to gorgonians in areas well exposed to slow but consistent water movement. *Analcidometra caribbea* is found from depths of 15 to at least 60 m and occurs widely in the West Indies. One individual was observed attached to the same gorgonian for a five-year period.

Swimming is carried out by bringing the arms downward in three or four coordinated groups with the pinnules extended and raising them with the pinnules retracted, resulting in movement in the aboral direction. About one second is required for each stroke of an arm. Specimens were observed to swim continuously for at least four minutes in aquaria, then after a period of rest continued for two minutes more.

References: Macurda 1973, Meyer 1973a, 1973b.

CLASS OPHIUROIDEA: BRITTLESTARS AND BASKET STARFISH

This class has arms which are long and slender, with the central body a flattened disk sharply set off from the arms. They lack an anus, with the mouth being used for ingestion of food and egestion of wastes.

Analcidometra caribbea is the only Caribbean crinoid that is able to swim with any ability. Normally, though, it is attached to gorgonians such as this *Ellisella* shown here. Bahamas, Crooked Island, 25 m depth.

Unidentified brittlestar. This gray species was photographed on a mat of algae. Jamaica, Discovery Bay, 3 m depth.

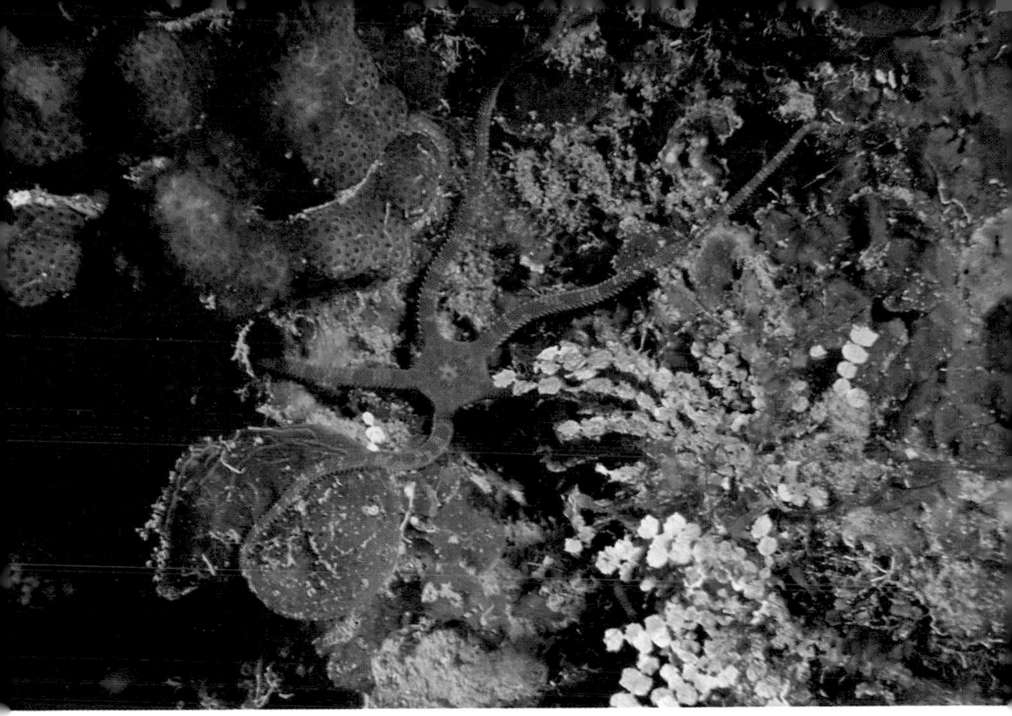

Unidentified brittlestar. The red species illustrated was found crawling among deep-reef algae and coral. Puerto Rico, Mona Island, 30 m depth, night.

Unidentified brittlestar. This specimen has the central disk in a small crevice in the rock and the arms extended, possibly to be used in filter feeding in the surrounding water. Puerto Rico, Mona Island, Playa Sardinera, 15 m depth, night.

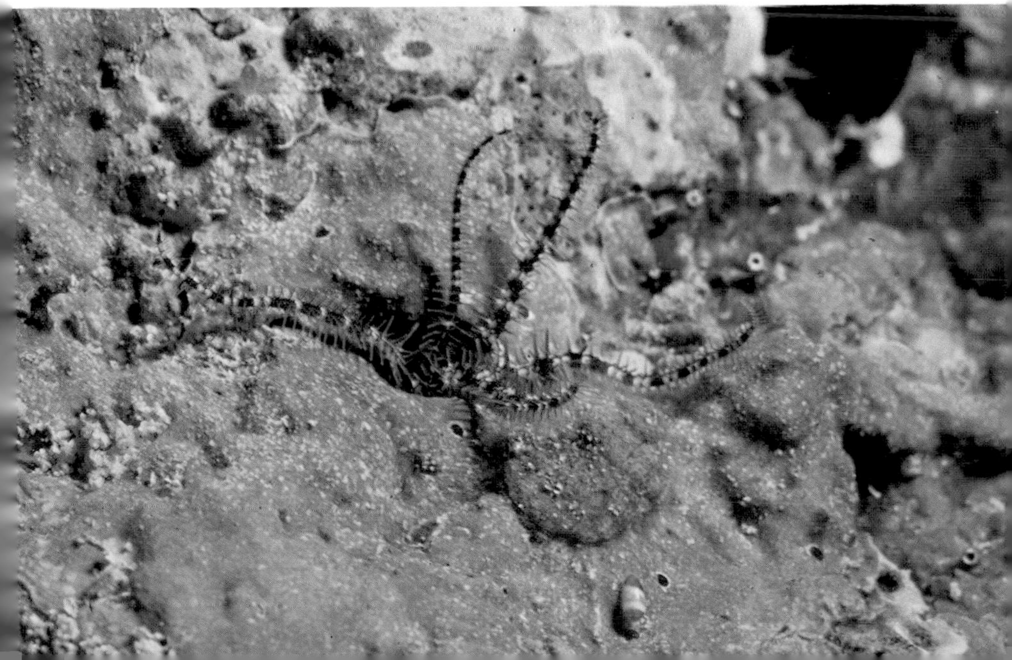

The arms are commonly used for locomotion and in many species are used for filter feeding. They are smooth or provided with spines of varying length. When handled, brittlestars tend to autotomize or break off their arms, hence their common name. The arms regenerate after a time. Often the filter-feeding species will have the central disk in the sediment or a rock crevice and the arms extended into the water at night. During the day the arms are retracted and the animal not visible. A light striking these filter-feeding ophiuroids causes them to quickly fold their outstretched arms.

Ophiuroids lack pedicellaria and are often found entwined or attached to organisms such as sponges or gorgonians.

ORDER LAEMOPHIURIDA

Family Ophiotrichidae

Ophiothrix suensonii Lutken
Sponge brittlestar

The sponge brittlestar is one of the most attractive of the ophiuroids, with the spines on the arms long, thin and glassy. The disk is flat with long spinelets and red, lavender or pink in color. The dorsal side of the arm has a dark purple, crimson or black stripe. The disk may be 2 cm in diameter and the arms up to 12 cm in length.

The species, both juveniles and adults, is found on sponges and gorgonians, climbing to exposed portions of these organisms at night and hiding during the day. *Ophiothrix suensonii* does not occur where sponges or gorgonians are lacking. It is known from Bermuda and southern Florida to Brazil and is common in the West Indies. Its depth range is from 2 to 458 m.

Reference: Clark 1933.

ORDER PHRYNOPHIURIDA

Family Gorgonocephalidae

Astrophyton muricatum (Lamarck)
Basket starfish

When fully expanded at night, the basket starfish is certainly the most striking of all the reef echinoderms. They reach nearly 1 m in diameter when expanded and occur at night on the top of prominent objects such as gorgonians or tall coral heads. During the day they hide, the large individuals in coral heads or rocks and small ones in the tangles of gorgonian arms, compressed into a compact ball the size of a fist. They may be tan to black in color.

The arms are divided many times so that the creature forms an efficient filtering net. The arms are oriented into the current, similar to what has been observed for many crinoids. The smallest arms at the edges fold over, rolling into a ball when touched by a planktonic animal; surprisingly large organisms (larval fishes, shrimp) can be captured this way. The food is then brought to the mouth.

When a light is directed at an expanded *A. muricatum* at night, it quickly begins to fold its arms and within a few minutes will be rolled into a confused mass. A caridean shrimp, *Periclimenes perryae* Chace, has been observed associated with *A. muricatum*, and large specimens are colored to match the basket starfish.

Astrophyton muricatum is known from southern Florida, the Bahamas and the Caribbean at depths from 6 to 30 m.

References: Davis, W. 1966, Clark 1933.

CLASS ASTEROIDEA: SEA STARS

The sea stars are not important animals of Caribbean coral reefs. They are star-shaped, the number of arms varying within and between species. The mouth is on the undersurface and the anus generally on the upper surface. Tube feet are used in locomotion.

A few species may occasionally be encountered on grass beds adjacent to reefs or on sandy areas of reefs. Two species are discussed further.

Family Oreasteridae

Oreaster reticulatus (Linnaeus)
West Indian sea star

The West Indian sea star is the largest and most massive

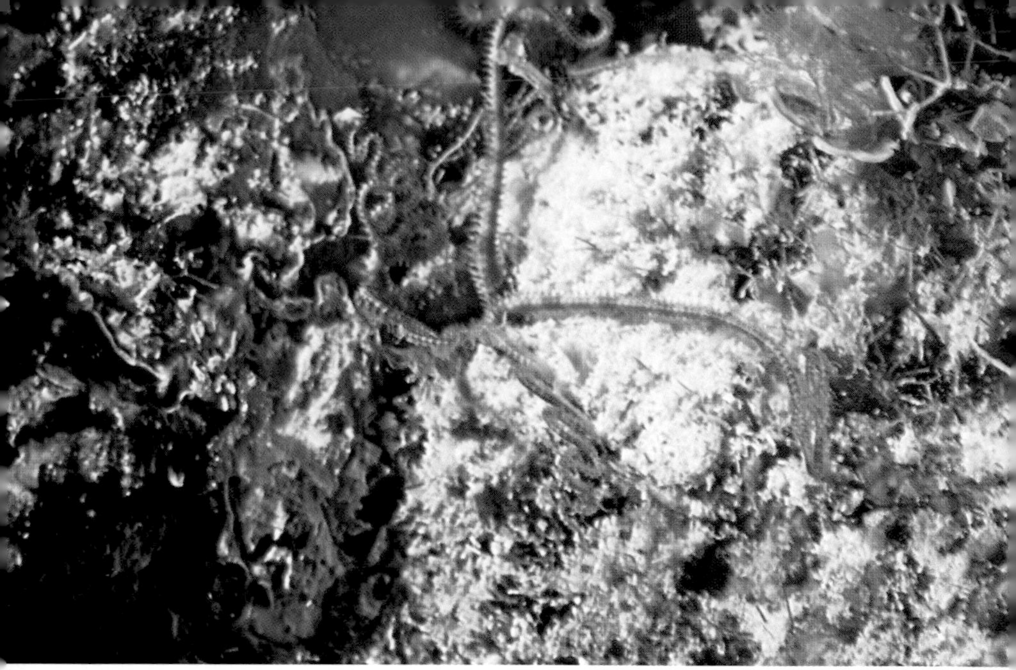

Unidentified brittlestar. This is a filter-feeding brittlestar which extends its arms into the water column. When a light is shined on this creature at night it rapidly folds the arms. Puerto Rico, Mona Island, 15 m depth, night.

The brittlestar *Ophiothrix suensoni* is found on sponges and gorgonians. It has long, thin spines on the arms and is shown here on the sponge *Haliclona rubens*. Jamaica, Discovery Bay, 15 m depth, night.

The fully expanded basket starfish, *Astrophyton muricatum,* is an impressive sight at night. It forms a circular fan for filtering material from the water passing through its arms at night. Puerto Rico, La Parguera, Caracoles Reef, 5 m depth, night.

Basket starfish, *Astrophyton muricatum,* climb up the stalks of gorgonians from their daytime refuges to filter plankton from the water at night. Jamaica, Discovery Bay, 15 m depth, night.

of the starfishes occurring in the region. It has four to seven arms which are broad with numerous blunt spines on the upper surface. They occur in various colors, including olive green, yellow, brown and reddish brown. The maximum size is about 50 cm in diameter.

It is exceedingly common in sandy and seagrass areas. Rarely does it occur on reefs. It is found in shallow water, generally less than 15 m, from Cape Hatteras and Bermuda to Brazil and the Cape Verde Islands.

References: Downey 1973, Ummels 1963, Thomas 1960, Clark 1933.

Family Astropectinidae

Astropecten duplicatus Gray

This flattened sea star has a row of large supramarginal plates around its edge and seldom has other than five acutely pointed arms. It occurs on sandy bottoms, often adjacent to reefs, and is the only member of the genus in the Caribbean to occur shallower than 20 m.

Astropecten duplicatus occurs throughout the Gulf of Mexico and the Caribbean as far north as Cape Hatteras at depths of nearly 0 to over 500 m.

Reference: Downey 1973.

CLASS ECHINOIDEA

This class contains the sea urchins, sand dollars and heart urchins. The body is enclosed with a test, spherical, oval or flattened in shape, of closely fitted calcium carbonate plates with the mouth on the surface normally directed toward the substrate (oral) and the anus on the surface away from the substrate (aboral). Hundreds of movable spines project from the test and are used for locomotion. Tube feet and pedicellaria, jaw-like structures set on movable stalks and often containing poison glands, occur on the test's surface. The mouth contains a masticatory apparatus, the Aristotle's lantern, containing five calcareous teeth used for grazing.

References: Hyman 1955, Boolootian 1966.

ORDER CIDAROIDA

Family Cidaridae

Eucidaris tribuloides (Lamarck)

The thick, flat-tipped spines of *Eucidaris tribuloides* distinguish this species from the other shallow-water Caribbean sea urchins. There are other members of this family in the Caribbean area, but they occur well below the depths of coral reefs. The spines are often densely overgrown with various organisms and are about equal in length to the diameter of the test. The test is light or dark brown, often mottled or marbled in color, with occasionally a greenish or red tinge.

The species occurs in or around rocks, often in the rubble of back reef or lagoonal areas, but does not aggregate like some other sea urchins. The spines nearest the mouth are used in locomotion. It is probably fairly slow growing.

The species is known from Bermuda, although reportedly rare, and is common in southern Florida, the Bahamas, the Caribbean to Brazil and offshore along the Carolina coast. It is also known from the Cape Verde Islands and Ascension Island, but not definitely from the west African coast. It occurs from low tide to at least 25 m.

References: Clark 1933, Cutress 1965, Kier 1966, 1975.

ORDER CENTRECHINOIDA

Family Echinidae

Lytechinus variegatus (Lamarck)

This species is found around or under rocks on reefs, in seagrass beds or on muddy bottoms. They can be quite abundant, with up to 15 adults per square meter. There are three distinct subspecies with discrete geographic ranges. The color of the test and spines of *Lytechinus variegatus* varies considerably between and within subspecies. The test may be purple, green, red or mottled with white. The spines, which are short, can be green, purple, red or white and one color at the tip (distally) with a second color at the base (proximally). The most common combination of colors in the Caribbean is a green (solid or mottled) test with green spines.

Small *Astrophyton muricatum,* including one black individual, are seen here at night on a gorgonian. Bahamas, Acklins Island, 18 m depth.

The West Indian sea star, *Oreaster reticulatus*, is seen here in a bed of the seagrasses *Thalassia testudinum* and *Syringodium filiforme.* Bahamas, Cat Island, 3 m depth.

The sea star *Astropecten duplicatus* is found on fine sediment bottoms on occasion next to reefs. Puerto Rico, La Parguera, Caracoles Reef, 5 m depth.

The slate-pencil urchin, *Eucidaris tribuloides,* has blunt spines which are usually densely covered with sedimentary material. Jamaica, Discovery Bay, 3 m depth.

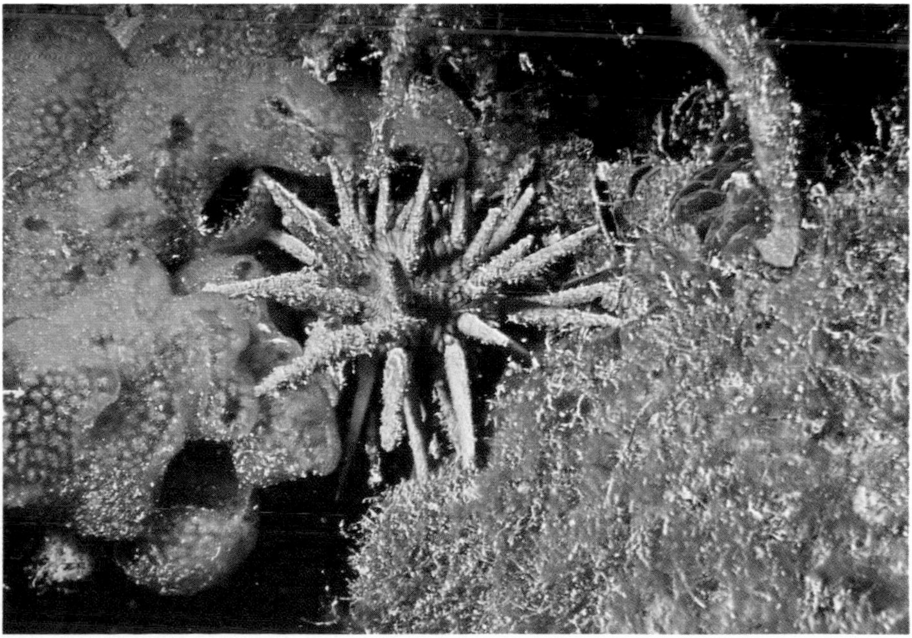

The pedicellaria are white, a character which distinguishes *L. variegatus* from *L. williamsi* (which has purple pedicellaria), and contain a mild toxin.

These urchins often cover the upper (aboral) surface with bits of seagrass or parts of bivalve mollusc shells. It is believed this provides protection from the bright sunlight in the shallows where these sea urchins abound.

Lytechinus variegatus is often found in aggregations. In some instances these are evidently for spawning purposes, in others this is not the case. It is believed to reach a maximum age of two to three years and is preyed on by some gastropod molluscs and birds.

The species occurs throughout the tropical western Atlantic. One subspecies, *L. variegatus variegatus,* occurs from Brazil, throughout the Caribbean to southern Florida and the Yucatan Peninsula. This subspecies is reported not to occur at the island of Barbados. A second subspecies, *L. variegatus carolinus,* occurs from North Carolina south through the Gulf of Mexico. These two subspecies occur together in southern Florida, Yucatan and the northern Bahamas. The third subspecies is endemic to Bermuda.

References: Serafy 1973, Harvey 1947, Clark 1933, Halstead 1965, Moore *et al.* 1963, Kier 1966, 1975, Kier and Grant 1965.

Lytechinus williamsi Chesher

This urchin seems to be most common on coral reefs. It can occur in beds of *Agaricia agaricites* or *Acropora cervicornis,* but evidently does not feed on the coral itself. The test is lightly colored with a brown or red stripe at the joints of the major plates. The spines are green or white with distinct ridges along their length which *L. variegatus* lacks. The pedicellaria of *L. williamsi* are purple and easily distinguished from the white ones of *L. variegatus.* There is no evidence of *L. williamsi* ever covering the aboral surface with bits of shell or other material.

The distribution of *L. williamsi* is not well known, with definite records only from Panama, Belize and Jamaica. It occurs between 5 and 12 m on reefs.

References: Chesher 1968a, Serafy 1973, Kier 1975.

Tripneustes esculentus (Leske)
West Indian sea egg

This species has the largest test of any West Indian sea urchin, occasionally 15 cm in diameter. The spines are short and white, contrasting with the normally black test. The eggs of this species are highly esteemed as food in a number of localities and its edibility has led to a severe decline in populations around some West Indian islands. Certain persons may experience gastrointestinal disturbances from eating the roe of the species, possibly due to an allergic reaction. A large part of the animal's energy is put into production of eggs and sperm during the summer spawning season, so much that the growth is nearly stopped during this period. During non-spawning seasons the species again grows normally. Individuals with ripening gonads tend to form aggregations. The species may live up to two or three years.

Tripneustes esculentus feeds almost exclusively on algae but does not consume much of the highly calcified species. They often cover the upper (aboral) surface of the test with bits of seagrass or pieces of mollusc shells held with the tube feet, probably to protect themselves from the intense sunlight in the shallow waters they occupy. The young live under rocks or other cover while the adults generally remain in the open. Various commensals, including a shrimp and a gastropod mollusc, may occur with this sea urchin.

The species is widely distributed from Bermuda and the Carolinas to Brazil, and also at Ascension Island and the west coast of Africa. It seldom occurs deeper than 6 m.

References: Lewis 1958, Clark 1933, Halstead 1965, Harvey 1947, Kier 1966, 1975, Kier and Grant 1965.

Family Echinometridae

Echinometra lucunter (Linnaeus)
Rock-boring sea urchin

This sea urchin reaches a test diameter of over 10 cm and has short, sharp-pointed spines with a thick base. They occur in shallow rocky areas, on coral reefs and rarely in

Lytechinus variegatus is a sea urchin which is found in a variety of habitats including coral reefs. Its pedicellaria are white and the test and spines are various colors. Puerto Rico, La Parguera, Mario Reef, 1 m depth.

Lytechinus williamsi occurs almost exclusively on coral reefs. Its purple pedicellaria distinguish it from similar sea urchins. Jamaica, Discovery Bay, 12 m depth.

The West Indian sea egg, *Tripneustes esculentus,* is an important food item in the Caribbean. Its dark test with white spines makes it easily identified. Cayman Islands, 3 m depth.

Lytechinus williamsi is not known to feed on corals, but occurs almost always in the vicinity of corals. Jamaica, Discovery Bay, 10 m depth.

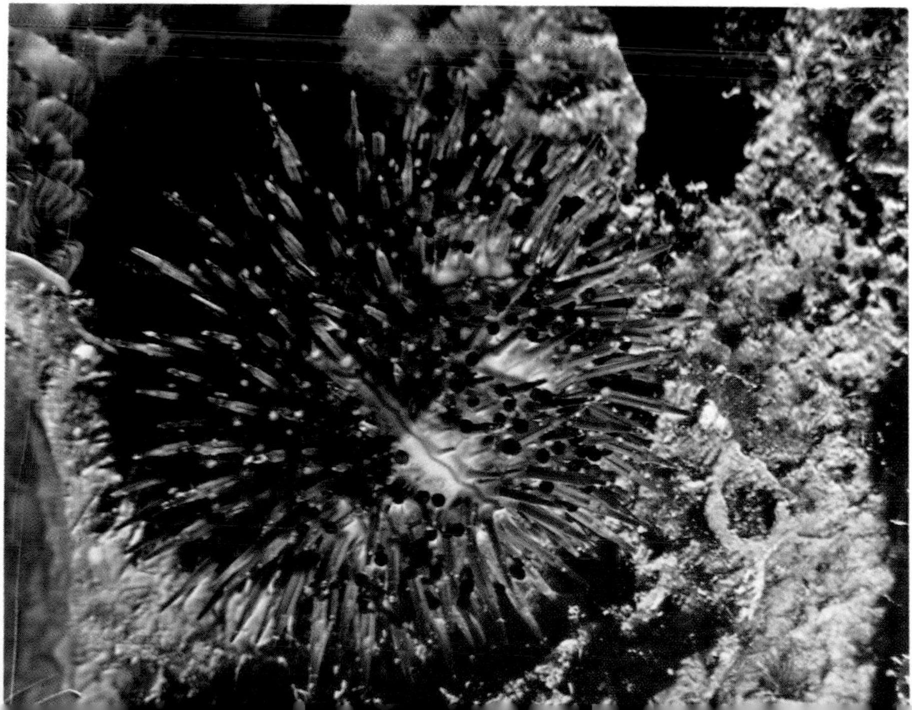

grassy areas. Usually they are encountered in rock holes just large enough for the urchin and may occur beneath slabs of rock or coral, particularly in areas of surf and wave surge.

The test is black to red, often particularly red on the upper (aboral) surface; the spines are normally black. The Aristotle's lantern, the movable toothed opening of the mouth, is used to erode the substrate away beneath the sea urchin; over a period of time this is enlarged into a spherical hole which the urchin occupies during the day. This hole is defended against other members of the species. *Echinometra lucunter* often leaves this hole at night to feed on algae in the area within a few centimeters of the opening.

The erosive effect of large numbers of these urchins is particularly important on reefs formed by coralline algae. The coralline algae produce shallow reefs in wave-swept areas where *E. lucunter* thrives. The large helmet shell, *Cassis tuberosa*, preys on the urchin. The fishes *Ginsburgellus novemlineatus*, *Gobiosoma multifasciatus* (Gobiidae) and *Arcos rubiginosus* (Gobiesocidae) occur with *E. lucunter*, often hiding beneath the test. One of the gobies, *G. novemlineatus*, feeds on the tube feet and pedicellaria of *E. lucunter*.

In the West Indies the species does not spawn year-round and shows peak spawning activity during the summer months.

Echinometra lucunter occurs on both sides of the Atlantic. On the western side it is known from Florida and Bermuda to Brazil. Interestingly, the largest individuals of the species have been collected near the northern and southern extremes of the range. Most specimens in the West Indies are about one-half of the size of these largest individuals. While most occur between 0 and 2 m, the species has been taken as deep as 45 m.

References: Abbott *et al.* 1974, Harvey 1947, McPherson 1969, Clark 1933, Kier 1966, 1975, Kier and Grant 1965.

Echinometra viridis Agassiz

This sea urchin is primarily a reef dweller, living in holes and crevices. The test is reddish and the spines lighter in color at their base, often with a white ring where they meet the test. The spines are tipped with a dark color, usually violet, which is the easiest method of distinguishing this species from *E. lucunter*. The spines of *E. viridis* are also longer relative to the test size than in *E. lucunter*.

Echinometra viridis may occur with *E. lucunter* as it does on the patch reefs off the Florida Keys, but does not occur in all the habitats that *E. lucunter* can occupy. This is particularly true of subtidal rocky shelves where *E. lucunter* forms its holes and is extremely abundant. *Echinometra viridis* feeds almost exclusively on algae.

The geographic distribution of *E. viridis* is more restricted than other Caribbean sea urchins. It occurs from southern Florida to Venezuela but is believed to be absent in the West Indies east of the Virgin Islands. It occurs from 0 to about 12 m.

References: Clark 1933, Adams *et al.* 1974, McPherson 1969, Adey 1974, Harvey 1947, Kier 1966, 1975, Kier and Grant 1965.

Diadema antillarum Phillipi
Long-spined sea urchin

The long-spined sea urchin is a member of nearly every reef community, except where the surf is heavy, occurring in such numbers as to be of prime importance in the ecology of reef communities. While the test of *D. antillarum* is fairly small (to 10 cm maximum diameter), the spines are long (up to 30-40 cm) and slender with points that easily penetrate human skin. Once they have penetrated some distance into the flesh of the unlucky person who steps or falls onto a *D. antillarum*, the spines break off, remaining embedded. Due to their brittle nature and microscopic serrations lining their surface, these spines are almost impossible to remove, but will eventually dissolve in the body without harmful effects. The spines are hollow and contain a mild toxin to which certain individuals seem to be sensitive. The danger of infection from the spines also exists.

It would be nearly impossible to mistake *D. antillarum* for any other Caribbean sea urchin. The test is black; the

Echinometra lucunter, the rock-boring sea urchin, produces holes in rock surfaces by the scraping action of its Aristotle's lantern. The hole is occupied during the day, but the urchin sometimes leaves it at night. Puerto Rico, La Parguera, Laurel Reef, 1 m depth.

Echinometra viridis has strong, but not particularly sharp, spines which are tipped with a dark color. Jamaica, Discovery Bay, 12 m depth.

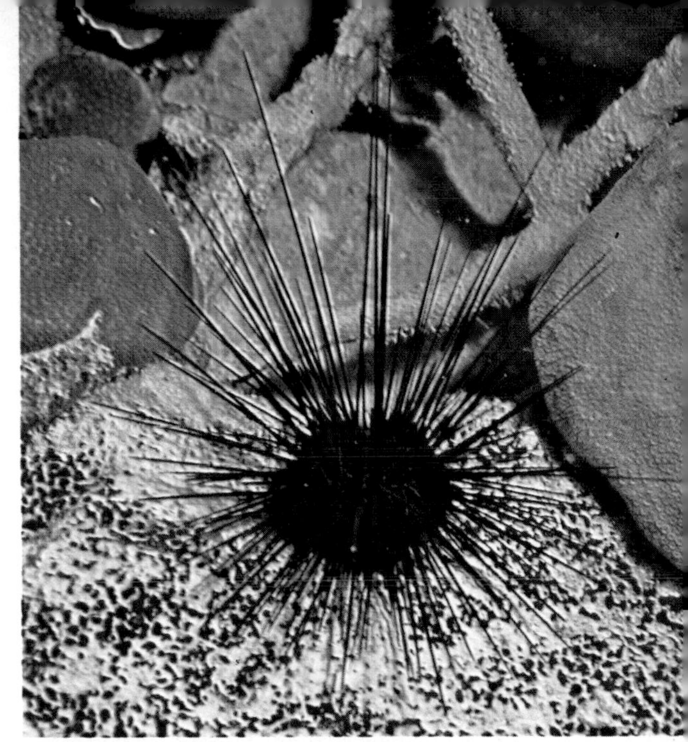

This is a typical *Diadema antillarum* photographed at night on a rock substrate where it is grazing. Heads of the coral *Montastrea annularis* lie on either side of it, but the surface on which it rests is nearly devoid of macroalgae. Jamaica, Discovery Bay, 15 m depth, night.

The spines of this young *Diadema antillarum* are black and white and it is sitting exposed at night on a rock surface with sponges. Puerto Rico, La Parguera, Laurel Reef, 12 m depth, night.

spines are black or white. In juvenile individuals the spines are often black-and-white-banded. During the day *D. antillarum* hides within crevices or around sheltered locations on the reef. At night they move into the open, often onto sandy areas adjacent to reefs where they feed on algae or seagrasses. The species also can be abundant in seagrass beds, rocky shelves, mangrove areas and even sandy bottoms.

Densities of *D. antillarum* can often be high on reefs, with as many as 13 per square meter having been reported. They are one of the agents, perhaps the major one, responsible for the formation of "halos," areas of bare sand without a covering of seagrass or macroalgae 2 to 10 m wide around West Indian patch reefs. The urchins migrate from the reef at night to feed in this sandy area, but are limited in the amount of distance they can travel. The seagrasses are grazed heavily to a certain distance from the reef, but no farther. *Diadema antillarum* remaining on reefs at night feed mainly on algae, rather than seagrasses.

Recent work indicates that *D. antillarum* is important, perhaps critical, in the maintenance of the populations of benthic reef invertebrates. This sea urchin regularly grazes the rocky surfaces of the reef, reducing the amount of macroalgae present and perhaps preying on newly settled larvae of benthic invertebrates. Experimental removal of all the *D. antillarum* from patch reefs has resulted in great increases in the amount of macroalgae present and a reduction in the coral cover. In some instances where *D. antillarum* was not present, algae were literally taking over the reef from the corals.

Quantities of calcium carbonate rock are also scraped from the reef surface and ingested during feeding, producing an erosive effect on the reef itself.

At least 15 species of fishes, such as the queen triggerfish, and some gastropod molluscs, including the helmet shell *Cassis tuberosa*, are known to prey on *D. antillarum*. Fishes that have been punctured by spines, probably during predation, have been noted by the characteristic purplish spot left on the skin by the embedded spines. Some juvenile fishes (particularly grunts) and mysid shrimp hide within the spines of this urchin. The spines are directed toward any movement in the vicinity or a shadow falling on the urchin. Two species of shrimp, black in color, are known to occur as commensals with *D. antillarum* on the test and spines.

These urchins often aggregate for spawning and appear to spawn throughout much of the year in the Caribbean. The geographic range includes Florida and Bermuda to Surinam, the islands of the tropical mid-Atlantic and the coast of western Africa. The species is known from the surface to 400 m.

References: Randall *et al.* 1964, Lewis 1964, 1966, Clark 1933, Sammarco *et al.* 1974, Ogden *et al.* 1973, Millott 1950, 1966, Kier 1966, 1975, Kier and Grant 1965.

ORDER SPATANGOIDA

Family Spatangidae

Meoma ventricosa Lamarck

This sand-dwelling member of the Spatangoida, the largest and most successful order of sea urchins, is extremely common but seldom observed on Caribbean reefs. Both young and adults remain buried in sand during the day; the adults emerge to rest on the sandy surface at night while the young remain hidden.

The biology of *M. ventricosa* has been studied. The urchin occurs in a wide variety of habitats (*Thalassia* beds, coral reef areas, deep reefs) in sediments from fine silty sand to coarse reef rubble and at depths from the intertidal to 200 m. It occurs throughout the Caribbean, the Bahamas, Florida and as far north as Bermuda.

In Florida spawning occurs from August to February with the peak from November to January. Unlike some other echinoids which have a simultaneous mass spawning of numerous individuals, *M. ventricosa* spawns individually without regard to what its neighbors are doing. The male and female gametes are shed directly into the water and the larvae are planktonic.

The larvae evidently settle in reef areas, as small individuals are almost exclusively found there buried 5 to 8 cm in the sediment beneath and adjacent to coral slabs. It may take

This specimen of *Diadema antillarum* has the spines white rather than black, an unusual occurrence. The periproct, a bulbous extension of the anus (on the aboral surface) of the urchin can be seen in the center of the photograph. Bahamas, Crooked Island, 10 m depth, night.

Meoma ventricosa, one of the heart urchins, buries in the sand during the day and is found exposed usually only at night. It has very short spines which pose little danger to someone handling it. Puerto Rico, La Parguera, Mario Reef, 8 m depth.

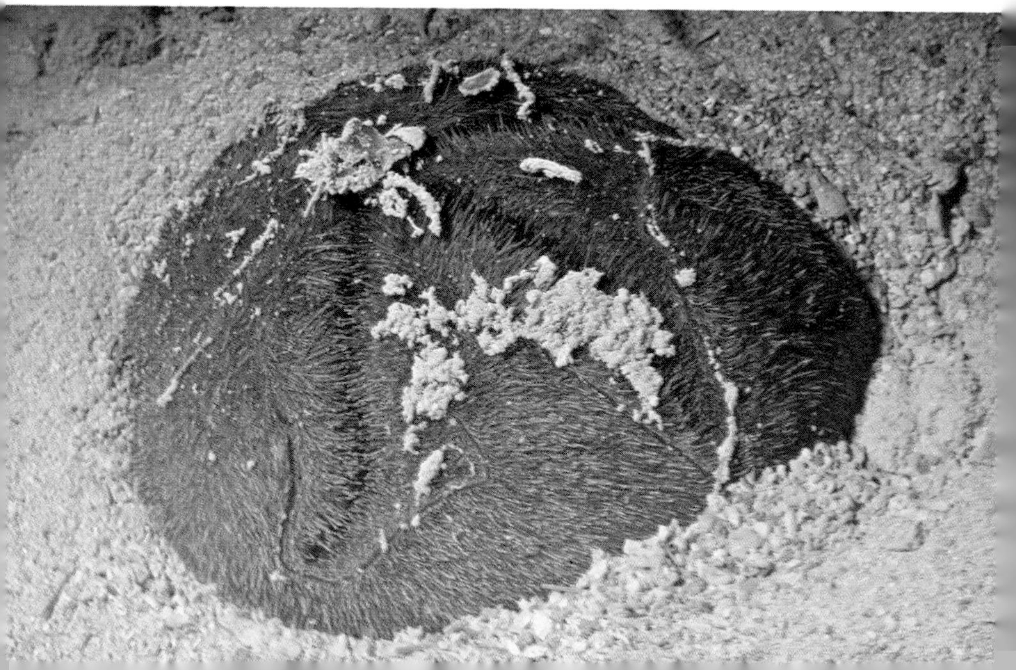

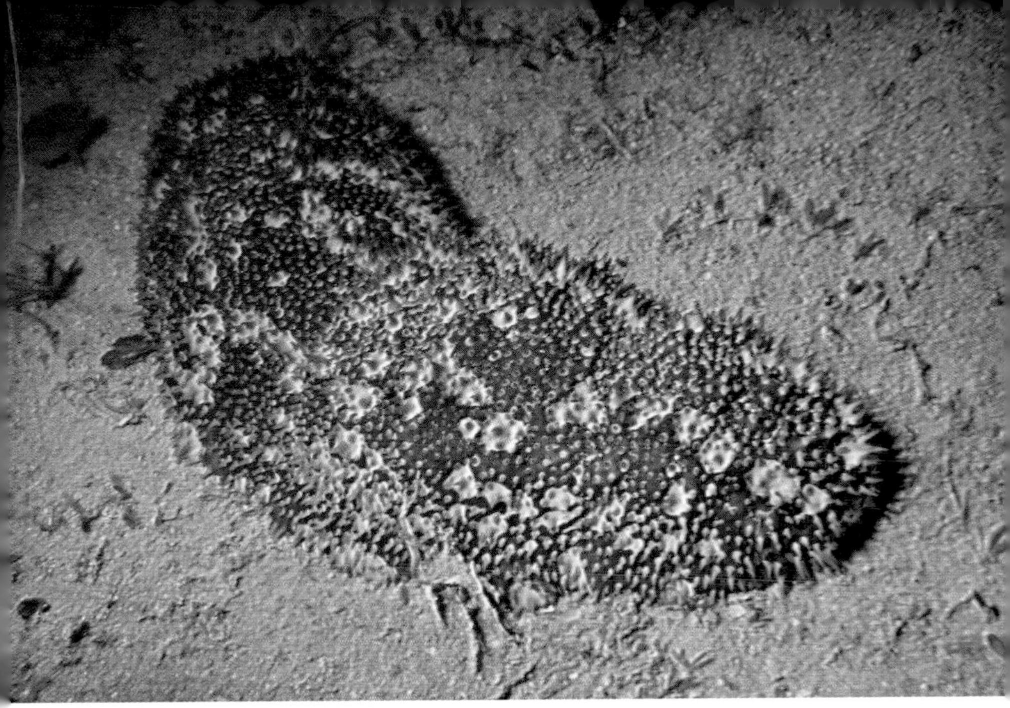

The large sea cucumber *Astichopus multifidus* is found on sediment bottoms adjacent to reefs. It is variable in color, some individuals being brown with white spots as shown here. Jamaica, Discovery Bay, 30 m depth.

Astichopus multifidus can also be white with brown spots. It moves rapidly for a sea cucumber, reaching speeds of 2 m per minute. Jamaica, Discovery Bay, 20 m depth, night.

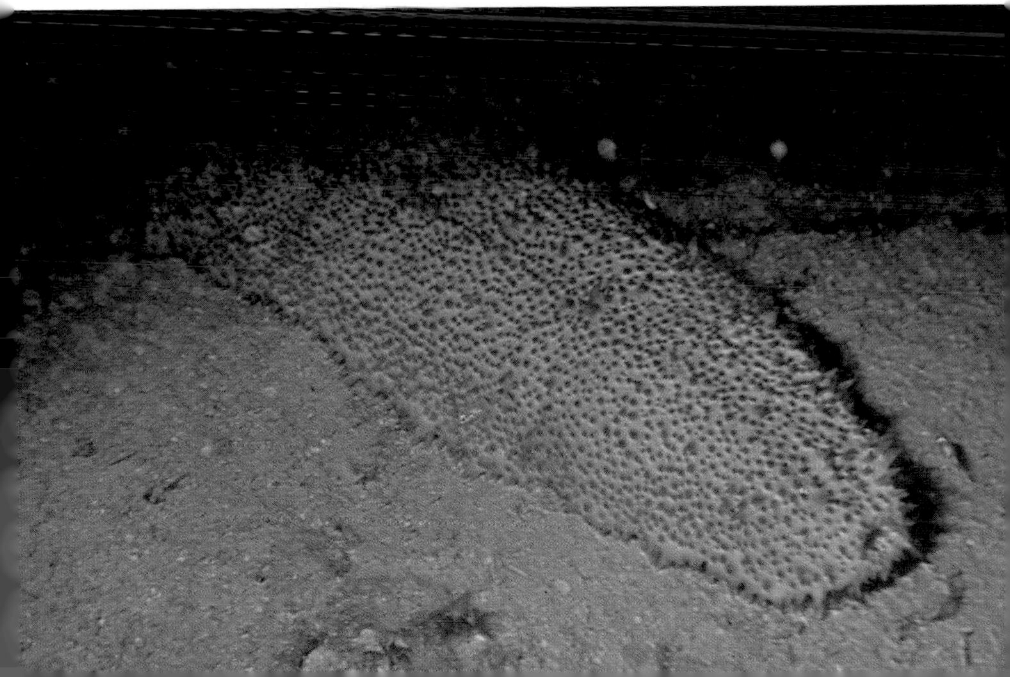

four years for an individual to reach a test length of 14 cm, but growth rates vary with different habitats. The food available to *M. ventricosa*, the organic material (algae, bacteria) found growing on sediment grains and digested by passing the sediment through the gut, is evidently not the only factor limiting growth. The oxygen concentration in the interstitial water of the sediment may be just as important in limiting respiration and feeding rate, therefore growth. The oxygen concentration can decrease substantially at night, and this may be the factor causing larger *M. ventricosa* to emerge from the sediment at night.

The urchins move slowly through the sand (3-6 cm per hour during the day, 7-8 cm per hour at night) and are often found in "herds" with two to three individuals per meter-square. They have few known predators (including stingrays and *Calappa* crabs) and secrete a yellowish substance when disturbed which is repellent to and able to kill small fish. A commensal pinnotherid crab occurs with *M. ventricosa*, with up to 12 crabs per urchin, usually on the oral surface.

References: Chesher 1968b, 1969, 1970.

CLASS HOLOTHUROIDEA: SEA CUCUMBERS

The sea cucumbers are unusual echinoderms in that the oral-aboral axis is greatly elongated and they lie on one side instead of on the oral surface. The skeleton is reduced to microscopic ossicles embedded in the body wall. They lack arms and pedicellaria, and some even lack tube feet.

About 25 species occur in shallow Caribbean water, but most are not easily distinguished. They occur on sediment bottoms or during the day beneath coral or rocks. Others burrow into the bottom. They feed by passing sediment through the gut and digesting any organic material contained in it or by "catching" detritus or small planktonic organisms on mucous-covered "tentacles" centered around the mouth. Breathing is by means of the skin, tube feet or internal respiratory trees. The body wall often contains a toxin, called "holothurin," which makes them distasteful. They move slowly, usually by means of the tube feet, and often have slender commensal fishes of the family Carapidae living within them during the day.

References: Deichmann 1963, Tikasingh 1963, Clark 1933.

ORDER ASPIDOCHIROTA

Family Holothuriidae

Astichopus multifidus (Sluiter)

Although larger than most, this sea cucumber is fairly similar in appearance to many Caribbean holothurians. It reaches 45 cm in total length and about 10 cm in diameter. The body is soft, with the dorsal and ventral surfaces uniformly covered with hundreds of tube feet. On the dorsal surface these tube feet are papillate.

The color is variable. The dorsum may be brown or gray, often a chocolate brown with small scattered white spots; some have three large chocolate brown spots on a light brown dorsum.

This species is very active for a holothurian. It moves by crawling, rolling or even "bounding." It can approach a speed of 2 m per minute or an astounding 0.12 km per hour! Like many holothurians, it feeds by ingesting sediment and digesting the organic material contained in it.

Astichopus multifidus is known from the West Indies, the Campeche Bank area and Dry Tortugas, Florida. They rarely occur on grass beds in shallow water but are most common on the sandy areas on and around reefs to a maximum depth of about 40 m.

References: Glynn 1965, Clark 1933, Deichmann 1963.

ORDER APODA

Family Synaptidae

Euapta lappa (Mueller)

This synaptid holothurian looks more like a giant worm than a typical sea cucumber, but it is one of the most characteristic inhabitants of West Indian reef flats. It is long (over 1 m), thin (about 4-8 cm) and flaccid, with the skin covered with large rounded knobs. The body can be contracted,

The synaptid sea cucumber *Euapta lappa* hardly resembles a typical holothurian, but rather seems to be a giant worm with a knobby body. The "mop" of tentacles is to the left in the photograph and is used for feeding on detrital material. Jamaica, Discovery Bay, 3 m depth.

Ascidia sydneiensis has only the tip of its siphon visible with most of the animal hidden in a crevice. It looks more like a tubular sponge than an ascidian. Jamaica, Rio Bueno, 18 m depth.

Unidentified ascidian, possibly *Clavelina picta.* This is a colonial ascidian found growing on gorgonian stalks, mangrove roots and other hard substrates. Puerto Rico, La Parguera, Laurel Reef, 3 m depth.

Unidentified ascidian, possibly *Clavelina* sp. Bermuda, south shore reefs, 8 m depth.

which they do when disturbed, to a fraction of its extended length. *Euapta lappa* is light gray to light brown in color, often variegated with small white patches or darker longitudinal stripes. A "mop" of tentacles occurs at the mouth end of the body, this mass being drawn in and out of the oral opening.

During the day *E. lappa* hides beneath coral slabs; at night it emerges and occurs on open bottoms. The species lacks tube feet and respiratory trees. It is found in the Caribbean and the Bahamas but not in Bermuda. It also reaches the Canary Islands in the east. It is limited to shallow depths and is seldom found below 10 m.

References: Clark 1933, Deichmann 1963.

Mixed zone at Discovery Bay, Jamaica. Visible are *Acropora palmata*, *A. cervicornis*, *Montastrea cavernosa* and *M. annularis*.

Phylum Chordata

SUBPHYLUM UROCHORDATA
CLASS ASCIDIACEA: ASCIDIANS OR SEA SQUIRTS

The ascidians are bottom-dwelling organisms on hard substrates generally in shallow water. They are sac-like or irregular in shape and vary from a few millimeters to several centimeters in length. They may occur singly (generally the larger, more complex species) or colonially, reproducing by budding, with each individual of the colony being termed a "zooid." The body is enclosed in a flexible outer covering, often called the "test," with openings for the incurrent and excurrent siphons. The test is the portion of the animal attached to the substrate. When touched or disturbed many species of ascidians can contract and close the openings of the siphons.

Most ascidians are hermaphroditic, producing a larva which resembles vertebrate larvae. It possesses a notochord which is lost after metamorphosis, and the larva eventually attaches to the substrate. It then transforms into the typical sea squirt.

Probably close to 100 species of Ascidiacea occur in the Caribbean, many of which may occur on reefs. Some are sizable, but mostly hidden in crevices, such as *Ascidia sydneiensis* Stimpson in which only the translucent yellow incurrent siphon is generally visible. *Ascidia sydneiensis* occurs in the tropics of all the oceans at moderate depths. Other large solitary species occur openly, but the test may be so covered with encrusting organisms that the ascidian is almost unrecognizable except for the siphons. Colonial species, such as members of the genus *Clavelina*, are found on rocky substrates, old gorgonian stalks or mangrove roots. *Distaplia*, a colonial genus, may be encrusting or upright, resembling a

Distaplia stylifer is a small ascidian whose colonies resemble berries on a short stalk. It is found widely on reefs. Puerto Rico, Mona Island, 15 m depth.

A species of *Distaplia.* The individual zooids of the colony are visible. Puerto Rico, La Parguera, Laurel Reef, 12 m depth.

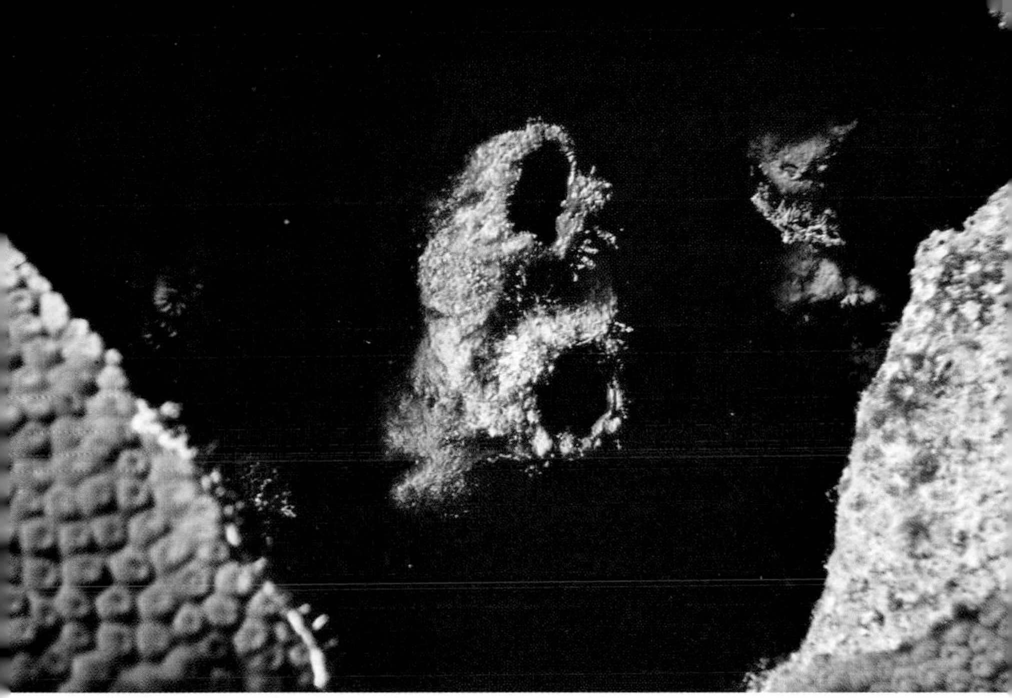

This and the following photograph show the rapid closing of the siphons of an ascidian when disturbed. The undisturbed ascidian has the two siphons open wide. Bahamas, Bimini, Turtle Rocks, 10 m depth.

One second after being touched the two siphons are closed by sphincters which effectively shut them off. Bahamas, Turtle Rocks.

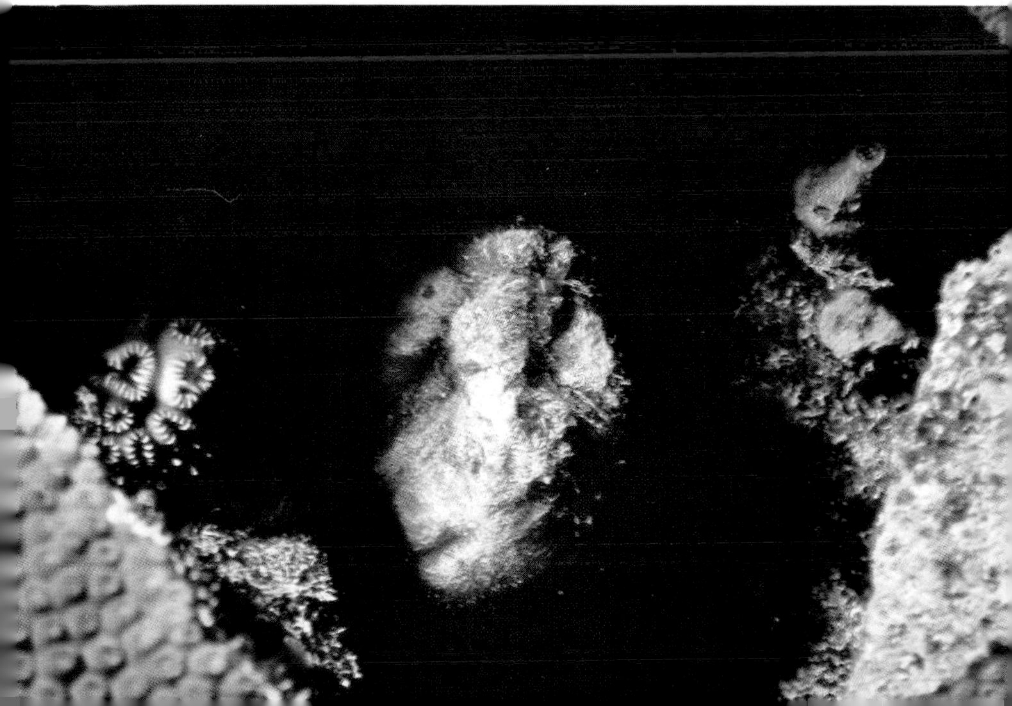

raspberry on a thick stalk. Ascidians are found at all depths on the reef and most species are widespread in their distribution.

References: Van Name 1921, 1930, Van Der Sloot 1969, Millar 1962, Millar and Goodbody 1974.

Deep-reef environment at 55 m depth, Discovery Bay, Jamaica. Several species of corals are visible.

Marine Plants

The plants of the sea encompass a wide spectrum of the plant kingdom. They fall generally into three groups: the flowering plants or spermatophytes, the algae and the fungi. Although there are relatively few species of marine spermatophytes, they are usually abundant and of great importance in shallow water communities. The algae are much more diverse and are divided into macro- and microscopic species on the basis of size. They are further divided by their pigments into the green, brown and red algae plus other groups such as the diatoms and dinoflagellates. Since microalgae are not apparent to underwater observers, they are not considered in the guide except for the symbiotic zooxanthellae and the endolithic algae. The microalgae are, however, important elements of reef communities. Similarly the fungi are also microscopic and are not considered, although they are also important.

Photosynthetic marine plants are limited in the depth they can inhabit by the penetration of light. In even the clearest tropical waters, macroalgae are essentially absent below approximately 100 m. Marine plants are intimately associated with coral reefs, so much so that reefs really owe their existence to the characteristic of symbiotic algae to enhance calcification of corals and the ability of coralline and calcareous algae to produce carbonate material.

The sizable plants of reefs are secured to the substrate in various manners. On hard substrates the plants are attached by a holdfast secured to the substrate itself. On sediment bottoms root-like structures or rhizoids serve to anchor the plant, often entwining particles of the substrate. Other plants may grow upon the surface of animals or plants as epiphytes.

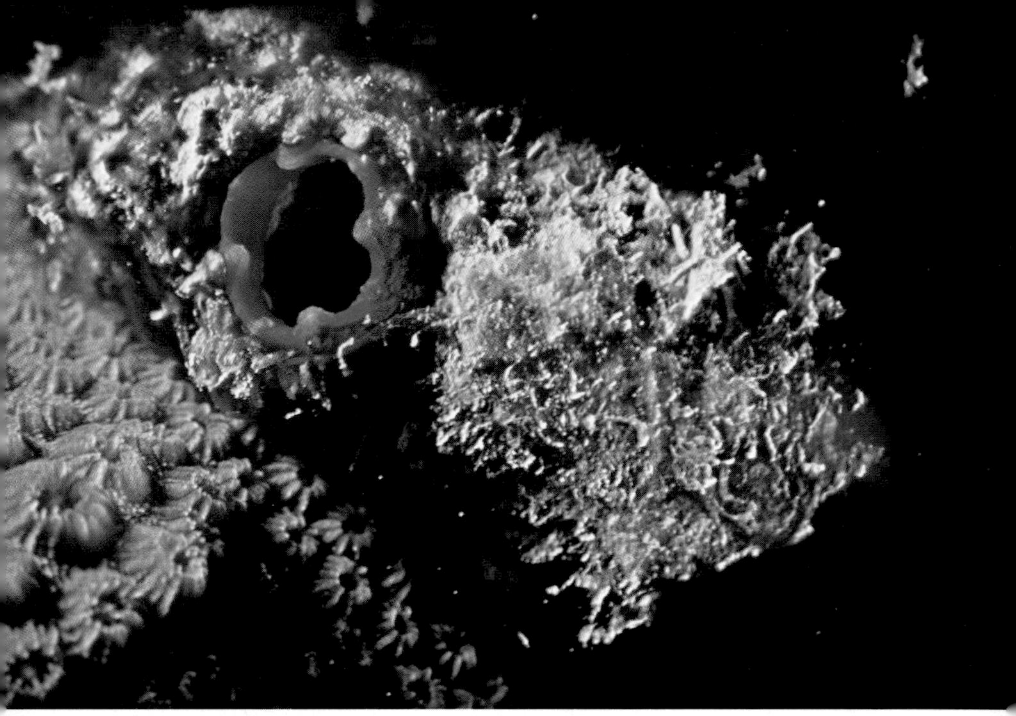

Unidentified ascidian. This species has the siphons lined in red but with the rest of the body covered by camouflaging material. Puerto Rico, La Parguera, Laurel Reef, 12 m depth.

Unidentified ascidian. This orange ascidian occurs in the photograph as a colony with an unidentified anemone in the upper left corner of the photograph. Puerto Rico, La Parguera, Laurel Reef, 12 m depth.

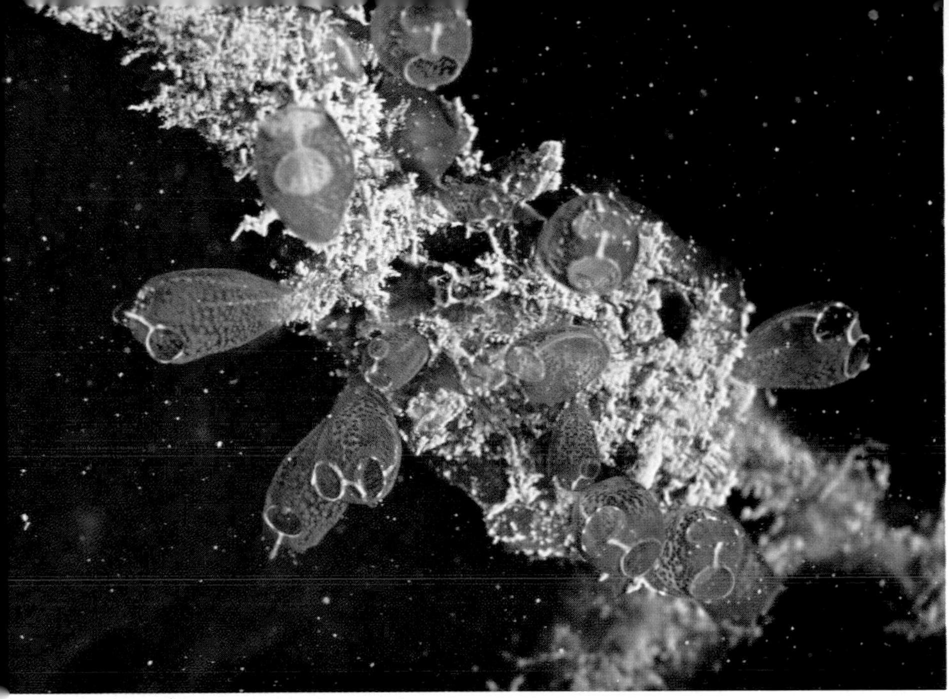

Unidentified ascidian. This lovely little blue ascidian occurs in groups along the stalks of dead gorgonians or on other hard substrates. It is nearly transparent except for the tiny spots of blue. Puerto Rico, La Parguera, Laurel Reef, 12 m depth.

The alga *Valonia ventricosa* can grow right among the fingers of branching corals. This individual is growing with the coral *Madracis mirabilis.* Jamaica, Discovery Bay, 25 m depth.

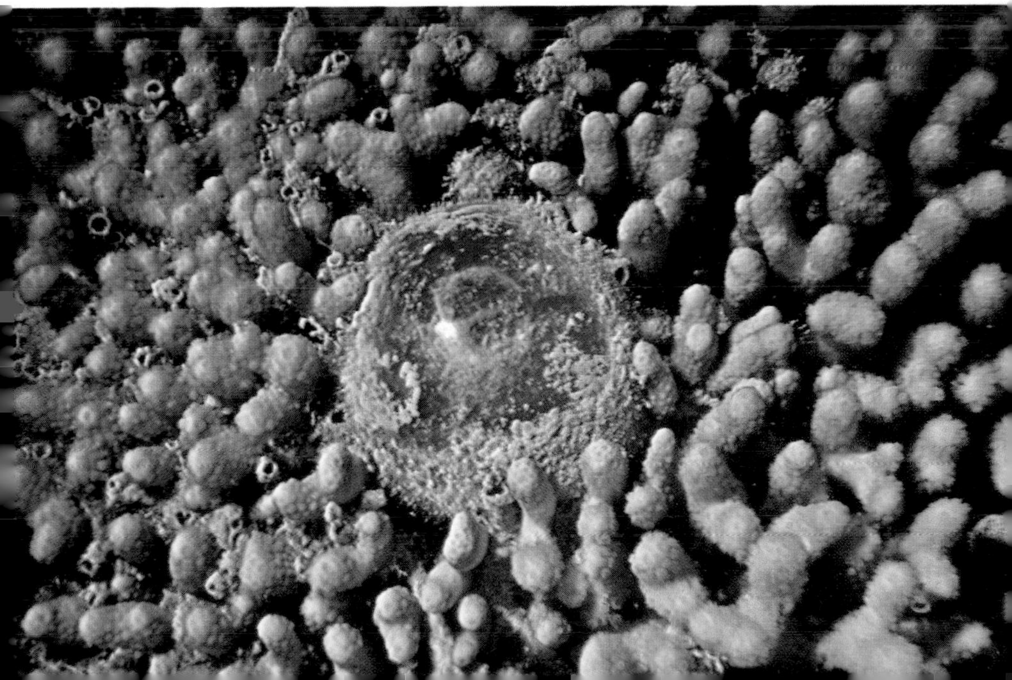

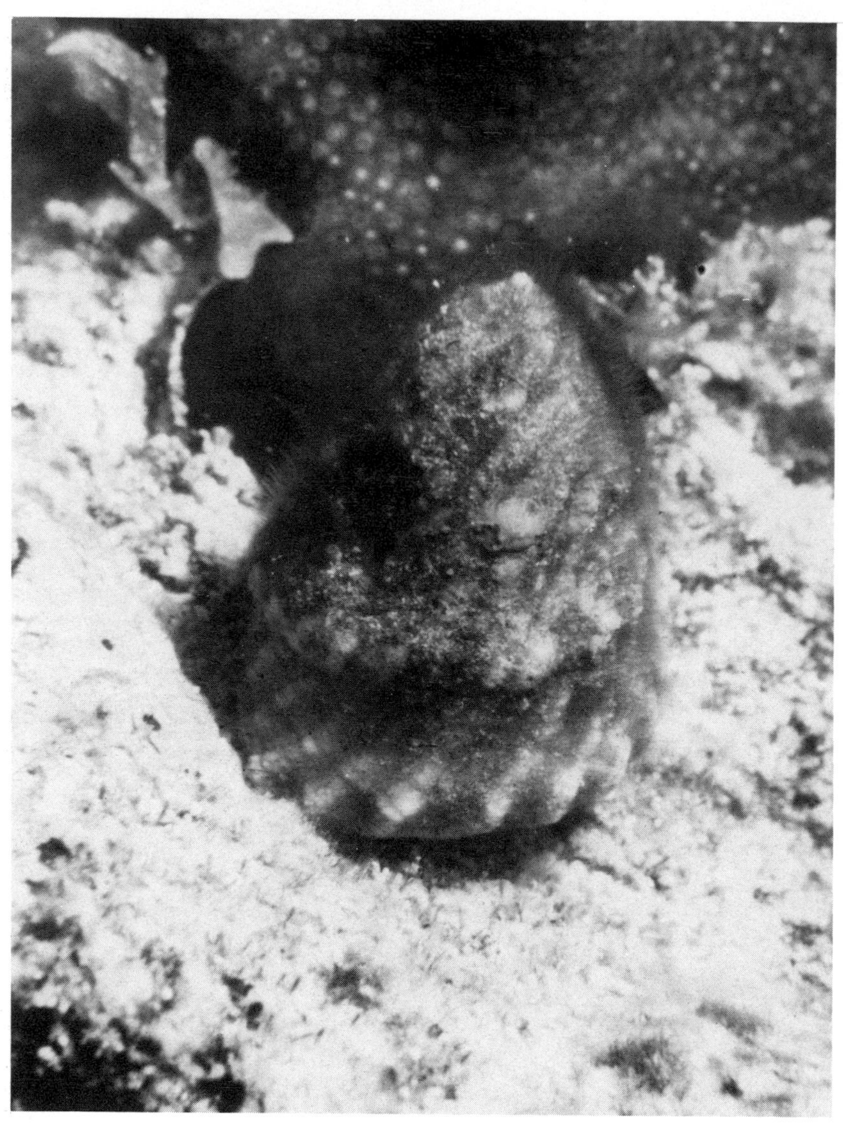

Astraea tecta with a fine coat of filamentous algae on its shell.

Marine Algae

The algae lack true roots, stems, leaves and flowers. Among macroalgae the vegetative portion of the plant is often divisible into root-like rhizoids, a stem-like stipe and leaf-like blades. They are photosynthetic. Several phyla are recognized, most of which are important in the marine environment. On reefs the Chlorophyta (green algae), Phaeophyta (brown algae) and Rhodophyta (red algae) are most apparent. The Cyanophyta (blue-green algae), Chrysophyta (diatoms and some microalgae) and Pyrrophyta (dinoflagellates) are present but not easily apparent.

Within the Caribbean there have been a number of studies published dealing with the algae occurring at specific geographic localities. Taylor (1960) summarizes these until about 1958, but no reference includes a summary of recent algal flora studies in the Caribbean and adjacent regions. Recent algal flora studies include Jamaica, Chapman 1961; Dominica, Taylor and Rhyme 1970; Curacao and Bonaire, Diaz 1964b; Puerto Rico, Almodovar 1962, 1970, Almodovar and Blomquist 1965, Diaz 1963; Virgin Islands, Earle 1972; Cuba, Diaz 1964a; Belize, Tsuda and Davies 1974; Yucatan, Taylor, 1972.

CLASS PYRROPHYTA: DINOFLAGELLATES

ZOOXANTHELLAE

The zooxanthellae are endosymbiotic unicellular algae belonging to the dinoflagellates which occur in reef-building corals and some other invertebrates. They carry out photosynthesis within the coral tissue, producing oxygen and carbon compounds. More oxygen may be produced than is used

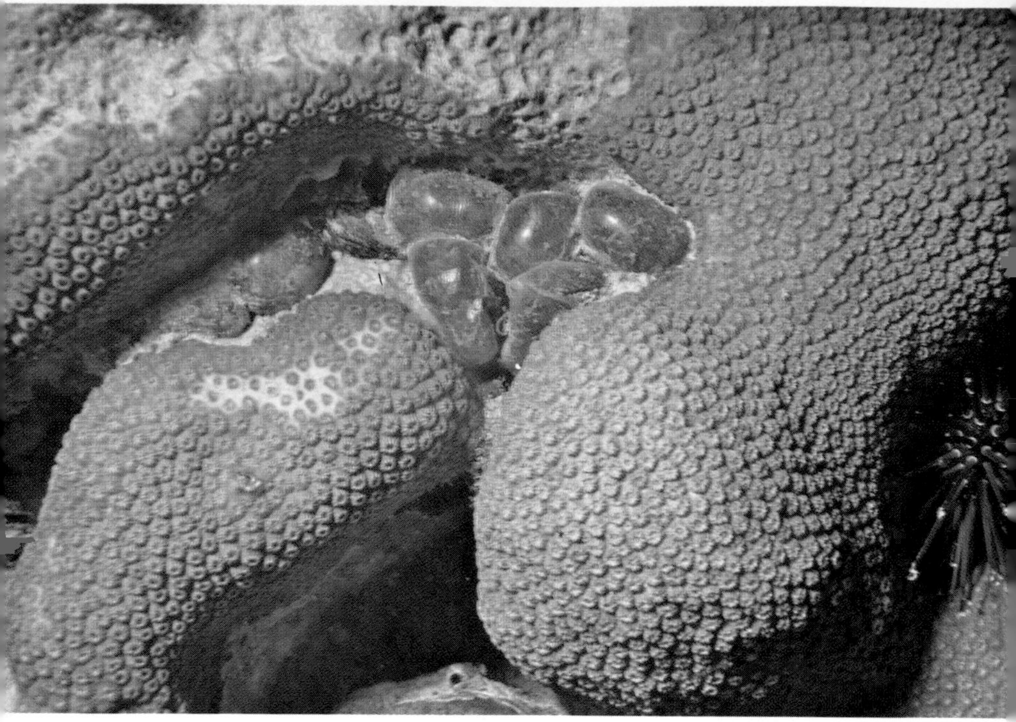

Valonia ventricosa can grow in the crevices of massive coral heads such as this *Montastrea annularis.* Each of the shiny-surface *V. ventricosa* represents only a single cell in the large vesicle. Jamaica, Discovery Bay, 5 m depth.

Valonia macrophysa has smaller vesicles than *V. ventricosa* but can cover cosiderably larger areas. This example is growing on the skeleton of a black coral. Jamaica, Rio Bueno, 20 m depth.

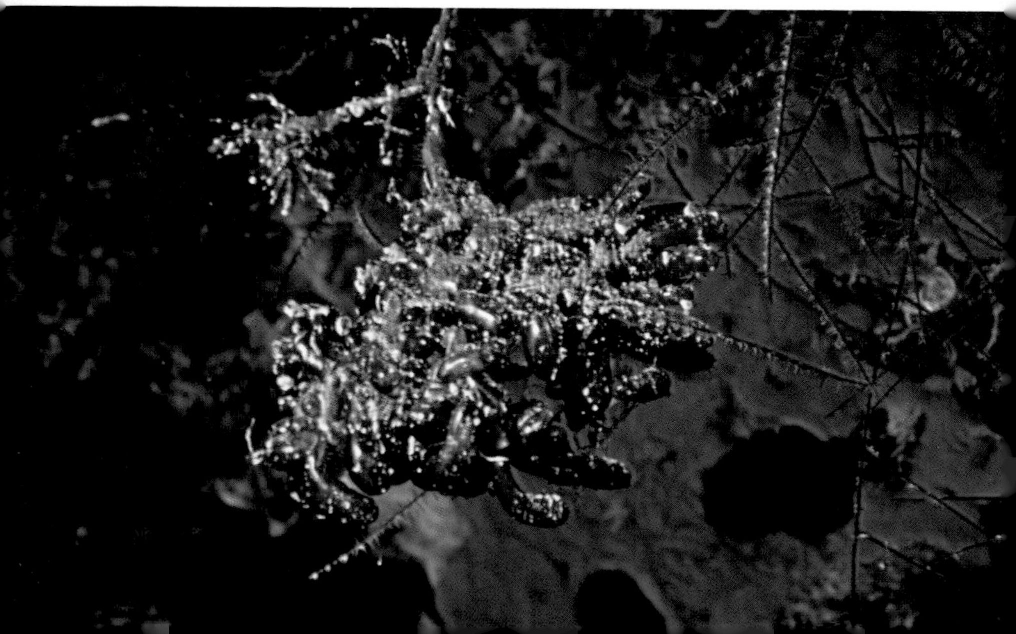

Dictyosphaeria cavernosa forms mats or layers of hollow, bubble-like cells. Jamaica, Discovery Bay, 3 m depth.

Dictyosphaeria cavernosa can grow adjacent to corals such as these spherical specimens next to the coral *Diploria labyrinthiformis.* Jamaica, Discovery Bay, 3 m depth.

by both the coral and zooxanthellae, but whether it is really needed by the coral in the high oxygen reef environment is uncertain. A portion of the carbon fixed by the zooxanthellae in photosynthesis is transferred to the animal tissues of the polyps, but the importance of these compounds in coral nutrition is not fully understood, as corals are also capable of capturing zooplankton for nutritional requirements.

Zooxanthellae are important in the production of the calcium carbonate skeleton of stony corals. All hermatypic reef corals possess zooxanthellae, while most non-reef building species do not. Light is necessary for rapid calcification in corals, and it is believed that during photosynthesis the zooxanthellae remove carbon dioxide, an end product in the calcification process (along with calcium carbonate), tipping the chemical equilibrium so that more calcium carbonate is produced than if the carbon dioxide was not removed. This explanation has certain problems, but that zooxanthellae play an important role in skeleton production is not disputed.

It is also believed that zooxanthellae help corals conserve essential nutrients by converting coral waste products into useful forms which are then recycled to the host coral.

Zooxanthellae are responsible for most of the color of reef corals. They have various complements of pigments producing the greens, browns, blues and reds of coral tissue. Some corals, such as *Colpophyllia natans*, will have differently colored zooxanthellae in different areas of the polyps, producing a two-color polyp. If a reef-building coral is maintained in darkness for a period of weeks or months, the zooxanthellae can be expelled from the coral tissues; the coral is then nearly colorless and termed "bleached." If normal conditions are restored, the coral can be reinfected with zooxanthellae from the environment. Natural expulsion may also occur after severe freshwater flooding of reefs or after passage of hurricane-strength storms.

The taxonomy of zooxanthellae is not well understood. The species which inhabit corals may also occur free in the marine environment, and different host organisms can be shown to have physiologically different zooxanthellae.

References: Muscatine 1973, Goreau 1964, Goreau and Goreau 1959, Trench 1973, Franzisket 1970.

CLASS CHLOROPHYCEAE: GREEN ALGAE

The Chlorophyceae or green algae occur in marine, fresh water and terrestrial habitats. They range from microscopic single cells to sizable species of centimeter to meter dimensions. Only about 10% of the Chlorophyceae are marine, and these are generally inhabitants of the tropics.

ENDOLITHIC ALGAE

Various species of the filamentous chlorophyte genus *Ostreobium* occur as a band or series of bands within the calcareous skeleton of stony corals near the living coral surface. The bands of *Ostreobium* are green or brown in color and may be as much as 35 mm beneath the growing surface of the coral. A limited amount of light penetrates the skeleton of the coral to the depth where these algae occur and the distance of the first band beneath the coral surface becomes less as the depth that the coral inhabits increases, probably due to the decrease of ambient light with increasing depth.

Red bands may also occur in some corals and are produced by a filamentous blue-green alga of the genus *Plectonema*.

Ostreobium evidently enters the coral skeleton at an early age, in some cases when the colony consists of a single polyp. The endolithic algae are not transmitted with the planula as are the zooxanthellae of stony corals. *Ostreobium* also occurs in carbonate substrates besides scleractinians and certainly is abundant in the reef environment, so the initial source of the algae is not a problem. The band(s) of *Ostreobium* grow outward at a rate equalling that of the coral, and the filaments actually bore their way through the skeleton. Flattened corals tend to have a single band while rounded heads generally have more than one. Branched corals lack endolithic algae in the tips, the site of most rapid growth, but have them in the larger branches and trunks.

References: Lukas 1973, 1974.

Caulerpa racemosa has ovate bodies on its vertical stalks which give the appearance of a bunch of grapes. Jamaica, Discovery Bay, 3 m depth.

Caulerpa mexicana, here among other algae, has flattened blades which marginal pinnae arising from stolons. Jamaica, Discovery Bay, 9 m depth.

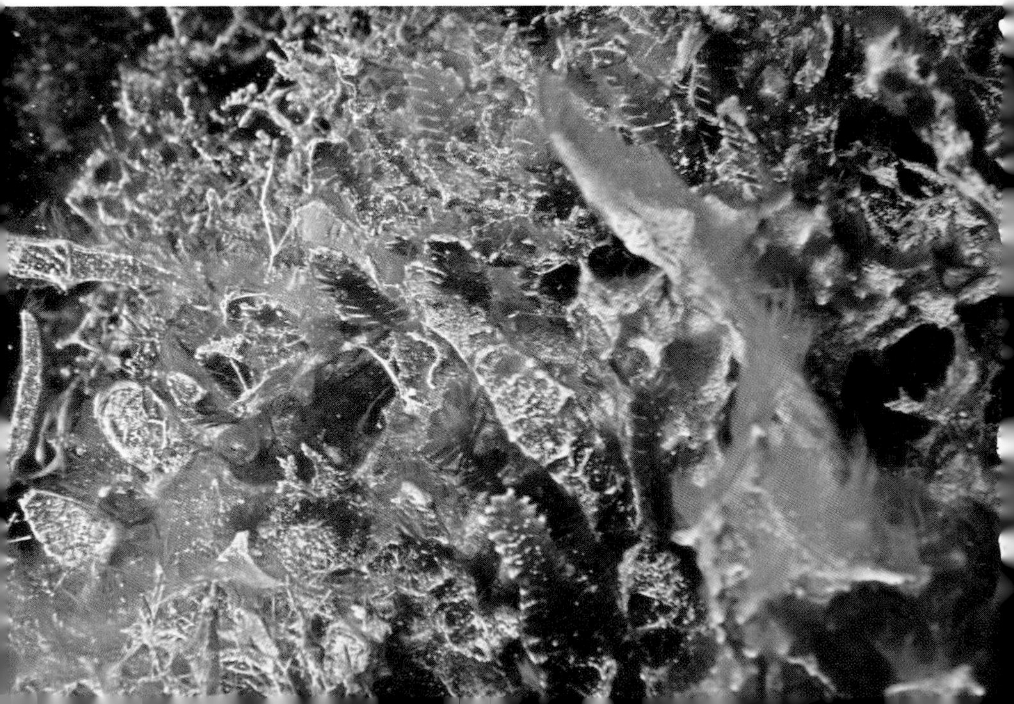

The calcareous green algae *Halimeda opuntia* forms dense clumps of its calcareous plates. These clumps form an excellent habitat for many other species of organisms. Puerto Rico, La Parguera, Mario Reef, 1 m depth.

Halimeda copiosa is a plant of the deep-reef environment. It is found on steep dropoff areas with the fronds of the algae hanging down vertically. Bahamas, Eleuthera Island, 30 m depth.

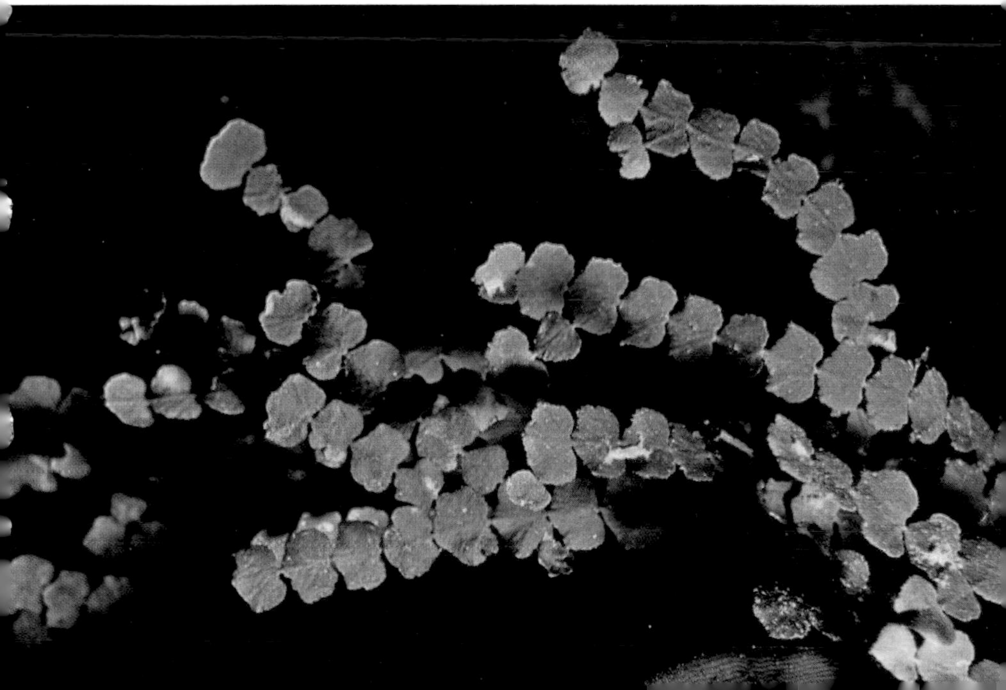

Family Valoniaceae

Valonia ventricosa J. Agardh.

This alga consists of an oval or spherical, thin-walled green body reaching over 5 cm in diameter. The large vesicle comprises only a single cell although other much smaller cells attaching this plant to the substrate occur. The shiny, balloon-like wall is fragile and easily ruptured, resulting in the death of the plant.

In the reef environment *Valonia ventricosa* is most common on rocky surfaces, in crevices of coral heads and between projections of finger corals. A few of these spherical plants may occur together, but generally they are solitary. Other microscopic algae (epiphytes) may cover much of the surface of this plant.

Valonia ventricosa is known from Bermuda and Florida to Brazil, generally in water shallower than 20 m and rarely below 30 m.

References: Taylor 1960, Earle 1972.

Several toadstool-like calcareous green algae at 30 m depth, Discovery Bay, Jamaica. These algae are probably *Penicillus dumetosus* with a greatly expanded "cap" and occur on rubble flats adjacent to reefs at moderate depths.

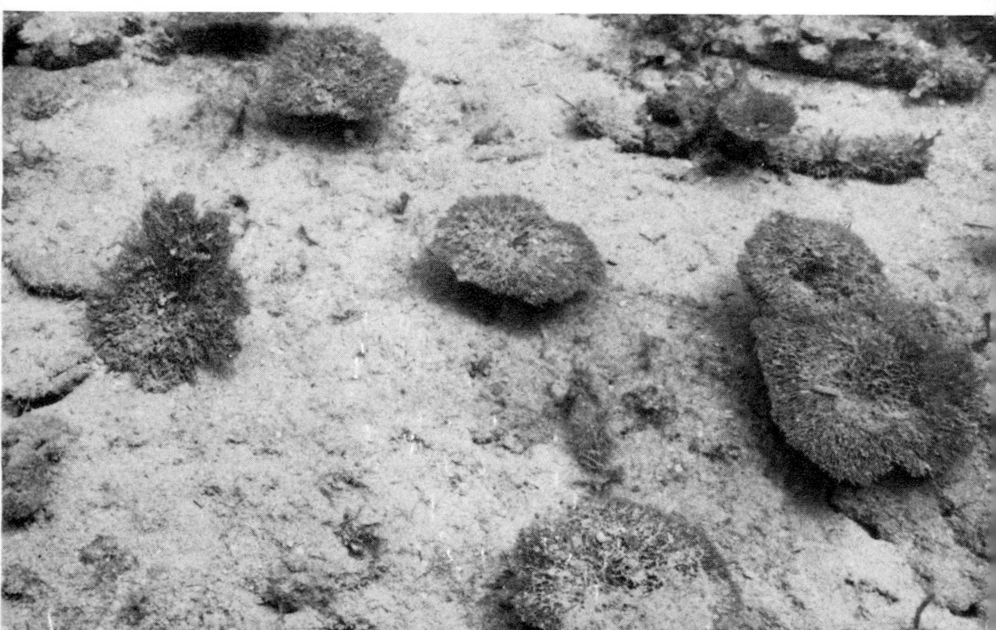

Valonia macrophysa Kuetzing

This species, like *Valonia ventricosa*, consists of large liquid-filled vesicles. *Valonia macrophysa* has smaller cells, generally not reaching over 5 by 15 mm in size, in large masses which may cover a considerable area with a thick mat. On reefs *V. macrophysa* may occur as an epiphyte on the skeletons of gorgonians or antipatharian corals. Its surface may also be heavily coated with epiphytic plants.

This algae is known from Bermuda and Florida to the Virgin Islands at depths to 30 m. It has also been reported from the Mediterranean Sea.

References: Taylor 1960, Chapman 1961.

Dictyosphaeria cavernosa Decaisne

This light green alga forms spherical to irregular lobes consisting of a single layer of cells that are hollow internally. The clearly visible individual cells are large, over one mm in dimension, and an individual plant may be 30 cm in diameter.

Dictyosphaeria cavernosa is generally found attached to a hard substrate, although records of it occurring epiphytically exist. It may occupy portions of live coral heads, seagrass beds or algal plains. It is eaten to a limited extent by herbivorous fishes.

The species is known from Bermuda and Florida throughout the West Indies to Brazil.

References: Chapman 1961, Taylor 1960, Earle 1972.

Family Caulerpaceae

Caulerpa racemosa (Forsskal)

This plant, like other species of *Caulerpa*, has erect branches arising from a horizontal stolon attached to the sediment at intervals by descending rhizomes. The form of the erect branches differs between species. In *Caulerpa racemosa* the branches arise every few centimeters, reaching as much as 30 cm in height. A large number of branchlets, resembling ovate or spherical bodies on stalks, arise from each

Halimeda goreaui is a deep-reef calcareous algae which forms dense mats of tiny plates. Bahamas, Cat Island, 30 m depth.

Halimeda discoidea has among the largest calcareous plates of any species of *Halimeda.* Jamaica, Discovery Bay, 25 m depth.

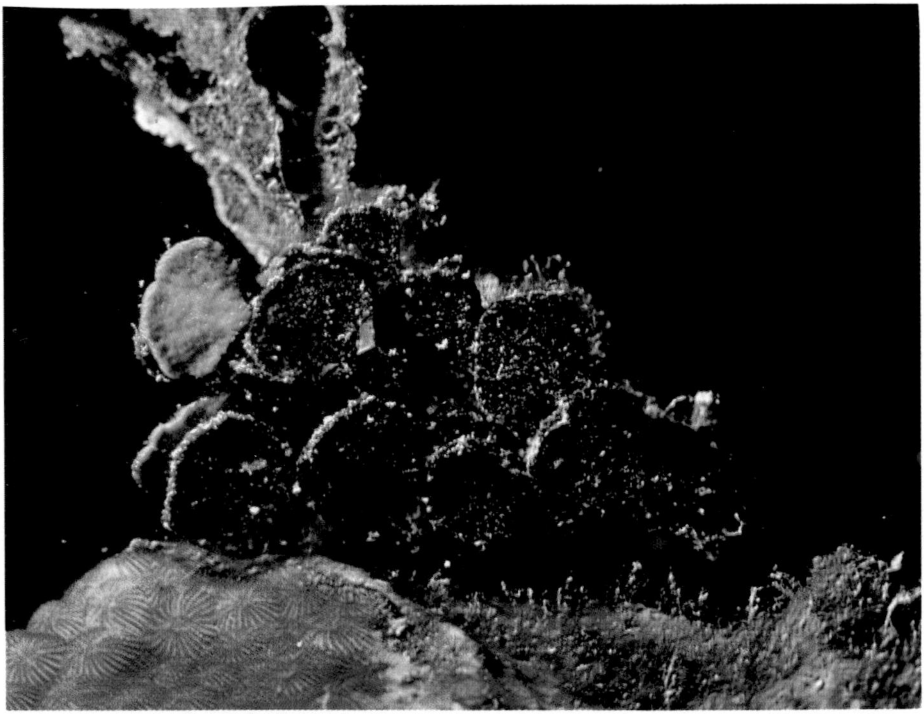

Udotea spinulosa is a calcerous green algae with a broad blade on a thin stalk. It is most common on algal plains adjacent to reefs, but on occasion is found right on the reef. Puerto Rico, Aguadilla, 12 m depth.

Udotea cyathiformis has the stipe formed into a funnel or cup-like shape. Jamaica, Discovery Bay, 30 m depth.

erect branch; where branches and stolons are close together the branchlets form a dense mat of seemingly spherical structures.

Caulerpa racemosa has been described as the most ubiquitous plant of the genus. It occurs from shallow muddy bays to clear water reef environments. It can occur adjacent to living corals and has even been observed growing on the coral *Acropora palmata.*

Its geographic range is from Bermuda and Florida to Brazil, including all of the Caribbean at depths from near the surface to 100 m.

Reference: Taylor 1960.

Caulerpa mexicana (Sonder)

The erect branches arising from the stolons of this plant are flattened with broad flattened marginal pinnae. The branches may reach 25 cm in height and 16 mm in width and are usually simple (not forked).

The stolons of this species are relatively short, seldom exceeding 30 cm in length, and the plant does not produce the tangle of branchlets and stolons as does *C. racemosa.* It occurs on sandy areas adjacent to reefs and more typically on inshore sandy or muddy bottoms. A heavy growth of epiphytes may occur on its surfaces. It is occasionally eaten by herbivorous fishes.

Caulerpa mexicana ranges from Bermuda and Florida to Brazil at depths to at least 70 m.

References: Taylor 1960, Earle 1972.

Family Codiaceae

Halimeda opuntia (Linnaeus)

The highly calcified segments of *Halimeda opuntia* are densely branched in multiple planes rather than nearly in a single plane as some other species are. Each segment may be up to only 12 by 20 mm, but this alga can cover larger areas with a dense mat 25 cm in thickness so that individual plants are indistinguishable. An abundance of invertebrates are sheltered in this mat, forming an identifiable community.

Halimeda opuntia is in some inshore locations the most important contributor of material to the sediments. It also occurs in abundance on reefs at a wide variety of depths and is the predominant species of *Halimeda* to about 20 m depth.

This species is found in all tropical oceans, and in the western Atlantic occurs from Florida to Brazil at depths of 0 to 55 m.

References: Taylor 1960, Goreau and Goreau 1973, Hillis-Colinvaux 1974.

Halimeda copiosa Goreau and Graham

As its name describes, this bright green alga occurs in large amounts in the proper habitat. The segments reach 3 cm across, occurring in elongate series essentially in one plane. A large number of these series of segments comprise a single plant reaching 70 cm in length and originating from a single holdfast. The segment series typically hang vertically or nearly so as a disorganized group of planar strings of segments.

Halimeda copiosa is a plant of the deep reef environment, seldom being found above 20 m depth. It grows on steep coral-overgrown slopes, often dangling from beneath sheet-like corals such as various species of *Agaricia.* Its contribution to deep reef sediments is extremely high, the large plates producing sediment of a much coarser nature than that occurring in many shallow environments. The ability of *H. copiosa* and other deep water species of *Halimeda* to calcify at depths of nearly 100 m is impressive, and their production of carbonate material at these depths may well exceed that produced by stony corals.

This alga is known from the West Indies and the Bahamas at depths of 10 to almost 100 m although it is rare shallower than 20 m.

References: Goreau and Graham 1967, Goreau and Goreau 1973.

Halimeda goreaui Colinvaux and Graham

The dark green segments of *Halimeda goreaui,* while only about one-quarter the size of those of *H. copiosa,* occur

Avrainvillea nigricans has a dark brown blade with a texture reminiscent of suede. The plants are flexible and sway back and forth in the current. Bahamas, Cat Island, 10 m depth.

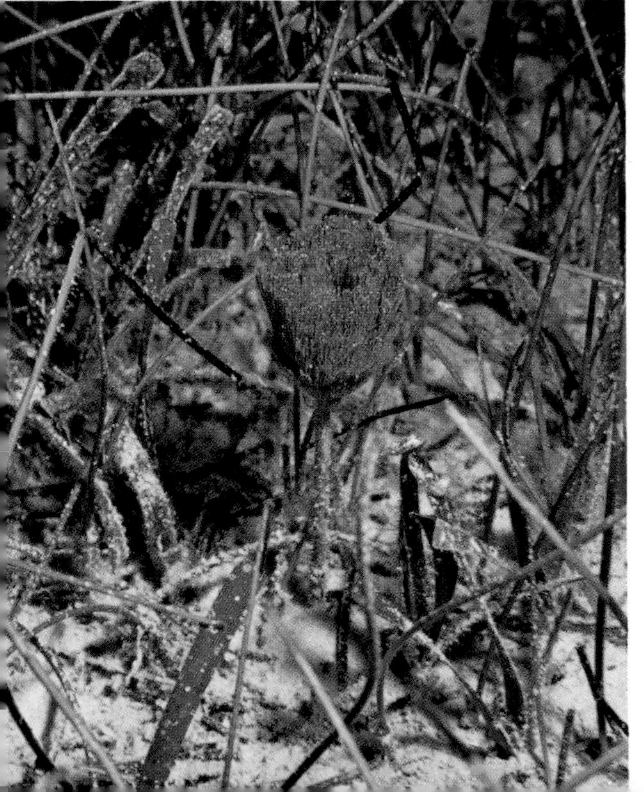

The shaving brush, *Penicillus capitatus,* is an easily recognized calcareous alga. This individual is among the seagrasses *Thalassia testudinum* and *Syringodium filiforme.* Bahamas, Cat Island, 3 m depth.

This calcareous alga is believed to be *Penicillus dumetosus* forma *expansus* which is somewhat different than the typical *P. dumetosus.* The "crown" is quite stiff and laterally expanded until it resembles a mushroom. Jamaica, Discovery Bay, 30 m depth.

Rhipocephalus phoenix is similar in general appearance to *Penicillus* but the crown of the alga is made up of flattened plates rather than filaments. Bahamas, Cat Island, 3 m depth.

in a dense mat of well organized series. This species is quite important on reef buttresses and in deep reef areas, occurring as dense clumps among living corals, but also occurs occasionally in shallow reef waters.

The species occurs in the Bahamas and West Indies at depths of 1 to at least 60 m.

References: (Hillis) – Colinvaux and Graham 1964, Goreau and Goreau 1973.

Halimeda discoidea Decaisne

This deep reef *Halimeda* has large segments (to 4 cm across) but is a sparsely branched low plant seldom reaching over 20 cm in height. It may occur on reef or algal plains and is known from Florida and the Bahamas to Brazil at depths to 60 m.

References: Taylor 1960, Earle 1972, Goreau and Goreau 1973.

Udotea spinulosa Howe

Several species of *Udotea* occur in the Caribbean. Most have a broadly flattened blade on a stalk, such as *U. spinulosa*. This species reaches about 8 cm, but other similar appearing members of the genus may reach 20 cm in height. All species of *Udotea* are heavily calcified and are stiff and erect plants. The elements making up the skeleton of this genus are relatively small and are not particularly important in the sediments of sloping fore reef areas.

Udotea spinulosa occurs on algal plains and sandy fore reef areas along with other members of the genus. It is generally found at depths greater than 9 m to at least 90 m and is known from southern Florida, the Bahamas and the West Indies.

References: Taylor 1960, Earle 1972.

Udotea cyathiformis Decaisne

This member of *Udotea* has a funnel- or cup-like form rather than a flattened blade. The funnel-like blade rests atop a stalk reaching 13 cm in height and may be open along one side. It is thin, papery and easily damaged.

Like many other members of this genus, *U. cyathiformis* is found on algal plains adjacent to reefs at 10 to 90 m and in shallower back reef areas. It does not occur on reef crests. It ranges from Bermuda and North Carolina to Brazil.

References: Taylor 1960, Earle 1972, Frederick 1964, Goreau and Goreau 1973.

Avrainvillea nigricans Descaine

The flattened blade or spatula-like growth form of this dark brownish to blackish plant may occur on a stalk reaching 20 cm in height. The surface is dull, its texture being almost like suede, and the plant as a whole is less stiff than the more heavily calcified members of the similar appearing genus *Udotea.* Several species of *Avrainvillea* occur in the West Indies which are somewhat similar in appearance.

Avrainvillea nigricans occurs on reefs, seagrass beds, sandy bottoms and algal plains, most commonly at depths below 10 m, and reaches over 50 m in depth. It will be eaten by herbivorous fishes where other algae are not available, but is not a preferred food. This species is known from Florida and Bermuda to Brazil.

References: Taylor 1960, Earle 1972, Frederick 1964, Chapman 1961.

Penicillus capitatus Lamarck
Shaving brush

This species, like other members of the genus, has a tuft of filaments atop a stalk. The plants are erect and well calcified, with the skeletal elements being relatively small. *Penicillus capitatus* reaches 15 cm in height. The filaments of the tuft are often whitened by calcification.

This alga occurs in a wide variety of habitats, from muddy or sandy bays to deep reef algal plains. The calcification of this and other species of *Penicillus* is believed to discourage herbivorous fishes from feeding extensively on them. It occurs to depths of 40 m, ranging from Bermuda and Florida to the entire West Indies.

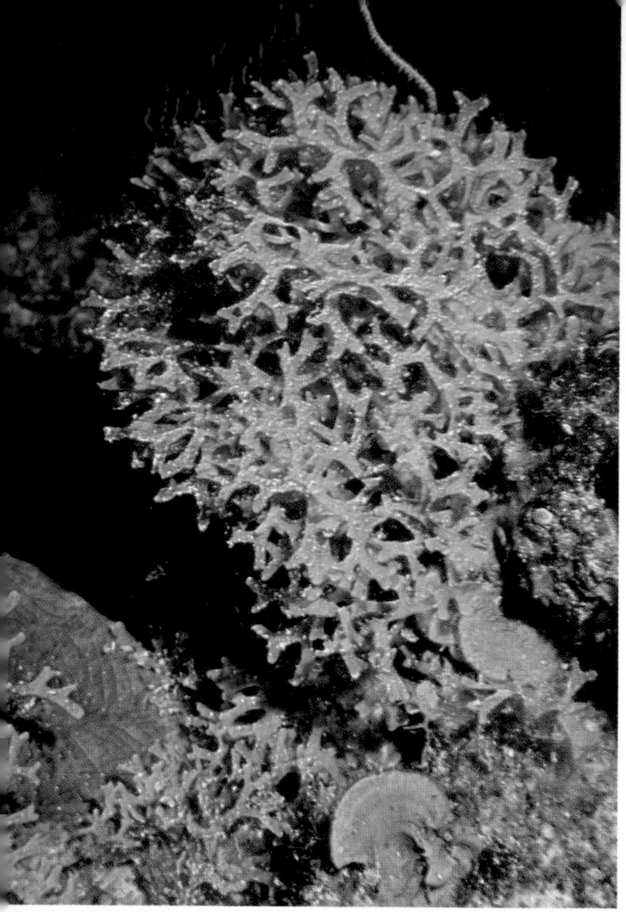

The brown alga *Dictyota* sp. is found abundantly on Caribbean reefs. They may possess chemicals which make them distasteful to fishes. Puerto Rico, La Parguera, 22 m depth.

A closeup view of the brown alga *Dictyota* sp. shows the bifurcate branching and the iridescent properties of the algae. Puerto Rico, La Parguera, Laurel Reef, 12 m depth.

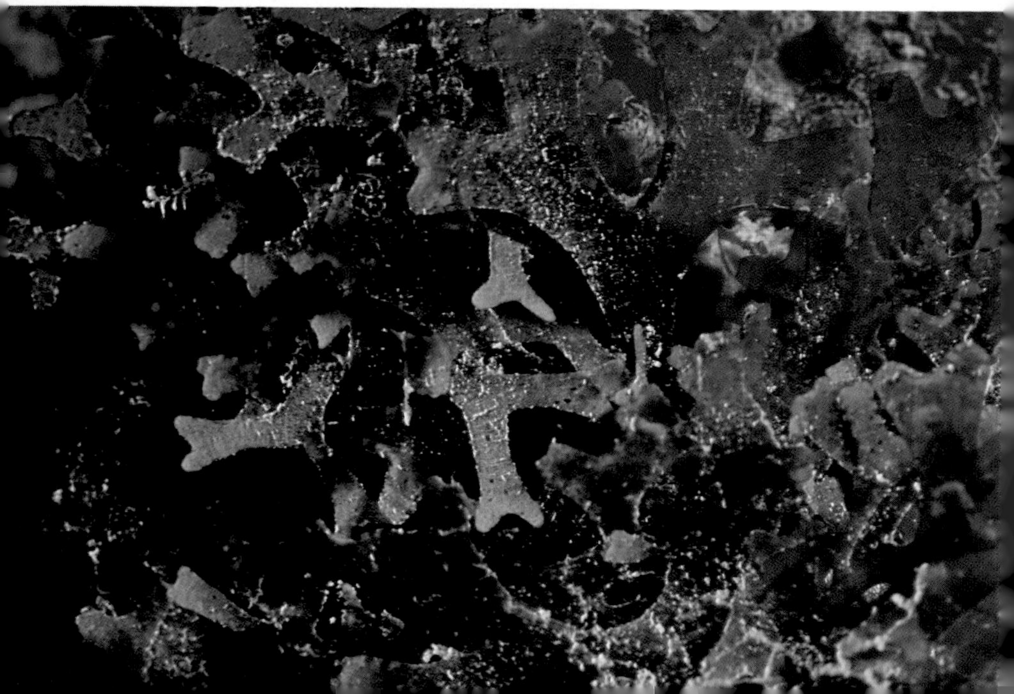

The brown alga *Stypopodium zonale* has brilliant greenish banding across its blades. The algae are tough but flexible and sway with the current. Puerto Rico, Mona Island, 15 m depth.

Lobophora variegata is a brown alga which is commonly found on rocky surfaces of reefs. There seem to be some sort of concentric growth markings on the fronds. Bahamas, Cat Island.

References: Taylor 1972, Earle 1972, Goreau and Goreau 1973.

Penicillus dumetosus

This species has a larger tuft with longer filaments than the similar *Penicillus capitatus.* It reaches 15 cm in height, and one form has the tuft of filaments (capitulum) reaching 15 cm in diameter.

While usually found in shallow water, the expanded form of *P. dumetosus* occurs to at least 30 m depth. The species is known from Bermuda and Florida to the West Indies.

References: Taylor 1960, Earle 1972.

Rhipocephalus phoenix (Ellis and Solander)

This stalked alga has the filaments arranged in upturned blades radiating around the stalk; it reaches 12 cm in height. It is not as heavily calcified as some other green algae it occurs with and is generally a dark green without white tinges.

Rhipocephalus phoenix is rare in back reef areas but is common in sandy areas of fore reef slopes at depths to 73 m. It is known from southern Florida, the Bahamas and the West Indies.

References: Taylor 1960, Goreau and Goreau 1973.

CLASS PHAEOPHYTA: BROWN ALGAE

The largely marine brown algae are more characteristic of temperate than tropical waters, but a number exist on Caribbean reefs and in certain cases are very abundant. Also within this group are the kelps of temperate Pacific coastlines, the largest of the algae. One member of the genus *Sargassum* is the most abundant plant in terms of biomass on the earth, and an entire area of the tropical North Atlantic is known as the Sargasso Sea.

The brown to olive-green coloration of the Phaeophyta is produced by the xanthophyll pigment fucoxanthin, which masks much of the color of the chlorophylls.

Family Dictyotaceae

Dictyota spp.

A number of species of the genus *Dictyota* occur on Caribbean reefs which, while distinctive species, each have certain characters which easily identify them as members of this genus. They have a flattened thallus branching dichotomously in the same plane as the flattening and are greenish to golden brown in color. The angle made by the dichotomous branches, the width of the branches, their relation to the distance between branchings and the condition of the edge of the branches are all important characters in determining species.

The members of *Dictyota* are important ecologically on reefs where they may form dense masses on rock surfaces and dead coral heads and are generally one of the abundant genera of algae on the reef. Although some fishes are reported to eat *Dictyota*, there is some evidence that possibly distasteful chemicals may make them unattractive to fishes.

Members of the genus *Dilophus* are quite similar to *Dictyota* and a specimen must usually be available to distinguish them. The various members of *Dictyota* are widely distributed in the tropical western Atlantic and occur from shallow depths to near 60 m.

References: Taylor 1960, Earle 1972.

Stypopodium zonale (Lamouroux)

This handsome alga has a fan-shaped clump broken into segments of strap-like branches. Reaching about 40 cm in height, the fan-like plants do not become separated until they reach about 10 cm in height. The ends of the blades separate when they reach a width of 1 to 5 cm, and the longer the plant, the more segments its outer margin is divided into. This greenish to golden brown plant has iridescent green lines every few centimeters parallel with the growing ends of the segments.

Stypopodium zonale occurs in high salinity environments from near the surface to 55 m. It ranges from Bermuda and Florida to Brazil. In most areas around Bermuda

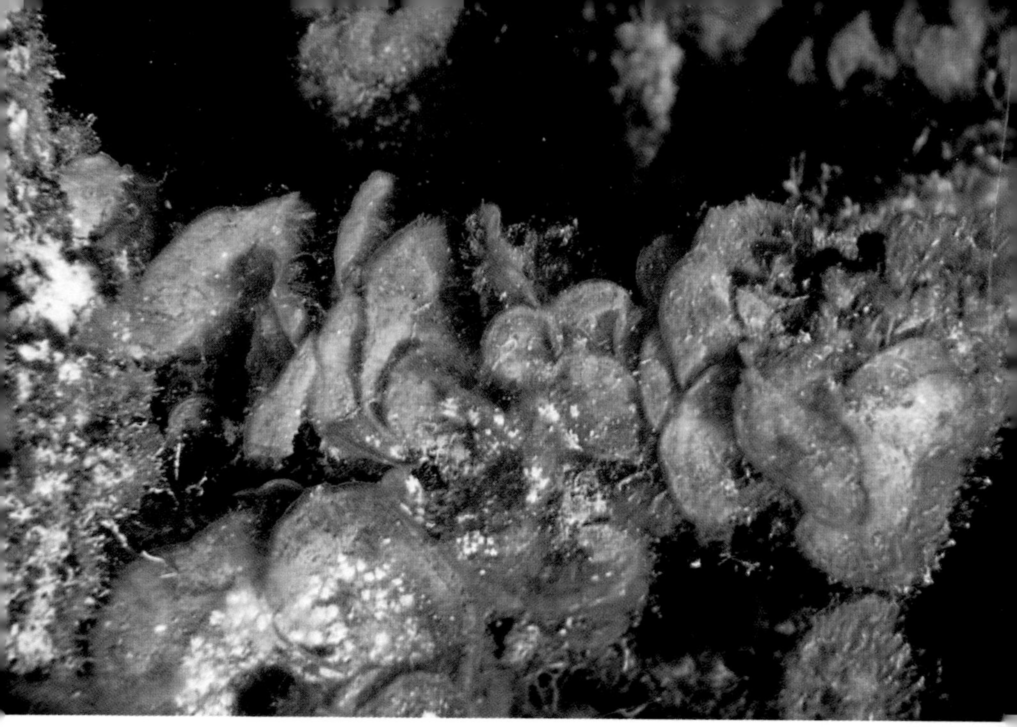

Lobophora variegata can occur in dense groups of fronds and often hangs down below the openings of reef caves. Bahamas, Eleuthera Island, 20 m depth.

Sargassum, usually thought of as a floating alga, grows on rocky substrates adjacent to reefs, particularly at fairly shallow depths. Bahamas, Exuma Islands, 4 m depth.

The blades of *Sargassum* can also serve as a substrate for other plants to grow on. Bahamas, Acklins Island, 20 m depth.

These specimens of *Sargassum* are on a rock substrate along with a variety of other algae. Puerto Rico, Mona Island, 15 m depth.

at 21 to 30 m S. *zonale* comprises the dominant benthic plant with extremely luxuriant growth.

Reference: Taylor 1960.

Lobophora (= *Pocockiella*) *variegata* (Lamouroux)

This alga has an erect form and a decumbent (on the substrate) form which occur on rocky surfaces, including dead coral heads. Often the fronds hang vertically on the lower edge of ledges and their surfaces are covered with epiphytes. It is one of the most common algae of hard surfaces on algal plains and is eaten by a variety of herbivorous fishes.

Lobophora variegata ranges from Bermuda and North Carolina to Brazil. It is common on intertidal rocks, but reaches to at least 55 m depth.

References: Taylor 1960, Earle 1972.

Family Sargassaceae

Sargassum spp.

The several species of *Sargassum* in the tropical Atlantic are unusual and important marine plants. The genus consists of both attached (benthic) and free-floating (pelagic) species. *Sargassum natans* (Linnaeus), which is known only in the pelagic condition, is the most abundant plant in biomass on either land or sea and has a large community of organisms associated with it.

Other species of *Sargassum* normally (at least initially) grow attached, although some can also live in the pelagic state. These plants may occur in reef environments on rocky substrates. The plants have a holdfast and the stalk is held erect by tiny gas-filled floats (which cause the pelagic forms to float) or bladders. They possess vascular tissue, an advanced characteristic, and are among the most complex of algae.

The species of *Sargassum* can be extremely abundant on limestone substrates, with over 30 individuals of S. *pteropleuron* occurring per square meter in the Florida Keys. Some attached species reach over 3 m in length, but large specimens are easily broken by wave action, the separated and unattached portion floating away. In southern Florida *S. pteropleuron* has a definite seasonality, dying back in the winter and reproducing in the fall.

It is difficult to distinguish between the over ten Atlantic species. Often plants must be fruiting for positive identification. Most species are known widely in the tropical Atlantic, and those growing attached on coral reefs have been observed to at least 30 m.

References: Taylor 1960, Earle 1972, Thompson 1971.

Turbinaria turbinata (Linnaeus)

The amazing leaves of this alga, resembling three- to six-sided pyramids attached in clusters to the stipe by their apexes, immediately identify members of the genus *Turbinaria.* Two similar species of *Turbinaria, T. turbinata* and *T. tricostata* Barton, occur in the West Indies. They are plants of shallow rocky areas exposed to some wave action and as such often occur on coral reefs and rocky shores. *Turbinaria turbinata* reaches about 40 cm in height and the leaves have an air vesicle which gives the erect branches added floatation. It ranges at depths less than 6 m from Florida, throughout the West Indies to Brazil; *T. tricostata* has a similar range but also occurs in Bermuda.

Reference: Taylor 1960.

CLASS RHODOPHYCEAE: RED ALGAE

The red algae are largely a marine class, living attached to the bottom. Their red color comes largely from phycoerythrin, a red pigment. They possess chlorophyll like other algae, but the ratio of the various types is different than in other classes.

Rhodophyceae constitute a large class with a wide range of diversity. Included are many species capable of producing calcium carbonate reef structures and also tiny filamentous species.

Turbinaria turbinata is a brown alga found in shallow rocky areas and is very distinctive, having leaves resembling small pyramids. It is often found in fairly turbid waters. Bahamas, Cat Island, 2 m depth.

Asparagopsis taxiformis is a red alga with the branches terminating in tufts. It grows very low to the substrate. Bermuda, south shore reef, 3 m depth.

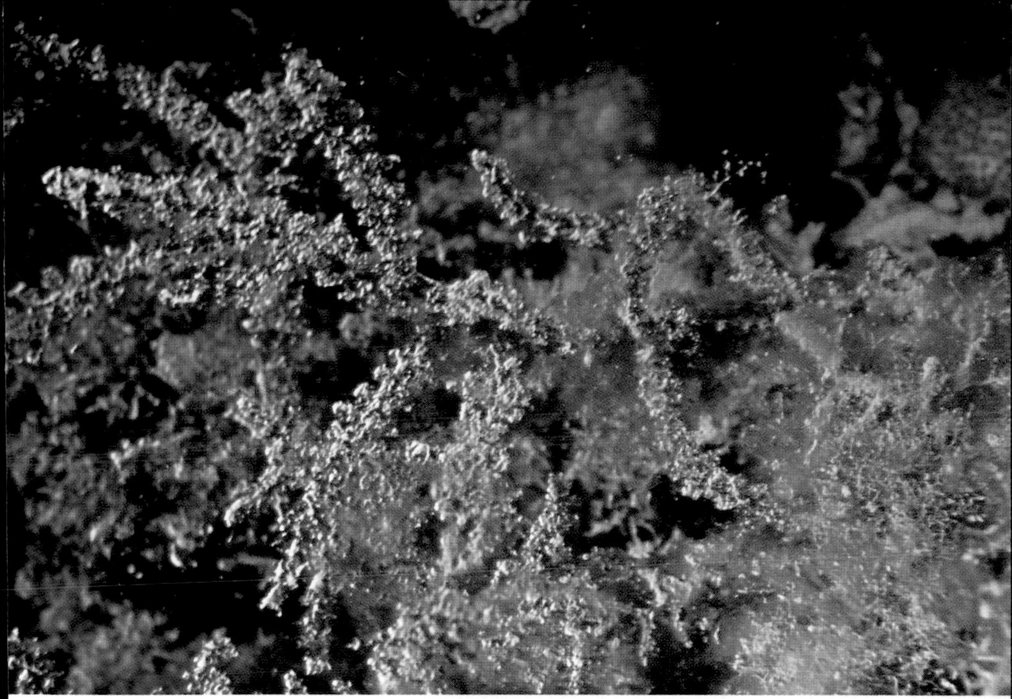

Acanthophora spicifera is a red alga which has markedly spinous branches. The spinations are usually hidden, however, by a dense coating of sediment and other organisms. Jamaica, Discovery Bay, 3 m depth.

A number of species of filamentous red algae grow on reefs. This particular example is growing from the stalk of a gorgonian or a sponge. Jamaica, Discovery Bay, 24 m depth.

Family Bonnemaesoniaceae

Asparagopsis taxiformis (Delile)

This short, bushy alga has the branches terminating in tufts. It occurs on rocky surfaces of reefs from near the surface to at least 30 m depth. *Asparagopsis taxiformis* is known from Bermuda through the Caribbean to Brazil.

Reference: Taylor 1972.

Family Rhodomelaceae

Acanthophora spicifera (Vahl)

This common inshore and reef alga is branched and markedly spinose. It is often densely covered with filamentous algae so that its true nature is not immediately apparent. *Acanthophora spicifera* forms dense clumps reaching at least 30 cm above the bottom. It occurs in a wide variety of habitats, from inshore bays and algal plains at moderate depths to true reef situations. It can be found adjacent to and on heads with living corals and is eaten by a number of herbivorous fishes.

This alga is known from Bermuda, Florida and throughout the West Indies to Brazil at depths of less than 1 to 25 m.

References: Taylor 1960, Chapman 1961, Earle 1972.

Family Corallinaceae

Coralline algae

The crustose coralline algae are quite important in coral reef ecology, producing calcareous structures on fixed substrates and on loose organic or mineral substrates. In areas exposed to heavy wave action an emergent ridge produced by crustose coralline algae may occur. These algae may also be found deeper on fixed substrates but do not form massive structures. They may also form oval nodules several centimeters in diameter, called rhodoliths, at depths of 50 m or more on level bottoms on offshore banks and insular margins. These rhodoliths are unattached and may cover large areas, forming a habitat particularly suitable for certain reef fishes.

The role of the crustose corallines on Atlantic reefs is just beginning to be understood. Emergent algal ridges were well known on Pacific reefs and atolls, but only recently have such structures definitely been recorded from the Atlantic. The coralline algae build the ridge at a rate of 3 to 6 mm per year where surf conditions restrict the grazing activities of parrotfishes and sea urchins, which causes the destruction of the calcareous material laid down as fast or faster than it is produced. Initially a dead coral or rock surface at 0 to 2 m is required. It is colonized by crustose corallines which produce a slight mound. The mounds grow into larger structures reaching to or near the surface, called "boilers," and by fusion of a line of boilers, a ridge is produced. In cases where other reefs have subsequently grown seaward of an algal ridge, reducing the wave action reaching the ridge, the fish and invertebrate grazers stop the growth and actually cause the erosion of the ridge by their feeding behavior of grazing algal material from the surface of the ridge.

Algal reefs ("boilers") have now been reported from a number of Atlantic locations including Yucatan and Bermuda. Emergent algal ridges are known from Panama, St. Croix and a number of the Lesser Antilles islands.

The taxonomy of Atlantic reef crustose coralline algae has only recently been examined in any detail. In the past the algae occurring on shallow ridges and reefs were simply referred to as *Lithothamnion*, a genus which occurs in deep water in the tropics. *Lithophyllum congestum* (Foslie), *Porolithon pachydermum* (Foslie) and species of *Neogoniolithon* are known to be important ridge constructors in St. Croix and Panama.

References: Adey 1974, 1975, Adey and Burke 1976, Adey and MacIntyre 1973, Ginsburg *et al.* 1971, MacIntyre and Glynn 1974, Boyd *et al.* 1963, Glynn 1973.

Jania sp. is an articulated coralline alga. Bahamas, Exuma Island, 10 m depth.

Galaxaurus sp. is an articulated coralline alga here seen growing among another algae, probably *Lobophora variegata*. Jamaica, Discovery Bay, 24 m depth.

Closeup view of an articulated coralline alga. Puerto Rico, La Parguera, 3 m depth.

A crustose coralline alga either of the family Corallinaceae or Squamariaceae. Bermuda, south shore reefs, 5 m depth.

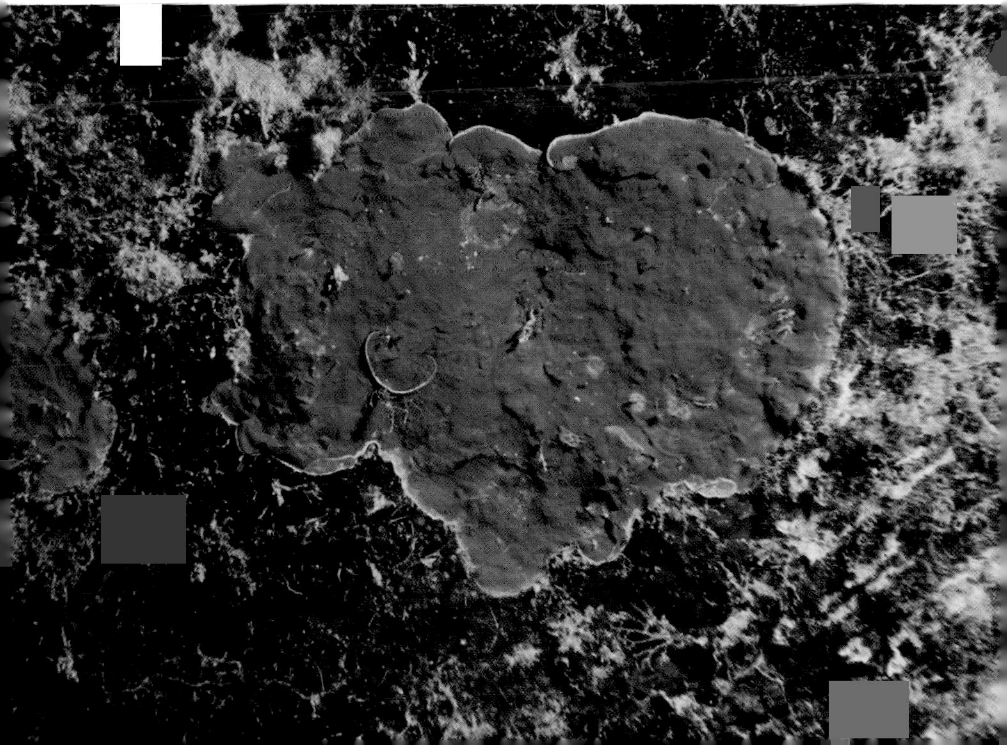

Flowering Plants

CLASS ANGIOSPERMAE: SPERMATOPHYTES

Family Hydrocharitaceae

Thalassia testudinum Koenig
Turtle grass

Turtle grass is the most ubiquitous plant of shallow water areas of the Caribbean, where it forms broad meadows. It is often mixed with various large algae and manatee grass, *Syringodium filiforme*, and its blades are usually coated with a wide variety of epiphytic and epizoic organisms.

Thalassia testudinum has a horizontal rhizome, buried as much as 25 cm deep in the sediment, which gives rise to erect, flattened green leaves (blades). A small percentage of the nodes on the rhizomes give rise to up to several leaves with several leafless nodes between the nodes with leaves. The older blades break off regularly but are quickly replaced. Expansion of *T. testudinum* beds occurs by lateral growth of the rhizomes at their ends. Except for the terminal part of the rhizome, they do not increase in length and if cut are incapable of rejoining or becoming lengthening rhizomes. Hence destruction of the rhizomes by mechanical excavation and damage is not easily repaired and *T. testudinum* beds may develop holes where the turtle grass cannot regenerate.

Turtle grass requires water of high salinity in areas sheltered from extreme wave action. It requies a moderate depth of sediment for its rhizome network. Where the sediments are minimal in thickness *T. testudinum* is usually stunted in its growth. Leaf kill occurs at temperatures below 20° C., but the rhizomes are able to survive somewhat lower temperatures. When suitable temperatures return, the blades are rapidly regenerated. Many herbivorous fishes and green turtles feed heavily on turtle grass, and the primary production of this plant is quite important in tropical marine Atlantic ecosystems.

Thalassia testudinum occurs throughout the Caribbean and the Gulf of Mexico. On the eastern coast of Florida it is not recorded north of Cape Canaveral. It reaches depths of 20 m in clear water and can stand slight exposure to air at low tide levels. Flowering occurs in Florida during the summer.

References: Phillips 1960, Moore 1963, Tomlinson and Vargo 1966, Tomlinson 1972, Zieman 1972, Thomas *et al.* 1961.

Syringodium filiforme Kuetzing
Manatee grass

Manatee grass is quite distinctive from other Atlantic marine spermatophytes in that its leaves are rounded, rather than flat, in cross section. It has a dense mat of rhizomes, about 5 cm deep, in the sediment and its beds are usually dense. Two to four leaves are present at each node, and individual leaves may reach 50 cm in length. *Syringodium filiforme* often occurs with *Thalassia testudinum* in mixed stands and has epiphytes growing on its leaves. It is eaten by various herbivorous fishes and the queen conch.

Manatee grass occurs from southern Florida throughout the Caribbean at depths of 1 to 22 m.

References: Phillips 1960, Earle 1972.

Halophila baillonis Ascherson

This spermatophyte occurs on silty, muddy substrates and may be found adjacent to reefs in areas of high sedimentation. It usually occurs as pure stands, but may be mixed with *Syringodium filiforme* occasionally. The small spatulate blades are green with definite midribs and rise up from a horizontal rhizome.

Halophila baillonis occurs deeper than other marine spermatophytes of the area, reaching from about 9 to 30 m,

An unidentified crustose coralline alga which is capable of fluorescence. Bahamas, Eleuthera Island, 30 m depth.

A typical view of a bed of the seagrasse *Thalassia testudinum* with its tall, flattened blades. Florida, Triumph Reef, 7 m depth.

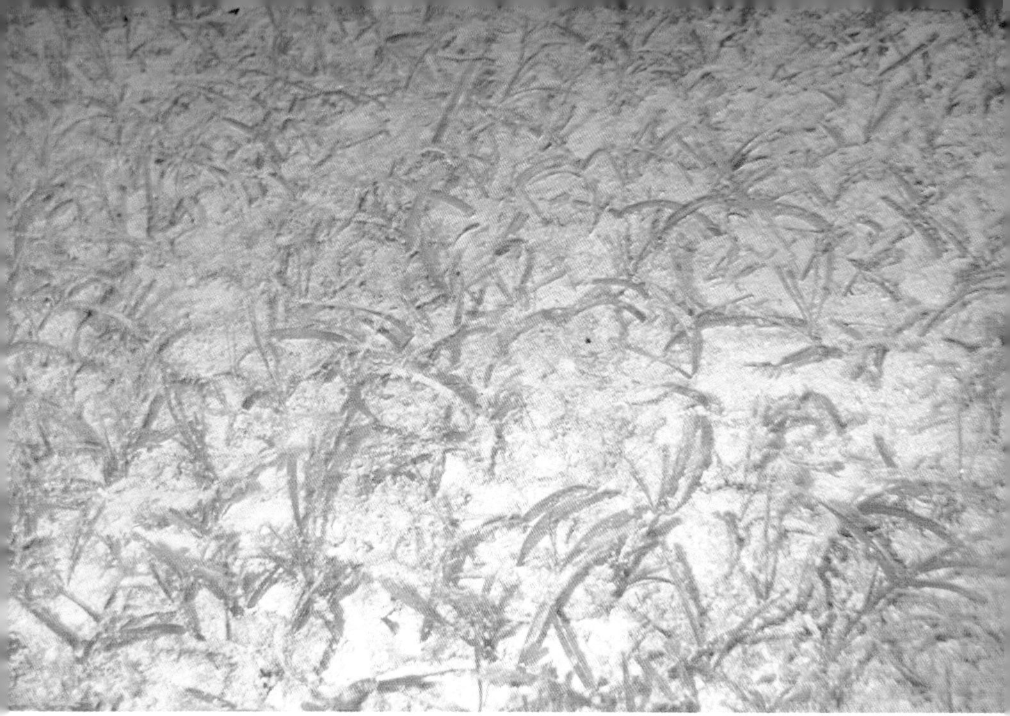

Area with scattered *Thalassia testudinum* near one of the halo zones around patch reefs. Florida, Triumph Reef, 7 m depth.

Syringodium filiforme has blades which are round in cross section. It is often found in mixed stands with *Thalassia testudinum*. Bahamas, Cat Island, 3 m depth.

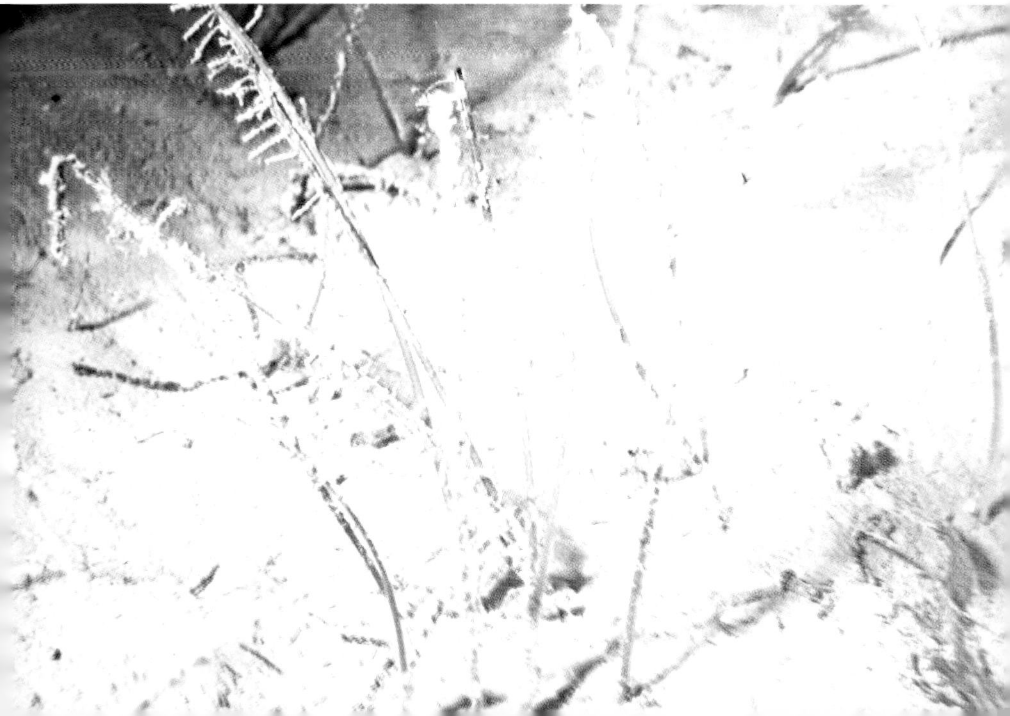

and is eaten by a variety of fishes and the queen conch. It ranges widely in the tropical western Atlantic. A similar species, *H. engelmanni* Ascherson, occurs in the Caribbean but in muddier conditions and is not found adjacent to reefs.

References: Earle 1972, Phillips 1960.

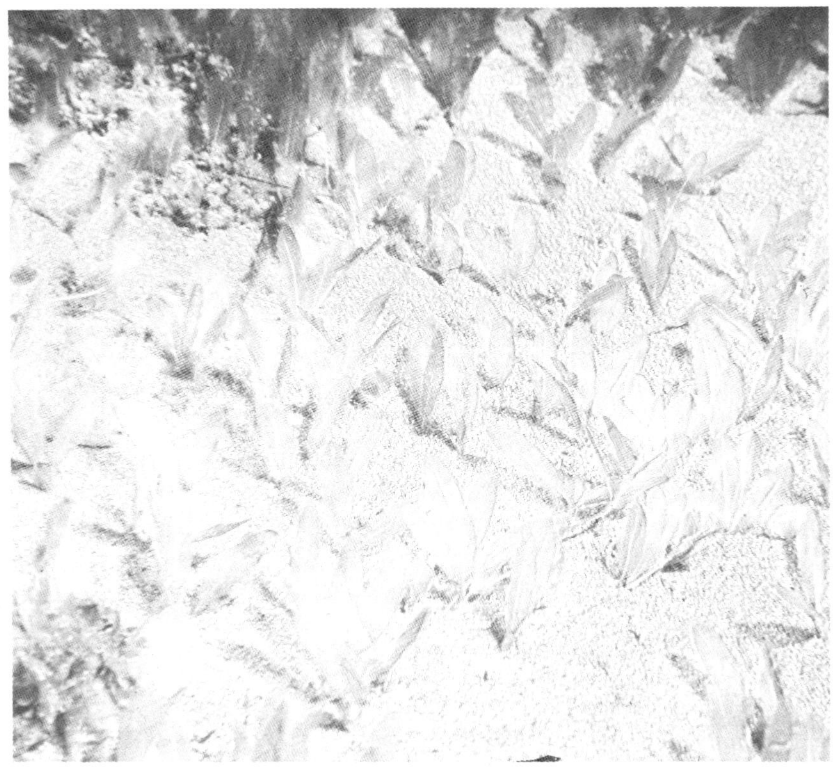

The seagrass *Halophila baillonis* has small, spade-like blades and is found quite deep for a seagrass, reaching to 30 m maximum. Puerto Rico, Aguadilla, 9 m depth.

Bibliography

Abbott, D.P., Ogden, J.C. and Abbott, J.A. (eds.). 1974. Studies on the activity pattern, behavior, and food of the echinoid *Echinometra lucunter* (Linnaeus) on beachrock and algal reefs at St. Croix, U.S. Virgin Islands. *Spec. Publ. 4, West Indies Lab., Fairleigh Dickinson Univ.*

Abbott, R.T. 1974. *American Seashells.* Van Nostrand Reinhold Co., New York, Second Edition, 663 pp.

Adams, R.D. 1968. The leeward reefs of St. Vincent, West Indies. *Journ. Geol.*, 76:587-595.

Adey, W.H. 1975. The algal ridges and coral reefs of St. Croix, their structure and Holocene development. *Atoll Res. Bull.*, 187:1-67.

________. 1974. A survey of red algal biology and ecology with reference to carbonate geology, and the role of reds in algal ridge and reef construction. In: Gerhard, L.C. and Multer, H.G. (eds.). *Recent Advances in Carbonate Studies, Spec. Publ. 6, West Indies Lab., Fairleigh Dickinson Univ.*: 3-6.

________ and Burke, R. 1976. Holocene bioherms (algal ridges and bank-barrier reefs) of the eastern Caribbean. *Geol. Soc. Amer. Bull.*, 87:95-109.

________ and MacIntyre, I.G. 1973. Crustose coralline algae: a re-evaluation in the geological sciences. *Geol. Soc. Amer. Bull.*, 84:883-904.

Almodovar, L.R. 1970. Deep-water algae new to Puerto Rico. *Quart. Journ. Fla. Acad. Sci.*, 33(1):23-38.

________. 1962. Notes on the algae of the coral reefs, off La Parguera, Puerto Rico. *Quart. Journ. Fla. Acad. Sci.*, 25(4):274-286.

________ and Blomquist, H.L. 1965. Some marine algae new to Puerto Rico. *Nova Hedwigia*, 9:63-71.

Almy, C.C. and Carrion-Torres, C. 1963. Shallow-water stony corals of Puerto Rico. *Carib. Journ. Sci.*, 3(2&3): 133-162.

Antonius A. 1972. Occurrence and distribution of stony corals (Anthozoa and Hydrozoa) in the vicinity of Santa Marta, Colombia. *Mitt. Inst. Colombo-Aleman Invest. Cient.*, 6:89-103.

Arnold, J.M. 1965. Observations on the mating behavior of the squid *Sepioteuthis sepioidea. Bull. Mar. Sci.*, 15(1):216-222.

Baisre, J.A. 1969. A note on the phyllosoma of *Justitia longimanus* (H. Milne Edwards) (Decapoda, Palinuridea). *Crustaceana*, 16(2):182-184.

Bakus, G.J. 1972. Marine studies on the north coast of Jamaica. *Atoll Res. Bull.*, 152:1-6 (abstracts).

Ball, M.M., Shinn, E.A. and Stockman, K.W. 1967. The geologic effects of hurricane Donna in south Florida. *Journ. Geol.*, 75(5):583-597.

Barnes, D.J. and Taylor, D.L. 1973. In situ studies of calcification and photosynthetic carbon fixation in the coral *Montastrea annularis. Helgolander wiss. Meeresunters*, 24:284-291.

Bayer, F.M. 1973. Coral reef project—Papers in memory of Dr. Thomas F. Goreau. 16. A new gorgonian octocoral from Jamaica. *Bull. Mar. Sci.*, 23(2):387-398.

________. 1961. The shallow-water Octocorallia of the West Indian region. *Stud. Fauna Curacao*, 12:1-373.

________ and Owre, H.B. 1968. *The Free-Living Lower Invertebrates.* Macmillan Co., New York, 229 pp.

Beebe, W. 1928. *Beneath Tropic Seas.* Halcyon House, New York, 234 pp.

Berg, C.L. 1975. Behavior and ecology of conch (superfamily Strombacea) on a deep subtidal algal plain. *Bull. Mar. Sci.*, 25(3):307-317.

________. 1974. A comparative ethological study of strombid gastropods. *Behavior*, 51:274-322.

Berrill, M. 1975. Gregarious behavior of juveniles of the spiny lobster, *Panulirus argus* (Crustacea: Decapoda). *Bull. Mar. Sci.*, 25(4):515-522.

Biffar, T.A. and Provenzano, A.J., Jr. 1972. Biological results of the University of Miami deep-sea expeditions. 94. A reexamination of *Dardanus venosus* (H. Milne-Edwards) and *D. imperator* (Miers), with a description of a new species of *Dardanus* from the western Atlantic. *Bull. Mar. Sci.*, 22(4):777-805.

Boolootian, R.A. (ed.). 1966. *Physiology of Echinodermata*. Interscience Publ., New York, 822 pp.

Boschma, H. 1955. The specific characters of the coral *Stylaster roseus*. *Pap. Mar. Biol Oceanogr., Deep-Sea Res.*, 73 (Suppl.):134-138.

———. 1953. On specimens of the coral genus *Tubastrea*, with notes on phenomena of fission. *Stud. Fauna Curacao*, 4:109-119.

———. 1948. Specific characters in *Millepora*. *Proc. Kon. Ned. Akad. Wetensch.*, 51(7):818-823.

Boyd, D.W., Kornicker, L.S. and Rezak, R. 1963. Coralline algae microatolls near Cozumel Island, Mexico. *Wyoming Univ. Contr. Geology*, 2:105-108.

Brattegard, T. 1973. Mysidacea from shallow water on the Caribbean coast of Colombia. *Sarsia*, 54:1-66.

———. 1970. Mysidacea from shallow water in the Caribbean Sea. *Sarsia*, 43:111-154.

———. 1969. Marine biological investigations in the Bahamas. 10. Mysidacea from shallow water in the Bahamas and southern Florida. Part 1. *Sarsia*, 39:17-106.

Bright, T.J. and Pequenat, L.H. (eds.). 1973. *Biota of the West Flower Garden Bank*. Gulf Publ. Co., Houston, Texas.

Bruce, A.J. 1974. On *Lysmata grabhami* (Gordon), a widely distributed tropical hippolytid shrimp (Decapoda, Caridea). *Crustaceana*, 27:107-109.

Bryan, E.H., Jr. 1953. Check list of atolls. *Atoll Res. Bull.*, 19:1-38.

Bullock, D.K. 1972. The corals of Bermuda. *Underwater Naturalist*, 7(3):18-21.

Bunt, J.C., Lee, C.C. and Heeb, M.A. 1972. The site off Freeport intended for Hydro-Lab. *Hydro-Lab Jour.*, 1(1):3-6, 53-56.

Caillouet, C.W., Beardsley, G.L. and Chitty, N. 1971. Notes on size, sex ratio and spawning of the spiny lobster *Panulirus guttatus* (Latreille), near Miami Beach, Florida. *Bull. Mar. Sci.*, 21(4):944-951.

Carlgren, O. and Hedgepeth, J.W. 1952. Actiniaria, Zoantharia and Ceriantheria from shallow water in the northwestern Gulf of Mexico. *Publ. Inst. Mar. Sci. Univ. Texas*, 2(2):142-172.

Chace, F.A., Jr. 1972. The shrimps of the Smithsonian-Bredin Caribbean expeditions with a summary of the West Indian shallow-water species (Crustacea: Decapoda: Natantia). *Smithsonian Contr. Zool.*, 98:1-179.

Chamberlain, K.C. 1967. Some octocorallia of Isla de Lobos, Veracruz, Mexico. *Brigham Young Univ. Stud.*, 13:47-54.

Chapman, V.J. 1961 and 1963. The marine algae of Jamaica. *Bull. Inst. Jamaica, Sci. Ser.*, 12, pts. 1 and 2.

Chesher, R.H. 1970. Biological results of the University of Miami deep-sea expeditions. 68. Evolution in the genus *Meoma* (Echinoidea: Spatangoida) and a description of a new species from Panama. *Bull. Mar. Sci.*, 20(3):731-761.

________. 1969. Contributions to the biology of *Meoma ventricosa* (Echinoidea: Spatangoida). *Bull. Mar. Sci.*, 19(1): 72-110.

________. 1968a. *Lytechinus williamsi*, a new sea urchin from Panama. *Breviora*, 305:1-13.

________. 1968b. The systematics of sympatric species in West Indian spatangoids. *Stud. Trop. Oceanogr., Miami*, No. 7:1-168.

Clark, H.L. 1933. A handbook of the littoral Echinoderms of Porto Rico and the other West Indian islands. *Sci. Sur., Porto Rico and Virgin Isl.*, 16(1):1-147.

Clark, W.D. 1955. A new species of the genus *Heteromysis* (Crustacea, Mysidacea) from the Bahama Islands, commensal with a sea anemone. *Amer. Mus. Novit.*, 1716: 1-13.

Clifton, E.H. and Phillips, R.L. 1972. Physical setting of the

Tektite experiments. *Nat. Hist. Mus., Los Angeles Co., Sci. Bull.*, 14:13-16.

Coe, W.R. 1951. The Nemertean faunas of the Gulf of Mexico and of southern Florida. *Bull. Mar. Sci. Gulf and Carib.*, 1(3):149-186.

Correa, D.D. 1973. Sobre anemonas-do-mar (Actiniaria) do Brasil. *Bol. Zool. e Biol. Mar., N.S.*, 30:457-468.

________. 1964. Corallimorpharia e Actiniaria do Atlantico oeste tropical. *Serv. Doc. R.U. Sao Paulo*, 1-139.

________. 1963. Nemerteans from Curacao. *Stud. Fauna Curacao*, 75:41-56.

________. 1961. Nemerteans from Florida and Virgin Islands. *Bull Mar. Sci. Gulf and Carib.*, 11(1):1-44.

Craft, L.L. 1975. Aspects of the biology of the crab *Percnon gibbesi* (Milne-Edwards) and its commensal association with the sea urchin *Diadema antillarum* Philippi. M.S. Thesis, Univ. Puerto Rico, Mayaguez.

Cutress, B.M. 1965. Observations on growth in *Eucidaris tribuloides* (Lamarck), with special reference to the origin of the primary oral spines. *Bull. Mar. Sci.*, 15(4): 797-834.

Cutress, C. and Ross, D.M. 1969. The sea anemone *Calliactis tricolor* (Lesueur) and its association with the hermit crab *Dardanus venosus* (H.M. Edwards). *Journ. Zool. (London)*, 158:225-241.

________ and Studebaker, J.P. 1973. Development of the cubomedusae, *Carybdea marsupialis. 9th Meeting, Proc. Assoc. Isl. Mar. Lab.:* 25.

D'asaro, Charles N. 1965. Organogenesis, development and metamorphosis in the queen conch, *Strombus gigas*, with notes on breeding habits. *Bull. Mar. Sci.*, 15(2):359-416.

Davis, C.C. 1966. A study of the hatching process in aquatic invertebrates, XXII. Multiple membrane shedding in *Mysidium columbiae* (Zimmer) (Crustacea: Mysidacea). *Bull. Mar. Sci.*, 16(1):124-131.

Davis, W.P. 1966. Observations on the biology of the ophiuroid *Astrophyton muricatum. Bull. Mar. Sci.*, 16(3): 435-444.

Deichmann, E. 1963. Shallow water Holothurians from the Caribbean waters. *Stud. Fauna Curacao*, 14:100-118.

Diaz-Piferrer, M. 1964a. Adiciones a la flora marina de Cuba. *Carib. Journ. Sci.*, 4(2-3):353-371.

________. 1964b. Adiciones a la flora marina de las Antilles Holandesas Curazao y Bonaire. *Carib. Journ. Sci.*, (4(4): 513-543.

________. 1963. Adiciones a la flora marina de Puerto Rico. *Carib. Journ. Sci.*, 3(4):215-235.

Downey, M.E. 1973. Starfishes from the Caribbean and the Gulf of Mexico. *Smithsonian Contr. Zool.*, 126:1-158.

Duarte-Bello, P.P. 1961. Corales de los arrecifes cubanos. *Acuario Nacional, Ser. Educ., Marianao, Cuba*, 2:1-85.

Duerden, J.E. 1904. The coral *Siderastrea siderea* and its post larval development. *Publ. Carnegie Inst. Wash.*, 20: 1-30.

________. 1901. Report on the actinians of Porto Rico. *Bull. U.S. Fish. Comm.*, 20(2):323-374.

Earle, S.A. 1972. The influence of herbivores on the marine plants of Great Lameshur Bay, with an annotated list of plants. *Nat. Hist. Mus., Los Angeles Co., Sci. Bull.*, 14: 17-44.

________. 1969. Phaeophyta of the eastern Gulf of Mexico. *Phycologia*, 7(2):71-254.

Ebbs, N.K., Jr. 1966. The coral-inhabiting polychaetes of the northern Florida reef tract. Part 1. Aphroditidae, Polynoidae, Amphinomidae, Eunicidae, and Lysaretidae. *Bull. Mar. Sci.*, 16(3):485-555.

Emery, A.R. 1968. Preliminary observations on coral reef plankton. *Limnol. Oceanogr.*, 13(2):293-303.

Emiliani, C. 1951. On the species *Homotrema rubrum* (Lamarck). *Contr. Cushman Foundtn. Foram. Res.*, 2: 143-147.

Fabricius, F. 1964. Aktive Lage-und Ortsveranderung bei der Koloniekoralle *Manicina areolata* und ihre palaookologische Bedeutung. *Senck. leth.*, 45(1-4):299-323.

Fisher, E. 1973. The shallow water Actiniaria and Corallimorpharia of Jamaica with special reference to the genus

Bundeopsis. M.S. Thesis, Univ. of the West Indies.

Forbes, M.L. 1971. Habitats and substrates of *Ostrea frons* and distinguishing features of early spat. *Bull. Mar. Sci.*, 21(2):613-625.

Franzisket, L. 1970. The atrophy of hermatypic reef corals maintained in darkness and their subsequent regeneration in light. *Int. Revue ges. Hydrobiol.*, 55(1):1-12.

Frederick, J.J. 1964. The marine algae of the Bermuda platform. Ph.D. Diss., Univ. of Michigan.

Garrett, P., Smith, D.L., Wilson, A.O. and Patriquin, D. 1971. Physiography, ecology and sediments of two Bermuda patch reefs. *Journ. Geol.*, 79:647-668.

Geister, J. 1972. Zur Okologie and Wucksform der Saulenkoralle *Dendrogyra cylindricus* Ehrenberg Beobachtungen in den Riffen der Insel San Andres (Karibisches Meer, Kolumbien). *Mitt. Inst. Colombo-Aleman Invest. Cient.*, 6:77-87.

Gemerden-Hoogeveen, G.C.H., van. 1965. Hydroids of the Caribbean: Sertulariidae, Plumariidae and Aglaopheniidae. *Stud. Fauna Curacao*, 22:1-87.

Geyer, O.F. 1969. Vorlaufige Liste der scleractinen Korallen der Bahia de Concha bei Santa Marta, Kolumbien. *Mitt. Inst. Colombo-Aleman Invest. Cient.*, 3:25-28.

Ghiselin, M.T. and Wilson, B.R. 1966. On the anatomy, natural history and reproduction of *Cyphoma*, a marine prosobranch gastropod. *Bull. Mar. Sci.*, 16(1):132-141.

Ginsburg, R.N. and James, N.P. 1973. British Honduras by submarine. *Geotimes*, 18:23-24.

________ and Lowenstam, H. 1958. The influence of marine bottom communities on the depositional environment of sediments. *Journ. Geol.*, 66(3):310-318.

________, Schroeder, J.H. and Shinn, E.A. 1971. Recent synsedimentary cementation in subtidal Bermuda reefs. In: Bricker, O.P. (ed.). *Carbonate Cements.* The Johns Hopkins Press, Baltimore: 54-58.

Glynn, P.W. 1973a. Ecology of a Caribbean coral reef. The *Porites* reef-flat biotope: Part 1. meteorology and hydrography. *Mar. Biol.*, 20:297-318.

________. 1973b. Ecology of a Caribbean coral reef. The *Porites* reef-flat biotope: Part II. Plankton community with evidence for depletion. *Mar. Biol.*, 22:1-21.

________. 1973c. Aspects of the ecology of coral reefs in the western Atlantic region. In: Jones, O.A. and Endean, R. (eds.). *Biology and Geology of Coral Reefs*, 2:271-324.

________. 1968. Mass mortalities of echinoids and other reef flat organisms coincident with midday, low water exposures in Puerto Rico. *Mar. Biol.*, 1:226-243.

________. 1965. Active movements and other aspects of the biology of *Astichopus* and *Leptosynapta* (Holothuroidea). *Biol. Bull.*, 129(1):106-127.

________. 1962. *Hermodice carunculata* and *Mithraculus sculptus*, two hermatypic coral predators. *Fourth meeting, Assoc. Isl. Mar. Labs.:* 16-17.

________, Almodovar, L.R., and Gonzalez, J.G. 1964. Effects of Hurricane Edith on marine life in La Parguera, Puerto Rico. *Carib. Journ. Sci.*, 4(2-3):335-345.

Goldberg, W.M. 1973a. The ecology of the coral-octocoral communities off the southeast Florida coast: geomorphology, species composition and zonation. *Bull. Mar. Sci.*, 23(3):465-488.

________. 1973b. Ecological aspects of salinity and temperature tolerances of some reef-dwelling gorgonians from Florida. *Carib. Journ. Sci.*, 13(3-4):173-177.

Gonzalez-Brito, P. 1970a. Algunos octocorales de la Isla de Margarita, Venezuela. *Bol. Inst. Oceanog. Univ. Oriente*, 9(1-2):79-92.

________. 1970b. Una lista de los octocorales de Puerto Rico. *Carib. Journ. Sci.*, 10(1-2):63-69.

Gore, R.H. 1970. *Pachycheles cristobalensis*, sp. nov., with notes on the porcellanid crabs of the southwestern Caribbean. *Bull. Mar. Sci.*, 20(4):957-970.

Goreau, T.F. 1969. Post Pleistocene urban renewal in coral reefs. *Micronesica*, 5(2):323-326.

________. 1967. Gigantism and abundance in the macrobenthos of Jamaican coral reefs. *Seventh meeting, Assoc. Isl. Mar. Labs.:* 26-27.

———. 1964. Mass expulsion of zooxanthellae from Jamaican reef communities after Hurricane Flora. *Science*, 145:383-386.

———. 1960. On the physiological ecology of the coral *Meandrina braziliensis* (Milne-Edwards and Haime) in Jamaica. *Third meeting, Assoc. Isl. Mar. Labs.:* 17-18.

———. 1959. The ecology of Jamaican coral reefs. I. Species composition and zonation. *Ecology*, 40(1):67-90.

——— and Goreau, N.I. 1973. Coral reef project—Papers in memory of Dr. Thomas F. Goreau. 17. The ecology of Jamaican coral reefs. II. Geomorphology, zonation, and sedimentary phases. *Bull. Mar. Sci.* 23(2):399-464.

——— and Goreau, N.I. 1959. The physiology of skeleton formation in corals. II. Calcium deposition by hermatypic corals under various conditions in the reef. *Biol. Bull.*, 117:239-250.

——— and Graham, E.A. 1967. A new *Halimeda* from Jamaica. *Bull. Mar. Sci.*, 17(2):432-441.

——— and Hartman, W.D. 1966. Sponge: Effect on the form of reef corals, *Science*, 151:343-344.

——— and Hartman, W.D. 1963. Boring sponges as controlling factors in the formation and maintenance of coral reefs. In: Sognnaes, R.F. (ed.). *Mechanisms of Hard Tissue Destruction. Amer. Assoc. Adv. Sci. Publ. 75, Washington, D.C.* :25-54.

——— and Land, L.S. 1974. Fore-reef morphology and depositional processes, north Jamaica. In: Laporte, L.F. (ed.). *Reefs in Time and Space. Soc. Econ. Paleontol. Mineralol. Spec. Publ. 18,* Tulsa, Oklahoma: 77-89.

——— and Wells, J.W. 1967. The shallow-water Scleractinia of Jamaica: Revised list of species and their vertical distribution range. *Bull. Mar. Sci.*, 17(2):442-453.

Graham, E.A. 1975. Fruiting in *Halimeda* (Order Siphonales) 1. *Halimeda cryptica* Colinvaux and Graham. *Bull. Mar. Sci.*, 25(1):130-133.

Haig, J. 1956. The Galatheidae (Crustacea, Anomura) of the Allan Hancock Expedition with a revision of the Porcellanidae of the western Atlantic. *Allan Hancock Atlantic Exped. Rep.*, 8:1-44.

Halstead, B.W. 1965. *Poisonous and Venomous Marine Animals of the World.* U.S. Gov. Printing Off., Wash., D.C., 3 volumes.

Hartman, O. 1959. Catalogue of the polychaetous annelids of the world. *Allan Hancock Found. Publ., Occ. Pap. 23.*

Hartman, W.D. 1973. Beneath Caribbean reefs. *Discovery,* 9:13-26.

________. 1969. New genera and species of coralline sponges (Porifera) from Jamaica. *Postilla,* 137:1-39.

________. 1967. Revision of *Neofibularia* (Porifera, Demospongiae), a genus of toxic sponges from the West Indies and Australia. *Postilla,* 113:1-41.

________ and Goreau, T.F. 1972. *Ceratoporella* (Porifera: Sclerospongiae) and the chaetetid "corals." *Trans. Conn. Acad. Arts Sci.,* 44:133-148.

________ and Goreau, T.F. 1970. Jamaican coralline sponges: their morphology, ecology and fossil relatives. In: Fry, W.G. (ed.). *The Biology of the Porifera. Symp. Zool. Soc. London* 25, Academic Press, London: 205-243.

________ and Goreau, T.F. 1966. *Ceratoporella,* a living sponge with stromatoporoid affinities. *Amer. Zool.,* 6: 563-564.

Hartnoll, R.G. 1965. The biology of spider crabs: a comparison of British and Jamaican species. *Crustaceana,* 9:1-16.

Harvey, E.B. 1947. Bermuda sea urchins and their eggs. *Biol. Bull.,* 93:217-218.

Hazlett, B. 1972. Shell fighting and sexual behavior in the hermit crab genera *Paguristes* and *Calcinus,* with comments on *Pagrus. Bull. Mar. Sci.,* 22(4):806-823.

________. 1966. Social behavior of the Pagurida and Diogenidae of Curacao. *Stud. Fauna Curacao,* 23:1-143.

________ and Rittschof, M. 1975. Daily movements and home range in *Mithrax spinosissimus* (Majidae, Decapoda). *Mar. Behav. Physiol.,* 1975(3):101-118.

Hechtel, G.J. 1969. New species and records of shallow water Demospongiae from Barbados, West Indies. *Postilla,* 132:1-38.

________. 1965. A systematic study of the Demospongiae of Port Royal, Jamaica. *Bull. Peabody Mus. Nat. Hist., Yale Univ.*, 20:1-103.

Hein, F.J. and Risk, M.J. 1975. Bioerosion of coral heads: inner patch reefs, Florida reef tract. *Bull. Mar. Sci.*, 25(1):133-138.

Hernnkind, W. 1970. Migration of the spiny lobster. *Nat. Hist.*, 79:36-43.

________. 1969. Queuing behavior of spiny lobsters. *Science*, 164:1425-1427.

________ and Cummings, W. 1964. Single file migration of the spiny lobster *Panulirus argus* (Latreille). *Bull. Mar. Sci. Gulf and Carib.*, 14:123-125.

________ and McLean, R. 1971. Field studies of homing, mass emigration, and orientation in the spiny lobster, *Panulirus argus. Ann. N.Y. Acad. Sci.*, 188:359-377.

Hillis-Colinvaux, L. 1974. Productivity of the coral reef alga *Halimeda* (Order Siphonales). *Proc. Sec. Int. Symp. Coral Reefs*, I:35-42.

(Hillis)-Colinvaux, L.H. and Graham, E.A. 1964. A new species of *Halimeda. Nova Hedwigia*, 8:5-10.

Hochberg, F.G. and Ellis, R.J. 1972. Cymothoid isopods associated with reef fishes (abstract). *Nat. Hist. Mus., Los Angeles Co., Sci. Bull.*, 14:84.

Hoffmeister, J.E. and Multer, H.G. 1968. Geology and origin of the Florida Keys. *Bull. Geol. Soc. Amer.*, 79:1487-1502.

Holthuis, L.B. 1965. On spiny lobsters of the genera *Palinurellus, Linuparus* and *Puerulus* (abstract). *Pap. Symp. Crustacea. Mar. Biol. Assoc. India*:1-2.

________. 1946. The decapod macrura of the Snellius Expedition, I. The Stenopodidae, Nephropsidae, Scyllaridae, and Palinuridae. Biol. Results Snellius Exped. XIV. *Temminckia*, 7:1-178.

________ and Eibl-Eibesfeldt, I. 1964. A new species of the genus *Periclimenes* from Bermuda (Crustacea, Decapoda, Palaemonidae). *Senckenbergiana Biol.*, 45(2):185-192.

Hubbard, J.A.E.B. 1973. Sediment-shifting experiments: a guide to functional behavior in colonial corals. · In: Boardman, Cheetham and Oliver (eds.), *Animal Colonies*. Dowden, Hutchinson and Ross, Inc., Stroudsburg, Pa.

Hughes, R.N. and Hughes, H.P.I. 1971. A study of the gastropod *Cassis tuberosa* (L.) preying upon sea urchins. *Journ. Exp. Mar. Biol. Ecol.*, 7:305-314.

Humm, H.J. and Taylor (Earle) S. 1961. Marine Chlorophyta of the upper west coast of Florida. *Bull. Mar. Sci. Gulf and Carib.*, 11(3):321-380.

Hyman, L.H. 1959. *The Invertebrates. Vol. 5. Smaller Coelomate groups.* McGraw-Hill Book Co. New York, 783 pp.

________. 1955. *The Invertebrates. Vol. 4. Echinodermata.* McGraw-Hill Book Co., New York, 763 pp.

________. 1940. *The Invertebrates. Vol. 1. Protozoa through Ctenophora.* McGraw-Hill Book Co., New York, 726 pp.

Jackson, J.B.C, Goreau, T.F. and Hartman, W.D. 1971. Recent brachipod-coralline sponge communities and their paleoecological significance. *Science*, 173:623-625.

Johnson, V.R., Jr. 1969. Behavior associated with pair formation in the banded shrimp *Stenopus hispidus* (Olivier). *Pac. Sci.*, 23:40-50.

Jones, M.L. 1962. On some polychaetous annelids from Jamaica, the West Indies. *Bull. Amer. Mus. Nat. Hist.*, 124(5):173-212.

Kanciruk, P. and Herrnkind, W. 1973. Preliminary investigations of the daily and seasonal locomotor activity rhythms of the spiny lobster, *Panulirus argus. Mar. Behav. Physiol.*, 1973(1):351-359.

Kier, P.M. 1975. The echinoids of Carrie Bow Cay, Belize. *Smithsonian Contr. Zool.*, 206:1-45.

________. 1966. Bredin-Archbold-Smithsonian biological survey of Dominica. 1. The echinoids of Dominica. *Proc. U.S. Nat. Mus.*, 121(3577):1-10.

________ and Grant, R.E. 1965. Echinoid distribution and habits, Key Largo Coral Reef Preserve, Florida. *Smith-*

sonian Misc. Coll., 149(6):1-68.

Kinzie, R.A. 1974. Experimental infection of aposymbiotic gorgonian polyps with zooxanthellae. *Journ. exp. mar. Biol. Ecol.*, 15:335-345.

________. 1973. Coral reef project—Papers in memory of Dr. Thomas F. Goreau. 5. The zonation of West Indian gorgonians. *Bull. Mar. Sci.*, 23(1):93-155.

Kissling, D.L. 1965. Coral distribution on a shoal in Spanish Harbor, Florida Keys. *Bull. Mar. Sci.*, 15(3):599-611.

Kornicker, L.S. and Boyd, D.W. 1962. Shallow-water geology and environments of Alacran Reef complex, Campeche Bank, Mexico. *Bull. Amer. Assoc. Petrol. Geol.*, 46:640-673.

Kornicker, L.S., Bonet, F., Cann, R. and Hoskin, C.M. 1959. Alacran Reef, Campeche Bank, Mexico. *Publ. Inst. Mar. Sci., Univ. Texas*, 6:1-22.

Kuhlmann, D.H.H. 1974. The coral reefs of Cuba. *Proc. Second Int. Symp. Coral Reefs*, 2:69-83.

________. 1971. Die korralenriffe Kubas II. Zur okologie der bankfriffe und ihrer korallen. *Int. Rev. ges. Hydrobiol.*, 56(2):145-199.

________. 1970. Die korralenriffe Kubas. I. Genese und evolution. *Int. Rev. ges. Hydrobiol.*, 55(5):729-756.

Kumpf, H.E. and Randall, H.A. 1961. Charting the marine environments of St. John, U.S. Virgin Islands. *Bull. Mar. Sci. Gulf and Carib.*, 11(4):543-551.

Laborel, J. 1974. West African reef corals, an hypothesis on their origin. *Proc. Second Int. Symp. Coral Reefs*, 1:425-443.

________. 1970. Madreporaires et hydrocorallia res recifaux des cotes bresiliennes, systematique, ecologie, repartition verticale et geographique. *Annls. Inst. oceanogr., Paris*, 47(1):171-229.

________. 1969. Les peuplements de madreporaires des cotes tropicales du Bresil. *Ann. de L'Univ. Abidjan, Ser. E.*, 2(3):1-260.

________. 1966. Contribution a l'etude des madreporaires des Bermudes (systematique et repartition). *Bull Mus. natn. Hist. nat. Paris, Ser. 2*, 38:281-300.

Lang, J.C. 1974. Biological zonation at the base of a reef. *Amer. Sci.*, 62(3):272-281.

________. 1973. Coral reef project—Papers in memory of Dr. Thomas F. Goreau. 11. Interspecific aggression by scleractinian corals. 2. Why the race is not only to the swift. *Bull. Mar. Sci.*, 23(2):260-279.

________. 1971. Interspecific aggression by scleractinian corals. 1. The rediscovery of *Scolymia cubensis* (Milne-Edwards and Haime). *Bull. Mar. Sci.*, 21(4):952-959.

________, Hartman, W.D. and Land, L.S. 1975. Sclerosponges: Primary framework constructors on the Jamaican deep fore-reef. *Journ. Mar. Res.*, 33(2):223-231.

LaRoe, E.T. 1971. The culture and maintenance of the loliginid squids *Sepioteuthis sepioidea* and *Doryteuthis plei*. *Mar. Biol.*, 9(1):9-25.

de Laubenfels, M.W. 1953. Sponges from the Gulf of Mexico. *Bull. Mar. Sci. Gulf and Carib.*, 2(3):511-557.

________. 1950. The Porifera of the Bermuda archipelago. *Trans. Zool. Soc. London*, 27(1):1-154.

________. 1949. Sponges from the western Bahamas. *Amer. Mus. Novitates*, 1431:1-25.

________. 1948. The order Keratosa of the phylum Porifera—a monographic study. *Allan Hancock Fndtn. Occ. Pap.* 3:1-217.

________. 1936. A discussion of the sponge fauna of the Dry Tortugas in particular and the West Indies in general, with material for a revision of the families and orders of the Porifera. *Pap. Tortugas Lab.*, 30:1-225.

________. 1936b. A comparison of the shallow-water sponges near the Pacific end of the Panama Canal with those at the Caribbean end. *Proc. U.S. Nat. Mus.*, 83:441-466.

Lewis, J.B. 1975. A preliminary description of the coral reefs of the Tobago Cays, Grenadines, West Indies. *Atoll. Res. Bull.*, 178:1-14.

________. 1974a. The settlement behavior of planulae larvae of the hermatypic coral *Favia fragum* (Esper). *Journ. Exp. mar. Biol. Ecol.*, 15:165-172.

________. 1974b. Settlement and growth factors influencing the contagious distribution of some Atlantic reef corals. *Proc. Second Int. Symp. Coral Reefs*, 2:201-206.

________. 1974c. The importance of light and food upon the early growth of the reef coral *Favia fragum* (Esper). *Journ. Exp. mar. Biol. Ecol.*, 15:299-304.

________. 1970. Spatial distribution and pattern of some Atlantic reef corals. *Nature*, 227(5263):1158-1159.

________. 1966. Growth and breeding in the tropical echinoid *Diadema antillarum* Philippi. *Biol. Bull.*, 132: 34-37.

________. 1965. A preliminary description of some marine benthic communities from Barbados, West Indies. *Canad. Journ. Zool.*, 43:1049-1074.

________. 1964. Feeding and digestion in the tropical sea urchin *Diadema antillarum* Philippi. *Canad. Journ. Zool.*, 42:549-557.

________. 1960. The coral reefs and coral communities of Barbados, W.I. *Canad. Journ. Zool.*, 38:1133-1145.

________. 1958. The biology of the tropical sea urchin *Tripneustes esculentus* Leske in Barbados, British West Indies. *Canad. Journ. Zool.*, 36(4):607-621.

Limbaugh, C., Pederson, H. and Chace, F.A., Jr. 1961. Shrimps that clean fishes. *Bull. Mar. Sci. Gulf and Carib.*, 11(2):237-257.

Little, C. 1965. Notes on the anatomy of the queen conch, *Strombus gigas. Bull. Mar. Sci.*, 15(2):338-358.

Lizama, J. and Blanquet, R.S. 1975. Predation on sea anemones by the amphinomid polychaete, *Hermodice carunculata. Bull. Mar. Sci.*, 25(3):442-443.

Logan, A. 1973. Life habits and taxonomy of the recent cemented bivalve *Spondylus americanus* from the Bermuda platform. *Geol. Soc. Amer.*, 5:189.

Logan, B.W. 1969. Coral reefs and banks, Yucatan shelf, Mexico. *Amer. Assoc. Petrol. Geol., Mem. 11:*129-198.

________, Harding, J.L., Ahr, W.M., Williams, J.D. and Snead, R.G. 1969. Carbonate sediments and reefs, Yucatan shelf, Mexico. Part 1. Late Quarternary sediments.

*Amer. Assoc. Petrol. Geol., Mem. 11:*1-128.

Lukas, K.J. 1973. Taxonomy and ecology of the endolithic microflora of reef corals with a review of the literature on endolithic microphytes. Ph.D. Diss., Univ. Rhode Island.

________. 1974. Two species of the chlorophyte genus *Ostreobium* from skeletons of Atlantic and Caribbean reef corals. *Journ. Phycol.*, 10(3):331-335.

Lyons, W.G. 1970. Scyllarid lobsters (Crustacea, Decapoda). *Mem. Hourglass Cruises*, 1(4):1-74.

Macurda, D.B., Jr. 1973. Ecology of Comatulid crinoids at Grand Bahama Island. *Hydro-lab Journ.*, 2(1):9-24.

MacIntyre, I.G. 1972. Submerged reefs of eastern Caribbean. *Amer. Assoc. Petrol. Geol. Bull.*, 56(4):720-738.

________. 1968. Preliminary mapping of insular shelf of the west coast of Barbados. *Carib. Journ. Sci.*, 8(1-2):95-100.

________. 1967. Submerged coral reefs, west coast of Barbados, West Indies. *Canad. Journ. Earth Sci.*, 4:461-474.

MacIntyre, I.G. and Glynn, P.W. 1974. Internal structure and developmental stages of a modern Caribbean fringe reef, Galeta Point, Panama. *Seventh Carib. Geol. Conf. Abs.*:40.

________ and Pilkey, O.H. 1969. Tropical reef corals: Tolerance of low temperatures on the North Carolina continental shelf. *Science*, 166:374-375.

Manhken, C. 1972. Observations on cleaner shrimps of the genus *Periclimenes*. *Nat. Hist. Mus., Los Angeles County, Sci. Bull.*, 14:71-83.

Manning, R.B. 1970. *Mithrax (Mithraculus) commensalis*, a new West Indian spider crab (Decapoda, Majidae) commensual with a sea anemone. *Crustaceana*, 19(2): 157-160.

________. 1969. Stomatopod crustacea of the western Atlantic. *Stud. Trop. Oceanogr.*, 8:1-380.

________. 1968. A revision of the family Squillidae (Crustacea, Stomatopoda) with the description of eight new species. *Bull. Mar. Sci.*, 18(1):105-142.

________. 1961. Notes on the Caridean shrimp, *Rhynchocinetes rigens* Gordon (Crustacea, Decapoda), in the western Atlantic. *Not. Nat.*, 348:1-7.

Marcus, E. and Marcus, E. 1967. American Opisthobranch mollusks. *Stud. Trop. Oceanogr.*, 6:1-256.

________ and ________. 1960. Opisthobranchs from American Atlantic warm water. *Bull. Mar. Sci. Gulf and Carib.*, 10(2):129-203.

Marsden, J.R. 1962. A coral-eating polychaete. *Nature, Lond.*, 193(4815):598.

________. 1960. Polychaetous annelids from the shallow waters around Barbados and other islands of the West Indies, with notes on larval forms. *Canad. Journ. Zool.*, 38(5):989-1020.

Mattraw, H.C. 1969. Variations of *Millepora* on the Bermuda platform. *Bermuda Biol. Sta. Spec. Publ.*, 6:29-38.

Matthews, R.K. 1966. Genesis of recent lime mud in southern British Honduras. *Journ. Sed. Petrol.*, 36:428-454.

Mayer, A.G. 1912. Ctenophores of the Atlantic coast of North America. *Carnegie Inst. Wash. Publ.*, 162:1-52.

________. 1910. Medusae of the world. *Carnegie Inst. Wash. Publ.*, 109 (3 vols.).

McLaughlin, P.A. 1975. On the identity of *Pagurus brevidactylus* (Stimpson) (Decapoda: Paguridae), with the description of a new species of *Pagurus* from the western Atlantic. *Bull. Mar. Sci.*, 25(3):359-376.

McPherson, B.F. 1969. Studies on the biology of the tropical sea urchins *Echinometra lucunter* and *Echinometra viridis*. *Bull. Mar. Sci.*, 19(1):195-213.

Meyer, D.L. 1973a. Coral reef project—Papers in memory of Dr. Thomas F. Goreau. 10. Distribution and living habits of comatulid crinoids near Discovery Bay, Jamaica. *Bull. Mar. Sci.*, 23(2):244-259.

________. 1973b. Feeding behavior and ecology of shallow-water unstalked crinoids (Echinodermata) in the Caribbean Sea. *Mar. Biol.*, 22:105-130.

Millar, R.H. 1962. Some ascidians from the Caribbean. *Stud. Fauna Curacao*, 13:61-77.

________ and Goodbody, I. 1974. New species of Ascidians from the West Indies. *Stud. Fauna Curacao*, 45:142-161.

Milliman, J.D. 1973. Caribbean coral reefs. In: Jones, O.A. and Endean, R. (eds.). *Biology and Geology of Coral reefs*, 1:1-50.

————. 1969. Four southwestern Caribbean atolls: Courtown Cays, Albuquerque Cays, Roncador Bank and Serrana Bank. *Atoll. Res. Bull.*, 129:1-41.

————. 1967. The geomorphology and history of Hogsty Reef, a Bahamian atoll. *Bull. Mar. Sci.*, 17(3):519-543.

———— and Supko, P.R. 1968. On the geology of San Andres Island, western Caribbean. *Geol. Mijn.*, 47:102-105.

Millott, N. 1966. Coordination of spine movements in echinoids. In: Boolootian, R.A. (ed.). *Physiology of Echinodermata*, Interscience Publ., New York: 465-485.

————. 1950. The sensitivity to light, reactions to shading, pigmentation, and colour change of the sea urchin *Diadema antillarum* Philippi. *Biol. Bull.*, 99:329-330.

Moore, D.R. 1963. Distribution of the sea grass, *Thalassia*, in the United States. *Bull. Mar. Sci. Gulf and Carib.*, 13 (3):329-342.

Moore, H.B., Jutare, T., Bauer, J.C. and Jones, J.A. 1963. The biology of *Lytechinus variegatus*. *Bull. Mar. Sci. Gulf and Carib.*, 13(1):25-53.

Multer, H.G. 1974. Some shelf-edge processes, Cane Bay, northwest St. Croix: a progress report. In: Multer, H.G. and Gerhard, L.C. (eds.). *Guidebook to the Geology and Ecology of Some Marine and Terrestrial Environments, St. Croix, U.S. Virgin Islands. Spec. Publ. West Indies Lab., Fairleigh Dickinson Univ.*, 5:101-114.

Munro, J.L. 1974. The biology, ecology and bionomics of Caribbean reef fishes: VI. Crustaceans (spiny lobsters and crabs). *Res. Rpt., Zool. Dept., Univ. West Indies, Kingston, Jamaica*: 1-57.

Muscatine, L. 1973. Nutrition of corals. In: Jones, O.A. and Endean, R. (eds.). *Biology and Geology of Coral Reefs*, 2:77-115.

Newell, N.D., Imbrie, J., Purdy, E.J. and Thurber, D.L. 1959. Organism communities and bottom facies, Great

Bahama Bank. *Bull. Amer. Mus. Nat. Hist.*, 117(4):177-228.

Noome, C. and Kristensen, I. 1976. Necessity of conservation of slow growing organisms like Black Coral. C.C.A. Ecology Conference, Bonaire, STINAPA, No. 11:76-77.

Ogden, J.C., Brown, R.A., and Salesky, N. 1973. Grazing by the Echinoid *Diadema antillarum* Philippi: formation of halos around West Indian patch reefs. *Science*, 182: 715-717.

________, Helm, D., Peterson, J., Smith, A. and Weisman, S. (eds.). 1972. An ecological study of Tague Bay Reef, St. Croix, U.S. Virgin Islands. *Spec. Publ. West Indies Lab., Fairleigh Dickinson Univ.*, 1:1-50.

Olivares, M.A. 1973. Estudio taxonomico de los madreporarios del Gulfo de Cariacao, Sucre, Venez. *Ninth Meeting, Assoc. Isl. Mar. Labs.*:9.

________ and Leonard, A.B. 1971. Algunos corals petreos de la Bahia de Mochima, Venezuela. *Bol. Inst. Oceanogr. Univ. Oriente*, 10(1):49-70.

Opresko, D.M. 1974. A study of the classification of the antipatharia. Ph.D. Diss., Univ. of Miami.

________. 1973. Abundance and distribution of shallow-water gorgonians in the area of Miami, Florida. *Bull. Mar. Sci.*, 23(3):535-558.

________. 1972. Biological results of the University of Miami Deep-Sea Expeditions. 97. Redescriptions and reevaluations of the antipatharians described by L.F. de Pourtales. *Bull. Mar. Sci.*, 22(4):950-1017.

Opresko, L., Opresko, D., Thomas, R. and Voss, G. 1973. Guide to the lobsters and lobster-like animals of Florida, the Gulf of Mexico and the Caribbean region. *Sea Grant Field Guide Ser., Univ. of Miami*, 1:1-44.

Osburn, R.C. 1940. Bryozoa of Porto Rico with a resume of the West Indian bryozoan fauna. *Sci. Sur. Porto Rico and Virgin Isl.*, 16(3):323-486.

Pang, R.K. 1973a. The systematics of some Jamaican excavating sponges (Porifera). *Postilla*, 161:1-75.

________. 1973b. Coral Reef Project—Papers in memory of

Dr. Thomas F. Goreau. 9. The ecology of some Jamaican excavating sponges. *Bull. Mar. Sci.*, 23(2):227-243.

Parrish, James D. 1972. A study of predation on tropical holothurians at Discovery Bay, Jamaica. In: Bakus, J.G. (ed.). *Marine studies on the north coast of Jamaica* (abstract). *Atoll. Res. Bull.*, 152:6.

Patton, W.K. 1967. Studies on *Domecia acanthophora*, a commensal crab from Puerto Rico, with particular reference to modification of the coral host and feeding habits. *Biol Bull.*, 132(1):56-67.

Pax, F. 1910. Studien an westindischen Actinien. *Zool. Jahrb., Suppl.*, 11:157-330.

Percharde, P.L. 1968. Notes on distribution and underwater observations of the molluscan genus *Strombus* as found in the water of Trinidad and Tobago. *Carib. Journ. Sci.*, 8(1-2):47-55.

Perkins, R.D. and Enos, P. 1968. Hurricane Betsy in the Florida-Bahamas area—geologic effects and comparison with Hurricane Donna. *Journ. Geol.*, 76:710-717.

Pfaff, R. 1969. Las scleractinia y Milleporina de las Islas de Rosario. *Mitt. Inst. Colombo-Aleman Invest. cient.*, 3: 17-24.

Phillips, R.C. 196;. Observations on the ecology and distribution of the Florida sea grasses. *Prof. Pap. Ser., Fla. Bd. Conserv.*, 2:1-72.

Porter, J.W. 1974a. Zooplankton feeding by the Caribbean reef-building coral *Montastrea cavernosa*. *Proc. Second Int. Symp. Coral Reefs*, 1:111-125.

________. 1974b. Community structure of coral reefs on opposite sides of the Isthmus of Panama. *Science*, 186: 543-545.

________. 1973. Ecology and composition of deep reef communities off the Tongue of the Ocean, Bahama Islands. *Discovery*, 9(1):3-12.

________. 1972. Ecology and species diversity of coral reefs on opposite sides of the Isthmus of Panama. In: Jones, M.L. (ed.). *The Panamic Biota: Some Observations Prior to a Sea-Level Canal. Bull. Biol. Soc. Wash.*, 2:89-116.

Pressick, M.L. 1970. Zonation of stony coral of a fringing reef southeast of Icacos Island, Puerto Rico. *Carib. Journ. Sci.*, 10(3-4):137-140.

Preston, E.M. and Preston, J.L. 1975. Ecological structure in a West Indian gorgonian fauna. *Bull. Mar. Sci.*, 25(2): 248-258.

Provenzano, A.J., Jr. 1968. The complete larval development of the West Indian hermit crab *Petrochirus diogenes* (L.) (Decapoda, Diogenidae) reared in the laboratory. *Bull. Mar. Sci.*, 18(1):143-181.

________. 1960. Note on *Paguristes cadenati*, a hermit crab new to Florida. *Quart. Journ. Fla. Acad. Sci.*, 23(4): 325-327.

________. 1959. The shallow-water hermit crabs of Florida. *Bull. Mar. Sci. Gulf and Carib.*, 9(4):349-420.

Randall, J.E. 1964. Predators of the queen conch (*Strombus gigas*). *Fifth meeting, Assoc. Isl. Mar. Labs.*: 17-18.

________, Schroeder, R.E. and Starck, W.A., II. 1964. Notes on the biology of the echinoid *Diadema antillarum*. *Carib. Journ. Sci.*, 4(2-3):421-433.

Rannfeld, J.W. 1972. The stony corals of Enmedio Reef off Veracruz, Mexico. M.S. Thesis, Texas A & M Univ.

Rathbun, M.J. 1933. Brachyuran crabs of Porto Rico and the Virgin Islands. *Sci. Sur. Porto Rico and Virgin Isl.*, 15(1): 1-121.

________. 1930. The cancroid crabs of America of the families Euryalidae, Portunidae, Atelecyclidae, Cancridae and Xanthidae. *Bull. U.S. Nat. Mus.*, 152:1-608.

________. 1925. The spider crabs of America. *Bull. U.S. Nat. Mus.*, 129:1-613.

Read, K.R.H., Davidson, J.M. and Twarog, B.M. 1968. Fluorescense of sponges and coelenterates in blue light. *Comp. Biochem. Physiol.*, 25:873-882.

Rees, J.T. 1973. Shallow-water octocorals of Puerto Rico: species account and corresponding depth records. *Carib. Journ. Sci.*, 13(1-2):57-58.

Rees, J.T. 1972. The effect of current on growth form in an octocoral. *J. Exp. Mar. Biol. Ecol.* 10:115-123.

————. 1969. Aspects of growth and nutrition in the octocoral *Telesto riisei*. M.S. Thesis, Univ. of Puerto Rico, Mayaguez.

Reiswig, H.M. 1973. Coral reef project—Papers in memory of Dr. Thomas F. Goreau. 8. Population dynamics of three Jamaican Demospongiae. *Bull. Mar. Sci.*, 23(2): 191-226.

————. 1971a. *In situ* pumping activities of tropical Demospongiae. *Mar. Biol.*, 9(1):38-50.

————. 1971b. Particle feeding in natural populations of three marine Demospongiae. *Biol. Bull.*, 141(3):568-591.

————. 1970. Porifera: sudden sperm release by tropical Demospongiae. *Science*, 170:538-539.

Rigby, J.K. and McIntire, W.G. 1968. The Isla de Lobos and associated reefs, Veracruz, Mexico. *Brigham Young Univ. Stud.*, 13:3-46.

Roberts, H.H. 1974. Variability of reefs with regard to changes in wave power around an island. *Proc. Second Int. Symp. Coral Reefs*, 2:497-512.

————. 1972. Coral reefs of St. Lucia, West Indies. *Carib. Journ. Sci.*, 12(3-4):179-190.

————. 1971. Environments and organic communities of North Sound, Grand Cayman Island, B.W.I. *Carib. Journ. Sci.*, 11(1-2):67-80.

Robertson, P.B. 1969a. The early larval development of the scyllarid lobster *Scyllarides aequinoctialis* (Lund) in the laboratory, with a revision of the larval characters of the genus. *Deep-Sea Res.*, 16(6):557-586.

————. 1969b. Phyllosoma larvae of a palinurid lobster, *Justitia longimana* (H. Milne Edwards), from the western Atlantic. *Bull. Mar. Sci.*, 19(4):922-944.

Rodriguez, G. 1959. The marine communities of Margarita Island, Venezuela. *Bull. Mar. Sci. Gulf and Carib.*, 9(3): 238-280.

Rooney, W.S., Jr. 1970. A preliminary ecologic and environmental study of the sessile foraminifera *Homotrema rubrum* (Lamarck). *Spec. Publ., Bermuda Biol. Sta.*, 6: 7-18.

Roos, P.J. 1971. The shallow-water stony corals of the Netherlands Antilles. *Stud. Fauna Curacao*, 37:1-108.

________. 1967. Growth and occurrence of the reef coral *Porites asteroides* Lamarck in relation to submarine radiance distribution. Acad. thesis, Univ. Amsterdam, 72 pp.

________. 1964. The distribution of reef corals in Curacao. *Stud. Fauna Curacao*, 20:1-51.

Rutzler, K. 1971. Bredin-Archbold-Smithsonian biological survey of Dominica: burrowing sponges, genus *Siphonodictyon* Bergquist, from the Caribbean. *Smithsonian Contr. Zool.*, 77:1-37.

Ryland, J.S. 1974. Bryozoa in the Great Barrier Reef province. *Proc. Second Int. Symp. Coral Reefs*, 1:341-348.

Sammarco, P.W., Levinton, J.S. and Ogden, J.C. 1974. Grazing and control of coral reef community structure by *Diadema antillarum* Philippi (Echinodermata: Echinoidea): a preliminary study. *Journ. Mar. Res.*, 32(1):47-53.

Sargent, R.C. 1975. Cleaning behavior of the shrimp, *Periclimenes anthophilus* Holthuis and Eibl-Eibesfeldt (Crustacea: Decapoda: Natantia). *Bull. Mar. Sci.*, 25(4):466-472.

Scatterday, J.W. 1974. Reefs and associated coral assemblages off Bonaire, Netherlands Antilles, and their bearing on Pleistocene and recent reef models. *Proc. Second Int. Symp. Coral Reefs*, 2:85-106.

Schopf, T.M. 1974. Ectoprocts as associates of coral reefs: St. Croix, U.S. Virgin Islands. *Proc. Second Int. Symp. Coral Reefs*, 1:353-356.

Serafy, D.K. 1973. Variation in the polytypic sea urchin *Lytechinus variegatus* (Lamarck, 1816) in the western Atlantic (Echinodermata: Echinoidea). *Bull. Mar. Sci.*, 23(3):525-534.

Shinn, E.A. 1963. Spur and groove formation on the Florida reef tract. *Journ. Sed. Petrol.*, 33:291-303.

Sims, W.H., Jr. 1966. The phyllosoma larvae of the spiny lobster *Palinurellus gundlachi* Von Martens (Decapoda, Palinuridae). *Crustaceana*, 11(2):205-215.

Smith, F.G.W. 1971. *Atlantic Reef Corals.* Univ. Miami

Press, 2nd ed., 164 pp.

Squires, D.F. 1958. Stony corals from the vicinity of Bimini, Bahamas, B.W.I. *Bull. Amer. Mus. Nat. Hist.*, 115:215-262.

Stearns, C.W. and Riding, R. 1973. Forms of the hydrozoan *Millepora* on a recent coral reef. *Lethaia*, 6:187-200.

Steinberg, J.C., Cummings, W.C., Brahy, B.D. and MacBain (Spires), J.Y. 1965. Further bio-acoustic studies off the west coast of North Bimini, Bahamas. *Bull. Mar. Sci.*, 15(4):942-963.

Steven, D.M. 1961. Shoaling behavior in a mysid. *Nature, Lond.*, 192(4799):280-281.

Stoddart, D.R. 1974. Post-hurricane changes on the British Honduras reefs: resurvey of 1972. *Proc. Second Int. Symp. Coral Reefs*, 2:473-483.

________. 1969. Post-hurricane changes on the British Honduras reefs and cays: resurvey of 1965. *Atoll Res. Bull.*, 131:1-35.

________. 1965. Re-survey of hurricane effects on the British Honduras reefs and cays. *Nature, Lond.*, 207:589-592.

________. 1963. Effects of hurricane Hattie on the British Honduras reefs and cays, October 30-31, 1961. *Atoll Res. Bull.*, 95:1-142.

________. 1962. Three Caribbean atolls: Turneffe Islands, Lighthouse Reef, and Glover's Reef, British Honduras. *Atoll Res. Bull.*, 87:1-151.

Storr, J.F. 1964. Ecology and oceanography of the coral reef tract, Abaco Island, Bahamas. *Geol. Soc. Amer., spec. Pap.*, 79:1-98.

Taylor, D.L. 1971. Photosynthesis of symbiotic chloroplasts in *Tridachia crispata* (Bergh). *Comp. Biochem. Physiol.*, 38A: 233-236.

Taylor, W.R. 1972. Marine algae of the Smithsonian-Bredin expedition to Yucatan—1960. *Bull. Mar. Sci.*, 22(1): 34-44.

________. 1960. *Marine Algae of the Eastern Tropical and Subtropical Coasts of the Americas.* Univ. of Michigan Press, 870 pp.

Taylor, W.R. and Rhyme, C.F. 1970. Marine algae of Dominica. *Smithsonian Contr. Bot.*, 3:1-16.

TenHove, H.A. 1970. Serpulinae (Polychaeta) from the Caribbean: I—the genus *Spirobranchus. Stud. Fauna Curacao*, 32:1-57.

Thomas, L.P. 1960. A note of the feeding habits of the West Indian sea star *Oreaster reticulatus* (L.). *Quart. Journ. Fla. Acad. Sci.*, 23:167-168.

________, Moore, R.N. and Work, R.C. 1961. Effects of hurricane Donna on the turtle grass beds of Biscayne Bay, Florida. *Bull. Mar. Sci. Gulf and Carib.*, 11(2):191-197.

Thompson, T.W. 1971. Gulf weed communities of south Florida. *Sea Frontiers,* 17(2):95-101.

Thorpe, J.E. and Stoddart, D.R. 1962. Cambridge Expedition to British Honduras. *Geog. Journ.*, 128:158-171.

Tikasingh, E.S. 1963. The shallow water Holothurians of Curacao, Aruba and Bonaire. *Stud. Fauna Curacao*, 14: 77-99.

Tomlinson, P.B. 1972. On the morphology and anatomy of turtle grass, *Thalassia testudinum* (Hydrocharitaceae). IV. Leaf anatomy and development. *Bull. Mar. Sci.*, 22(1):75-93.

________ and Vargo, G.A. 1966. On the morphology and anatomy of turtle grass, *Thalassia testudinum* (Hydrocharitaceae). I. Vegetative morphology. *Bull. Mar. Sci.*, 16(4):748-761.

Treadwell, A.L. 1939. Polychaetous annelids of Porto Rico and vicinity. *Sci. Sur. Porto Rico and Virgin Isl.*, 16(2): 151-319.

Tresslar, R.C. 1973. Corals. In: Bright, T.J. and Pequenat, L.H. (eds.). *Biota of the West Flower Garden Bank.* Gulf Publ. Co., Houston: 116-139.

Trench, R.K. 1973a. Coral reef project—Papers in memory of Dr. Thomas F. Goreau. 13. Further studies on the mucopolysaccharide secreted by the pedal gland of the marine slug *Tridachia crispata* (Opisthobranchia, Sacoglossa). *Bull. Mar. Sci.*, 23(2):299-312.

________. 1973b. Zooxanthellae—a problem in taxonomy. *Eighth meeting, Assoc. Isl. Mar. Labs.*:13.

________. 1969. Chloroplasts as functional endosymbionts in the mollusc *Tridachia crispata* Bergh (Opisthobranchia, Sacoglossa). *Nature, Lond.*, 222:1071-1072.

Tsuda, R.T. and Dawes, C.J. 1974. Preliminary checklist of the marine benthic plants from Glover's Reef, British Honduras. *Atoll Res. Bull.*, 173:1-13.

Turner, R.D. and Rosewater, J. 1958. The family Pinnidae in the western Atlantic. *Johnsonia*, 3:285-326.

Ummels, F. 1963. Asteroids from the Netherlands Antilles and other Caribbean localities (Oreasteridae, Ophidiasteridae, Asterinidae, Luidiidae). *Stud. Fauna Curacao*, 15:72-101.

U.S. Naval Oceanographic Office. 1967. Environmental atlas of the Tongue of the Ocean, Bahamas. *Spec. Publ. 94*, Washington, D.C.

Van Der Sloot, C.J. 1969. Ascidians of the family Styelidae from the Caribbean. *Stud. Fauna Curacao*, 30:1-57.

Van Name, W.G. 1921. Ascidians of the West Indian region and southeastern United States. *Bull. Amer. Mus. Nat. Hist.*, 44:283-494.

________. 1930. The ascidians of Porto Rico and the Virgin Islands. *Sci. Sur. Porto Rico and Virgin Isl.*, 10(4): 405-511.

Vervoort, W. 1968. Report on a collection of Hydroida from the Caribbean region, including an annotated checklist of Caribbean hydroids. *Zool. Verh.*, 92:1-124.

________. 1962. A redescription of *Solanderia gracilis* Duchassaing and Michelin, 1846, and general notes on the family Solanderiidae (Coelenterata: Hydrozoa). *Bull. Mar. Sci.*, 12(3):508-542.

Villalobos, A. 1974. Estudios ecologicos en un arrecife coralino en Veracruz, Mexico. *Symp. Invest. Resources Carib. and Adj. Regions, UNESCO:* 531-545.

Voss, G.L. 1956. A review of the cephalopods of the Gulf of Mexico. *Bull. Mar. Sci. Gulf and Carib.*, 6(2):85-177.

________, Opresko, L. and Thomas, R. 1973. The potentially commercially valuable species of octopus and squid of Florida, the Gulf of Mexico and the Caribbean area. *Sea Grant Field Guide Ser.*, Univ. of Miami, 2:1-33.

________ and Phillips, C. 1957. A first record of *Octopus macropus* Risso from the United States with notes on its behavior, color, feeding and gonads. *Quart. Journ. Fla. Acad. Sci.*, 20(4):223-232.

________ and Solis, M.J. 1966. *Octopus maya*, a new species from the Bay of Campeche, Mexico. *Bull. Mar. Sci.*, 16 (3):615-625.

Wainwright, S.A. and Dillon, J.R. 1969. On the orientation of sea fans (genus *Gorgonia*). *Biol. Bull.*, 136:130-139.

Walsh, G.E. 1967. An annotated bibliography of the families Zoanthidae, Epizoanthidae, and Parazoanthidae (Coelenterata, Zoantharia). *Tech. Rpt., Hawaii Inst. Mar. Biol.*, 13.

Warburton, F.E. 1958. The manner in which the sponge *Cliona* bores in calcareous objects. *Canad. Journ. Zool.*, 36:555-562.

Warmke, G.L. and Abbott, R.T. 1961. *Caribbean Seashells.* Livingston Press, Wynnewood, Pa., 348 pp.

Wells, J.W. 1973a. Coral reef project—Papers in memory of Dr. Thomas F. Goreau. 2. New and old scleractinian corals from Jamaica. *Bull. Mar. Sci.*, 23(1):16-58.

________. 1973b. Two new hermatypic scleractinian corals from the West Indies. *Bull. Mar. Sci.*, 23(4):925-932.

________. 1972. Some shallow water ahermatypic corals from Bermuda. *Postilla*, 156:1-10.

________. 1971. Note on the scleractinian corals *Scolymia lacera* and *S. cubensis* in Jamaica. *Bull. Mar. Sci.*, 21(4): 960-963.

________ and Lang, J.C. 1973. Systematic list of Jamaican shallow-water Scleractinia (appendix). In: Wells, J.W. New and old Scleractinian corals from Jamaica. *Bull. Mar. Sci.*, 23(1):55-58.

West, D.A. 1971. The symbiotic zoanthids of Puerto Rico with observations on their biology. M.S. Thesis, Univ. of Puerto Rico, Mayaguez.

Williams, A.B. 1965. Marine decapod crustaceans of the Carolinas. *Fish. Bull.*, 65(1):1-298.

Work, R.C. 1969. Systematics, ecology, and distribution of the mollusks of Los Roques, Venezuela. *Bull. Mar. Sci.*, 19(3):614-711.

Yonge, C.M. 1973a. Observations of the pleurotomarid *Entemnotrochus adamsoniana* in its natural habitat. *Nature, Lond.*, 241:66-68.

———. 1973b. Coral reef project—Papers in memory of Dr. Thomas F. Goreau. 1. The nature of reef-building (hermatypic) corals. *Bull. Mar. Sci.*, 23(1):1-15.

Zieman, J.C. 1972. Origin of circular beds of *Thalassia* (Spermatophyta: Hydrocharitaceae) in south Biscayne Bay, Florida, and their relationship to mangrove hammocks. *Bull. Mar. Sci.*, 22(3):559-574.

INDEX

Page numbers printed in **bold** refer to photographs.